PROJECT HYDROGEN '91

"LAUNCHING OUR SUSTAINABLE ENERGY FUTURE"

Technical Proceedings

Editors

T.N. Veziroglu

R.E. Billings

AMERICAN ACADEMY OF SCIENCE

Independence, Missouri
USA

PROJECT HYDROGEN '91

"LAUNCHING OUR SUSTAINABLE ENERGY FUTURE"

Technical Proceedings

16-18 September 1991
Adam's Mark Hotel
Kansas City, Missouri, USA

Presented by

International Association for Hydrogen Energy

American Academy of Science

Missouri Department of Natural Resources'
Division of Energy

Environmental Protection Agency

Organizing Committee

Dr. Roger E. Billings, Conference Chairman	American Academy of Science, USA
Dr. J. O'M. Bockris	Texas A&M Univ. USA
Dr. S. Furuhama	Musashi Institute of Technology, Japan
Walt Pyle	Chevron Research & Technology Co., USA
Dr. John W. Sheffield	University of Missouri-Rolla, USA
Olof Tegström	Tebetron, Sweden
Dr. William D. Van Vorst	UCLA, USA
Dr. T.N. Veziroglu, President	International Association for Hydrogen Energy, USA

Library of Congress Catalog Card Number: 92-70026

ISBN: 0-9631634-1-8

First Edition 1992

In order to make this volume available as economically and as rapidly as possible, the authors' typescripts have been reproduced in their original forms. This method unfortunately has its typographical limitations but it is hoped that they in no way distract the reader.

Printed in the United States of America.

FOREWORD

Project Hydrogen '91, the first World Hydrogen Conference held in an off-year, is notable as a milestone in the presentation of developments of hydrogen technologies that are, at last, practical and economically feasible for implementation in today's existing infrastructure.

The Conference presented major breakthroughs in the research and application of hydrogen fuel cells, solar production of hydrogen, and hydrogen-fueled internal combustion engines. Technical reports were also given on specific components of hydrogen applications and of on-going international hydrogen utilization programs.

In addition to the customary Technical Sessions where researchers reported on their latest work, Project Hydrogen '91 featured a unique Tutorial Session, in which experts in the various aspects of the hydrogen energy economy instructed and brought up-to-date those who are new to hydrogen technology. These first-time participants included industry administrators and technical decision-makers, who are anxious to become involved in alternative energy applications.

The continuing depletion of our natural resources and pollution of the environment by the burning of fossil fuels is beginning to have measurable physical and political consequences world-wide. The resulting changes to global weather patterns are starting to cause extremes in temperature and precipitation, and the nations of the world are resorting to war in an attempt to control or to stabilize important oil-producing regions. If this trend is to be reversed, we must begin now to change our energy consumption habits. It is our hope that this proceedings will further the goal of the Project Hydrogen '91 Conference -- "to launch our sustainable energy future" -- by showing that the technologies necessary to do so are now available.

--T. Nejat Veziroglu
--Roger E. Billings

ACKNOWLEDGMENTS

The Organizing Committee wishes to acknowledge the support and assistance of Dr. Roger E. Billings and the American Academy of Science, Dr. T. Nejat Veziroglu and the International Association for Hydrogen Energy, and Robert Hangebrauch and Brock Nicholson of the Environmental Protection Agency for their sponsorship of Project Hydrogen '91.

Acknowledgement is also gratefully extended to the session organizers, and to the lecturers and authors whose notable contributions are presented in this proceedings.

In addition, the Organizing Committee recognizes the contribution of Dr. John O'M Bockris of Texas A&M University, for his inspiration and vital support of the conference, to Dr. T. Nejat Veziroglu and his team at the Clean Energy Research Institute of the University of Miami, and to Jan Eyre at the American Academy of Science for the preparation of this proceedings.

Organizing Committee
Project Hydrogen '91

PROJECT HYDROGEN '91

CONTENTS

CONTENTS

CONTENTS

CONTENTS

CONTENTS

CONFERENCE GOALS AND CHALLENGES*

T. Nejat Veziroglu

President, International Association for Hydrogen Energy
Director, Clean Energy Research Institute, University of Miami
Coral Gables, Florida 33124, U.S.A.

Distinguished guests, colleagues, ladies and gentlemen: It is a great honor to be speaking at the opening of the Project Hydrogen '91 Conference. I would first like to thank the organizers and the staff members for putting together what obviously will prove to be the best Hydrogen meeting of the year 1991.

Ladies and Gentlemen: We are in the midst of challenging times. The life supporting environment of our planet Earth and its inhabitants, peoples, flora and fauna, are under growing attack by pollution, acid rains and the greenhouse effect - all caused by the fossil fuels. They are affecting our health, our environment and causing irreparable damage to ecology.

So far as we know, the planet Earth is the only one hospitable to life. Observations by telescopes, satellites, radars, electromagnetic listening devices have not come up with any other heavenly body which supports life. For all we know, our planet may be the only one with life. You see, the universe is 15 to 20 billion years old. Our solar system is much younger at 5 billion years of age. Some astronomers claim that there might be an even distribution of planets with life in the universe; then about two-thirds of these planets should have civilizations much older than ours - billions and billions of years older. And think of the progress they should have made in those additional billions of years as compared to ours. Think of the progress made by the human society on this planet in the past 100 years: radio, television, computers, air travel, space travel, men on the moon. If our young civilization could make such mind boggling progress in only 100 years or less, think of the progress other civilizations could have made, which would be billions and billions of years senior to ours. Of course, if they existed, we could have been found long ago. So it seems that ours is the only and lonely planet supporting life in this vast universe.

Ladies and gentlemen, we have nowhere else to go but to live and let live on the planet Earth. Therefore, we must take extremely good care of this planet. It has a well balanced, but fragile ecology. We must not disturb it. We must protect it. We must preserve it.

---

*Opening Address presented at the Project Hydrogen '91 Conference.

There is a fuel which can achieve this. It is hydrogen - hydrogen produced from clean and renewable primary energy sources. Hydrogen will produce little or no pollution, no acid rains, no greenhouse gases. It is the cleanest fuel we know.

Hydrogen is making inroads in several directions. Because of its unequalled advantages, it has become the primary fuel of the space programs around the world. In this country, in the Soviet Union, in Europe, in China, in Japan, in India, it is the principal space energy carrier.

Hydrogen is now moving from space to the atmosphere. The United States has decided to go ahead and build an aerospace plane named X30, which will have speeds of Mach 6 to 25. It is scheduled to fly in the second half of this decade. Its engine will obtain oxygen from the atmosphere while flying in it, and will use stored liquid oxygen outside the atmosphere. The only fuel capable of powering its engines is hydrogen. In addition to providing power, slush hydrogen (a mixture of liquid hydrogen and frozen hydrogen) will also provide cooling to reduce frictional heating of the plane's surfaces and also for the engine nozzles.

Great Britain has decided to move ahead with the HOTOL project. This is similar to the aerospace plane - though somewhat smaller - and will be used for carrying goods to space stations and satellites in orbits around the Earth. It will also run on hydrogen.

When we consider the fuels for the land vehicles - cars, buses, trucks, etc. - the trend is the same, towards hydrogen. The number of land vehicles running on propane, natural gas and methane is increasing throughout the world. The historical and natural progression is towards increasing the hydrogen-to-carbon ratio in the fuels - the ultimate being pure hydrogen.

Fuel cells are now close to becoming commercialized. A 4.5 MW fuel cell has been running in Tokyo for the past 6 years, with excellent results. Tokyo Electric Utility has now a new fuel cell of 11 MW capacity. In the United States, 50 fuel cells of 40 KW each have been tested for different applications, ranging from homes to buildings, from small businesses to shopping centers. They provide electricity and heat with an overall efficiency of 80 percent, running on hydrogen produced from natural gas or naphtha. It is planned to commercialize the phosphoric acid fuel cells soon - as early as next year; the higher efficiency molten carbonate fuel cells by the middle of the 1990s, and the simpler (but higher temperature and higher efficiency) solid oxide fuel cells by the turn of the century. These will all run on natural gas or naphtha, as hydrogen is not yet available in large quantities and in comparable prices. When hydrogen becomes available and when the fuel cells run directly on hydrogen, their capital costs will go down, as there will be no need for the reformer/shift reactors to produce hydrogen from hydrocarbons.

An important question is "When will hydrogen be available at competitive prices for non-aerospace applications?". Well, it is not too far away. Alkaline bi-polar electrolysers using inorganic ion exchange membranes have now been commercialized with very high efficiencies of about 90 percent for converting electricity to hydrogen. Countries having hydropower potential or cheap off-peak electricity can now use these to produce hydrogen at a price that can compete with hydrocarbons. For example, Canada is offering to sell to the United States electrolytic hydrogen produced by hydropower at costs comparable to those produced from natural gas.

There are important and encouraging developments on the photovoltaic front. Their conversion efficiencies continue to go up and the prices continue to go down. It is expected that early in the next century solar hydrogen will be able to compete with the fossil fuels.

Ladies and Gentlemen, we are living in very exciting times. The dawning of the hydrogen age has already begun. We are seeing the efforts of the hydrogen community bearing fruit. Hydrogen has become the undisputed fuel of the space sector and is on the threshold of being commercialized in other sectors. The production of fossil fuels has about peaked, and that of hydrogen is on the upswing. The question is no longer "whether or not hydrogen?", but how fast will hydrogen replace the fossil fuels?".

The International Association for Hydrogen Energy has an energy-environment resolution which states that every good should carry a surcharge to remedy its environmental damage. There is no doubt that if adopted, if implemented, the principles embodied in this resolution will speed up the changeover. Hence, it falls on all of us as scientists and engineers and as supporters of the Hydrogen Energy System, to ensure the implementation, and accelerate the conversion.

This is a challenge to all of us. We must bring the Energy-Environment Resolution to the attention of environmentalists, educators, doctors, lawyers, legislators and the public-at-large. We must educate them; show them that environmental problems and energy problems are inter-related and can be solved intelligently, efficiently and economically.

Environmentalists are our natural allies. We must go to their meetings. We must speak up and explain that energy does not have to be an environmental foe; they can be compatible. Hydrogen, the universe's primary energy source, is also the most "natural" energy carrier.

We must talk to the fossil fuel producers, show that they don't have to fear hydrogen, that the world needs fossil fuels for the changeover, that we must conserve as much of the fossil fuels as possible, and that eventually as a chemical feed-stock, the fossil hydrocarbons will command higher prices, and for a longer period.

We must urge the legislators to pass laws embodying the principle exposed in the Resolution. We must explain that it will protect the public health; it will reduce wasteful spending; it will provide built-in incentives for both the goods and production methods to be environmentally compatible; and it will help reduce taxes, since the unproductive spending for environmental damage and pollution-related health care will progressively phase out.

In our mission to convert to the clean and renewable energy system, even the efforts of one single person, one single dedicated person, can achieve a lot. Let us look at a shining example:

Society consists of "thinkers" and "doers." Thinkers dream up the solutions. Doers take the practical steps to implement them. It is rare to have these two qualities in one person. There was a young student in the 1960s in a western university. He was thinking of the environmental problems facing our society; facing our planet. He saw that most of them were caused by pollution coming from gas guzzling automobiles. He believed that if he ran these automobiles with a clean fuel, which did not produce pollutants, acid rain ingredients, then the environmental problems could be solved. He was one of the first men to convert a car to run on hydrogen. When the energy crisis started, he saw that hydrogen would be the answer to our energy and environmental woes. He was the first man to build a home running on hydrogen. He became the first man to establish a company dedicated to hydrogen energy. And now, he has established an organization dedicated to the education, research and training of young persons, covering all aspects of the hydrogen energy system, at the graduate and postdoctoral levels. It is designed to prepare them to be the leaders, who will be establishing the environmentally most compatible, clean and abundant, and the most friendly energy system for our lone life-supporting planet Earth - the Hydrogen Energy System.

Ladies and Gentlemen, on behalf of the International Association for Hydrogen Energy and the entire hydrogen energy community, I am taking this opportunity to express our gratitude and thanks to this young man, Dr. Roger Billings.

I would also like to add that it has been a pleasure to work with Dr. Billings on this important conference; and I wish all the participants a very informative and productive meeting. When you return to your home towns, states and countries, I urge you to carry the flag of the Hydrogen Energy System everywhere, just like Roger Billings has done, and continues to do. Thank you very much.

CHAPTER ONE

Solar Energy Collection and Conversion

BRIGHTER THAN THE SUN: NEW SOLAR CONCENTRATOR TECHNOLOGY

Roland Winston, David Cooke and Joseph J. O'Gallagher

Department of Physics and The Enrico Fermi Institute
The University of Chicago, 5640 Ellis Ave.,
Chicago, Illinois 60637

ABSTRACT

Sunlight at the surface of the sun has an intensity of ~63 W/mm^2 but by the time the light reaches the surface of the earth the intensity has fallen to about 0.8 to 1.0 mW/mm^2. In a recent set of experiments we have concentrated terrestrial sunlight by a factor of 84,000 ± 3,000 to intensities of 72 W/mm^2. This establishes a new world record for solar flux concentration, exceeding our previous result by a factor of 1.5. Such high concentrations cannot be attained with a conventional imaging device alone; our approach relies on a device designed according to principles of the new field of nonimaging optics. Nonimaging concentrators are essentially "funnels" for light, and the enormous intensities of sunlight we have attained with them actually exceed the intensity at the surface of the sun itself by about 15 percent. In other words, we have produced the highest concentration of sunlight in the solar system. Our approach provides techniques for developing high fluxes and/or high temperatures from concentrated sunlight. These in turn can be used not only for direct solar pumping of lasers, which has been demonstrated, but also to drive a variety of photochemical reactions including several that show promise for the direct production of hydrogen.

INTRODUCTION

Phase space conservation[1] and thermodynamic arguments[2] place a limit on the concentration of sunlight any optical device can attain, namely

$$C_{max} = n^2/\sin^2\theta, \quad (1)$$

where n is the index of refraction at the target surface and θ is the semiangle subtended by the sun. At the surface of the earth $\theta = 0.27^0$, and so with an ordinary refractive material (n ~ 1.5) one obtains an upper limit on concentration of about 100,000.

The concentrations attained with conventional imaging devices fall far short of this limit, because of abberations[3]. A parabolic mirror, for example, produces a perfect image on axis, but blurs and broadens the image off axis. To attain high concentrations we relax the requirement of forming images. Simply put, a laser crystal or a solar furnace does not care about receiving a picture-perfect image of the sun; all either care about is receiving the maximum power per area. By dispensing with image-forming requirements in applications where no image is required, we can attain concentrations that approach the theoretical limit.

NONIMAGING CONCENTRATION

The relatively new field of nonimaging optics provides a framework for designing such powerful concentrators.[3] We design our concentrator by an approach called the "edge ray" method[3], in which all light rays entering the device at the maximum collection angle are directed after one reflection at most to the rim of the exit aperture[4] (Fig. 1). In this way the other rays are reflected within the aperture itself.

In principle we could attain high concentrations of sunlight with a nonimaging concentrator alone, but such a device would be quite large and unwieldy. Our approach for concentrating sunlight utilizes a two-stage system incorporating a parabolic mirror and a nonimaging concentrator (Fig. 2). The overall concentration for a two-stage system is simply the product of the concentration of the parabolic mirror with that of the nonimaging concentrator:

$$C = n^2\cos^2\Phi/\sin^2\theta, \qquad (2)$$

which falls short of the theoretical limit (equation (1)) only by the factor of $\cos^2\Phi$, where Φ is the rim angle of the mirror. The effects can be minimized by making the rim angle small; i.e., choosing a mirror with a relatively large f number.

(a) Dielectric-filled secondary concentrators.

Employing a technique based on this concept, our group previously achieved a record concentration of sunlight of 56,000 times the intensity at the surface of the earth.[5] We have now modified and improved that apparatus to reach still higher concentrations. We employ a 40.6-cm-diameter silver coated telescope mirror with rim angle f = 11.5o (f = 2.5). Sunlight is concentrated to a one centimeter spot one meter away from the mirror. The light is further compressed to a one millimeter spot by a nonimaging concentrator made out of sapphire. Sapphire was chosen for its low absorption and high index of

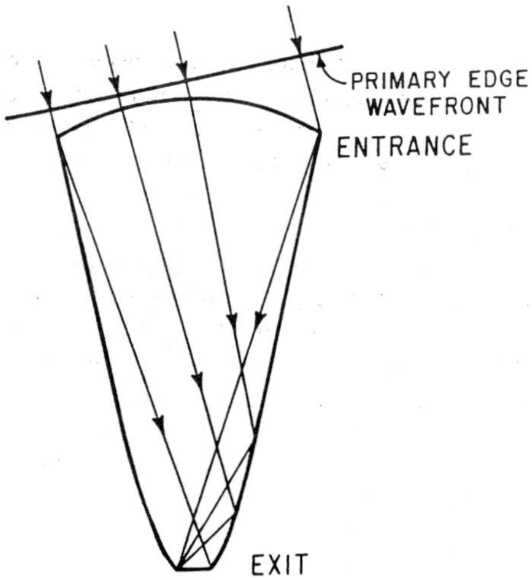

Fig. 1. In a nonimaging concentrator designed by the edge ray method all light rays entering the device at the maximum collection angle are directed after one reflection at most to the rim of the exit aperture.

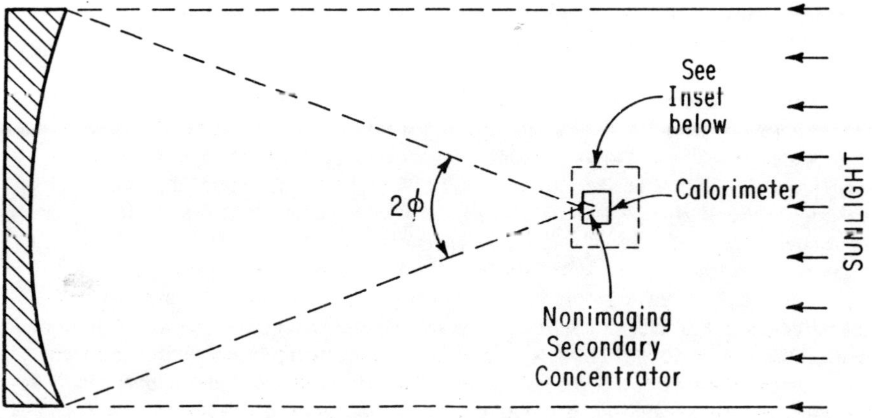

Fig. 2. Two-stage approach to concentrating sunlight is illustrated.

refraction, $n = 1.76$. (The concentrator in our previous experiment was a silver-coated vessel filled with an immersion oil having an index of refraction of 1.53.) Substituting these values into equation (2) gives the theoretical limit to the concentration of our new two-stage system:

$$C = [1.76\cos(11.5^0)/\sin(0.27^0)]2 = 137{,}000 \quad (3)$$

This is a 31 percent increase over the theoretical upper limit of 104,000 in our earlier device. Another modification we have made is that our concentrator now works by total internal reflection[6]. In the previous experiment the walls of the concentrator were silvered, and as such there were losses due to absorption. Total internal reflection, on the other hand, is nearly loss-free, and thereby increases the efficiency of the concentrator, allowing a greater fraction of the theoretical limit to be attained[7].

(b) Air-filled secondary concentrators

Since 1984, the University of Chicago and the Solar Energy Research Institute (SERI) have been pursuing an effort to develop techniques and applications for generating as high a flux of concentrated solar radiation as possible[8]. In the past, furnaces based on imaging optical designs have produced fluxes in the range 5000 to 16000 "suns". The angle subtended by the sun, together with physical conservation principles, sets a limit on the maximum achievable geometric concentration of sunlight at a factor of about 46,000 (or 46,000 "suns"), if the index of refraction at the target surface is unity (i.e. in air). SERI recently designed and constructed a 10 kilowatt solar concentrating furnace facility, which uses a modified long focal length off-axis design, so that when combined with a nonimaging secondary element the resulting configuration can approach this allowed limit[9]. A ray-trace analysis of a variety of nonimaging secondary designs was used to select a prototype secondary for a series of high flux experiments. The selected design is a Compound Parabolic Concentrator (CPC) which has an acceptance angle 14^0 and is truncated to 80% of its full height. The entrance and exit aperture diameters are 6.0 cm. and 1.48 cm. respectively corresponding to a geometric concentration ratio C = 16.4 X. Two mechanically identical water-cooled prototypes were constructed using an alloy of tellurium and copper. In a series of experiments conducted at SERI during the summer of 1990, the performance of both secondary units was measured using a small cold-water calorimeter[10]. The optical models for the full system predict that an average concentration in excess of 20,000 suns at a net power of about 3.5 kilowatts should be delivered at the CPC exit aperture. Net flux values corresponding to intensities of 17,000 to over 19,000 suns were obtained with the first unit exceeding the previous highest values achieved in air. Greater care was taken in preparing the reflecting surface of the second unit and on several occasions, flux concentrations at multikilowatt levels in excess of 22,000 suns were achieved, surpassing all previous values for the concentration of sunlight in air by more than 30%.

CONCLUSIONS

The levels of concentration reported here represent a new domain of high intensity solar fluxes which can be made available through nonimaging optical designs for delivery to a variety of targets and absorbers with relatively little technical difficulty. Possible uses for such high levels of solar flux are just beginning to be explored. Some potential applications include promoting various photochemical reactions including several for the production of hydrogen, processing of a variety of materials and pumping of high powered lasers.

ACKNOWLEDGEMENT

We thank the US Department of Energy, in particular the Solar Energy Research Institute and the Office of Basic Energy Sciences, for support of this work.

REFERENCES

1. Winston, R. J. Opt. Soc. Am. 60, 245 (1970).
2. Rabl, A. Sol. Energy 18, 93 (1976).
3. Welford, W. T. & Winston, R. High Collection Nonimaging Optics (Academic, New York, 1989).
4. The edge-ray principle is used in a slightly different form for the concentrator described here, to ensure that light is reflected by total internal reflection. Hence, the maximum exit angle of the light is reduced slightly, from 90o to 86o, which reduces concentration by one half percent.
5. Gleckman, P. Appl. Opt. 27, 4385-4391 (1988).
6. Winston, R. Appl. Opt. 15, 291-292 (1976).
7. Gleckman, P., O'Gallagher, J. & Winston, R. Nature 339, 198-200 (1989).
8. Lewandowski, A., C. Bingham, J. O'Gallagher, R. Winston, D. Sagie, "Performance Characterization of the Solar Energy Research Institute High Flux Solar Furnace." Proceedings Int'l. Energy Agency, 5th Symposium on Solar High Temperature Technologies, August 27-31, 1990, Davos, Switzerland.
9. Lewandowski, A.. "The Design of an Ultra High Flux Solar Test Capability." Presented at the 1989 Intersociety Energy Conversion Engineering Conference, August 1989, Davos, Switzerland.

10. O'Gallagher, J. , R. Winston, C. Zmola, L. Benedict, D. Sagie and A. Lewandowski, "Attainment of High Flux-High Power Concentration Using a CPC Secondary and the Long Focal Length SERI Solar Furnace," 1991 American Society of Mechanical Engineers, Solar Energy Division Conference, Reno, Nevada, March 17-22, 1991.

RESEARCHES CONCERNING THE PHOTOVOLTAIC EFFECTS AND SOME PHYSICAL INTERFACE PROCESSES AT MIXED SEMICONDUCTOR HETEROSTRUCTURES WITH PHOTOSYNTHETIC PIGMENT

MAGDA GHERGHEL
SOFIA MELINTE
The Polytechnical Institute of Jassy, Physics Department, 63 Splai Bahlui Str., Jassy 6600 ROMANIA

ABSTRACT

This paper presents the experimental results concerning the study of the photoelectric behaviour of a multijunction type multilayer mixed heterostructure (solar cell system tandem) having the form: SnO_x/CdS:Na-Cu_xS:Cu-Al-Chl/Cu. On some previous papers, we have underlined the fact that, to improve the photovoltaic performances of CdS-Cu_xS type heterostructures accomplished by the liquid phase reaction method, we have directed our researches on the use of some elements from the I_a and I_b groups as dopants in the CdS and Cu_xS layers, respectively. On taking into account, as well, the considerable photoconducting properties noticed at the photosynthetic pigments, we have tried to insert between the dopped layers from the photojunction, some thin layers of microcrystalline chlorophyll obtained by electrodeposition, sputtering and dipping; finally there resulted a mixed multijuction of SnO_x/CdS : Na-Cu_xS : Cu-Al-Chl/Cu type.

Our experimental results have led to the conclusion that both the photosynthetic pigments (chlorophyll) and the I_a and I_b group impurities (Na,Cu,Li,K) are able to improve the photoeffects at CdS-Cu_xS junction, by enhancing its photovoltages and photocurrents/cm^2, i.e. U_{CD}=(350-575)mV and J_{SC}=(10.5-15)mA/cm^2.

At the same time, by using the "progressive-analogical" modelling method, we have studied the influence of electric fields with different intensities and of the addition of the ascorbic acid on the photoconduction microcrystalline chlorophyll thin films included in a mixed heterostructure of the SnO_x/CdS:Na-Cu_xS:Cu-Al-Chl/Cu type. We have also studied the methods in which the temperature, the concentration of the dopants solutions amd the external electric fields intervene in the interface physical processes of this mixed heterostructure: the diffusion of the oxygen at the CdS/Cu_xS interface, the decrease of the Schottky barrier height at the Al/Chl interface: $\Delta\emptyset$=(0.012-0.025eV, the quantum tunnelling effects at the Al/Chl and Chl/Cu interfaces, the effects of the interface images forces.

At the same time, we have studied the optical absorption mechanism as well at the phonons scattering mechanism at this type of heterostructure, the knowledge of this phenomenon being extremely important for the understanding of the physical processes at the metal-semiconductor interface with a view to improve the photovoltaic qualities at these heterostructures.

I. INTRODUCTION

One of the developing directions of the research program on non-conventional energy sources is that including the improvement of the non-conventional energy sources performances based on the photovoltaic effect. In this context, a special attention is given to the technologies for the accomplishement of the heterojunctions and heterostructures with a Schottky barrier made up of thin photoconductive layers. One knows the fact that, whith a view to enter competition with other energy sources, the photovoltaically - bases devices should satisfy some main requirements, namely, a reduced cost price an as long as possible service life, an as high as possible coefficient of solar conversion, usable non-deficit materials; the last condition is closely connected with the first one.

Thus, the CdS-Cu_xS type heterostructures obtained by the method of liquid phase reaction [4,5] present the advantage that their constituent photoconductive consumption, in a relativoly short time, has a good adherence to the substrate and a controllable stoichiometry [4-8] .Their shortcoming consists in their photoconstructive and photosensitive properties worse than those of the films produced by other methods (vacuum evaporation, sintering and dipping) [10] . Yet, their performances can be improved by controlling some physico-chemical parameters (temperature, concentration of the solution, impurities doping degree), that are directly influencing the chemical reaction kinetics and, hence, the film generation on the substrate.

On some previous papers, we have underlined the fact that to improve the photovoltaic performances of CdS-Cu_xS type heterostructures accomplished by the method of the liquid phase reaction, we have directed our researches on the use of some elements from the I_a and I_b groups as dopants in the CdS and Cu_xS layers, respectively, on taking into account that impurities are cheaper and casier to find.

At the same time, on considering that the photosynthetic pigments (chlorophyll) show an intense absorption in the visible range (where the solar radiated energy is maximum),having a semiconduction activation energy of 1-2 eV, which corresponds to maximum efficiencies of the photovoltaic devices[4,6] as well as the fact that the microcrystalline chlorophyll films show considerable semiconduction properties[1-3,6], we have proposed to study the influence of an Al-Chl-Cu type heterostructure on the CdS-Cu_xS photojunction, with the aim of using the photosynthetic pigment thin films in producing mixed photovoltaic devices based on anorganic semiconductor materials. Simultaneously, we have folowed the influence of both electric fields of different intensities and the addition of the ascorbic acid (C_6H8O_6) on the energy transfer and electron transport through the photosynthetic membrane of the chlorophyll molecules, with the object of improving their photoconduction performances [8].

At the same time, we have decided to study both the process of the interface optical absorption and the influence that the scattering mechanism on the phonons of the structural lattice could have upon the photovoltaic efficiency at this type of mixed multijunction.

II. EXPERIMENTAL RESULTS AND DISCUSSIONS

The CdS-Cu_xS heterojunction was produced by the method of liquid phase reaction [3,4] on glass substrates of (3 x 7)cm on which semitransparent SnO_x thin films were previously formed by the same method, these films representing the negative electrodes of the heterojunction [3-5]. At the positive electrode, a copper grid covered later with a 0.2μm thin Al film for avoiding the oxidation process was made by vacuum evaporation. Under the optimum conditions that we have established in [3,4], we have now obtained the open circuit voltage V_{CD}=190-250 mV and short circuit current J_{SC}=8-8.5 mA/cm^2, for the heterojunction with CdS and Cu_xS films, (Fig.1).

During our experiments, we have found out that an improvement of the CdS-Cu_xS heterojunction photovoltaic parameters can be obtained by an optimum correlation between the sputtering temperature and the concentration of the solution, these parameters being able to influence the equilibrium of the chemical reaction for the CdS and Cu_xS film formation.

For the purpose of improving the photoeffects of this type of photoelement, and increasing the heterojunction photovoltages and photocurrents, we have sputtered above the CdS films, at a constant temperature of 120°C, a Na_2SO_4 solution with the concentration of 0.5-2%, thus obtaining a CdS:Na-Cu_xS structure. At the same time, we have sputtered above the Cu_xS film obtained by the same method, at a constant temperature of 150°C, a $CuSO_4$ solution with the concentration of 0.5-2%, thus producing the Cu_xS:$CuSO_4$:Cu structure.
A considerable improvement of the photovoltaic properties have been seen at the CdS-Cu_xS heterojunction with the CdS and Cu_xS films doped at the optimum concentration of Na_2SO_4 and $CuSO_4$ solutions, respectively, an amplification of both U_{CD} and J_{SC} being noticed: U_{CD}=300 mV and J_{SC}=8.7-9 mA/cm^2, respectively.

On considering that the photosynthetic pigments show an intense absorption in the visible range, where a maximum radiated solar energy exists, and that the chlorophyll microcrystalline thin films show remarkable photoelectric properties [6], an Al-Chl-Cu type heterostructure (Fig.2), was made above the CdS:Na-Cu_xS:Cu photojunction, thus finally resulting a mixed SnO_x/CdS-Cu_xS-Al-Chl-/Cu type multijunction. Thin microcrystalline films have been prepared by sputtering, electrodeposition and dipping methods, described in details in [3,8], having a thickness of 10-20μm and a mean resistivity of 3.10^{11} Ωcm. For this mixed photoelement, we have noticed an increase of both photocurrents and photovoltages, i.e. U_{CD}=350-450 mV and J_{SC}=10-10.5 mA/cm^2, [15].

On taking into account that in the vegetal tissues there exist physiological gradients connected with the electrical gradients (hence stationary electric fields) and that the photoelectric voltages of these biological materials depend on the intensity of the metabolical processes, we have decided to study the influence that an external electric field of different intensities (applied to biological materials from the photosynthetic pigments have been obtained) could have on the photoelectric behaviour of the Al-Chl-Cu heterostructure. Thus, we have applied to the vegetal samples that we have used (from which we have afterwards extracted the microcrystalline a and b chlorophylls), an artificial electrostatic field E directed downwards, with intensities ranging between 75 and 60; 60-35; 35-7.5 V/cm. Aluminium plates of (35x65)cm and 10 μm thick multipolar platinum and wolfram electrodes have been used.

During our experiments we have noticed that thin chlorophyll films obtained from vegetal samples subjected to an average electric field E=7.5 V/cm, have higher photoeffects as compared to pigment films extracted from vegetal materials that have not been subjected to electric fields. At the same time, on considering that the chlorophyll molecules are complexed "in vivo" with proteins, enzymes and vitamins (mainly the ascorbic acid that is always present in green tissues), we have supposed that a "progressive-analogical correlation" exists between the following parameters: electric fields E + the concentration C% of the ascorbic acid → proteic metabolism → photoconduction (volt-ampere characteristic). We could also notice an increase of the photoeffects in the microcrystalline chlorophyll films when adding in the pigment extract an ascorbic acid solution with the mean concentration of (0.5-2.5%).

In **Fig.**3 we give the volt-ampere characteristic J=f(V) for the Al-Chl-Cu heterostructure, where the microcrystalline chlorophyll films have been obtained by treating the pigment in the electric field E=7.5 V/cm with an addition of ascorbic acid in a concentration of 1.5%. For these heterostructures, the mean values for U_{CD} and J_{SC} are U_{CD}=575 mV and J_{SC}= 25 na/cm^2, respectively (Fig.3).

During our experimental tests to plot this characteristic, we could find that at the direct polarization, the darkness current had values much higher than at the inverse polarization. In exchange, the photocurrent J_L=J_{SC} presented an inverse behaviour, having small values at the direct polarization and high values at the inverse one, the characteristic J=f(V) at illumination passing through the IV quadrant, a property which is pelicular for the photovoltaic devices (see Fig.3). On the entire load interval taken into account during our experimental tests, the direct darkness current at this type of heterostructure presented an experimental tension dependence while the inverse darkness current depended exponentially on $\sqrt{V}$ (see Fig.4). Consequently, we have noticed that instead of the saturation phenomenon of the inverse current, there appears its slight variation with the applied tension (Figs.3 and 4); this fact suggests the existence of both a Schottky barrier at the Al/Chl interface and of a "p"-type semiconduction in the thin chlorophyll layer at the Al-Chl-Cu heterostructure.

Having considered this conclusion, we present in Fig.5 the diagram of the energy bands for the metal/chlorophyll (M/chl) structure in which we take into account the fact that the difference between the flow of electrons is toward the "p"-type semiconductor (chlorophyll film). Accordingly,we consider that a barrier layer which contains a negative spatial load (electrons)is formed at the Al/Chl interface, a lead due to the ionized acceptors, remained uncompensated by the disappearence of the holes in the Chl film on a distance l from the Al/Chl interface in the Al-Chl-Cu heterostructure. Thus, in the energetical diagram shown in Fig.5, the energy bands are curved downwards with the value $\Psi(x)=\phi_m-\phi_s = eV_c <0$, $\phi_m < \phi_s$, as for the passing from the metal to the Chl film the electrons should surpass the supplementary potential barrier $\Psi(x)=eV_c$ (ϕ_m and $\phi_s = \phi_{Chl}$ represent the energetical barrier for metal and chlorophyll, respectively, while V_c is the difference of contact potential metal - semiconductor M-S).

The expressions of the contact potential, of the contact electric field intensity $\mathcal{E}(x)$ as well as that of the tickness l of the deprived layer in carriers from the semiconductor at the Al/Chl interface, can be determined by solving the Poisson equation [15]:

$$\frac{d^2V}{dx^2} = -\frac{\rho}{\varepsilon}, \quad \varepsilon = \varepsilon_0 \varepsilon_s .$$

(ρ being the density of the spatial loads at the interface)and by resorting to modelling by applying the Bethe theory of the diode and the Schottky theory of the diffusion [12,15].

On taking the absolute value of the external tension,positive for a direct connecting direction, the expression for the characteristic J=f(V) of the Schottky contact [9,12,15] at the Al/Chl interface has the form:

$$J = A^* T^2 e^{-\frac{\phi_B}{kT}} \cdot \left(e^{\frac{eV}{kT}} - 1\right) = J_S \left(e^{\frac{eV}{kT}} - 1\right) \qquad (1)$$

where $\phi_B = \phi_{m-s} = eV_B = eV_c$ reprezents the effective barrier to be passed by the electrons which go from the metal to the Chl film (Schottky barrier), while A^* reprezents Richardson effective constant, [12-14]; T is the working temperature.

For $V > 0$, the relation (1) depicts the direct branch (from the first quadrant) of the characteristic J=f(V) of the Schottky contact at the Al/Chl interface. For $V<0$ in the relation (1), the exponential value becomes negligible, so that there results:

$$J = -J_S \qquad (2)$$

This expression reprezents the inverse branch ($V<0$) of the characteristic J=f(V), i.e. the inverse current of the Schottky diode or the saturation current.

On applying the Schottky electronic diffusion theory [9, 12,13], we notice the fact that in this case the characteristic J=f(V) for Al/Chl is similar to that from the p-n junction, but both the significance and the expression of the saturation current J_S are different. Thus, in this case, the inverse current J_S will depend on the applied tension by the agency of the contact electric field [15], $\mathcal{E} = -dV/dx$, from the Al/Chl interface:

$$J = e\, n_0 \mu\, \mathcal{E} \left(e^{\frac{eV}{kT}} - 1\right) = J_S \left(e^{\frac{eV}{kT}} - 1\right) \qquad (3)$$

where n_o is the equilibrium concentration of the carriers in the contact region, while u reprezents their mobility [15]. Finally, in the discussed case [15], the expression (3) becomes:

$$J = 2e\, n_o \mu \left[\frac{en_o}{2\varepsilon} (V_c - V) \right]^{\frac{1}{2}} \cdot (e^{\frac{eV}{kT}} - 1) = J_S (e^{\frac{eV}{kT}} - 1) \ . \quad (4)$$

Here we can see the dependence of the inverse current, function of the applied tension $J_S = f(V^{1/2})$, a phenomenon we have noticed experimentally, as well, during our tests concerning the plotting of the characteristic J=f(V) at the Al-Chl-Cu heterostructure (Fig.4).

The value of the Schottky barrier height at the Al/Chl interface has been evaluated experimentally both by measuring the characteristics J=f(V) at the Al-Chl-Cu heterostructure and by photoelectrical measurements [12,14]. So, on extrapolating the curve of the direct characteristic described by relation (1),[15], for $V \approx 0$, only the current J_S given by the following relation will pass through the Al/Chl interface:

$$J_S = \frac{4\pi e}{h^3} \cdot m^* k^2 T^2 \cdot e^{\frac{-\phi_B}{kT}} = A^* T^2 e^{-\frac{\phi_B}{kT}} , \quad (5)$$

where $A^* = \frac{4\pi e}{h^3} \cdot m^* k^2$ is Richardson constant [12,13],

h is Planck constant, k is Boltzmann constant, m^* is the effective mass of the carriers (electrons and holes).

On making the logarithm (5), we obtain for $\phi_B = eV_B$ the following relation:

$$\phi_B = kT \ln \left(\frac{A^* T^2}{J_S} \right) = eV_B \ . \quad (6)$$

On expressing the Schottky barrier by the value of the Schottky potential, we get[15]:

$$V_B = \frac{kT}{e} \ln \left(\frac{A^* T^2}{J_S} \right) = \frac{kT}{e} \ln \left(\frac{A\, S\, T^2}{J_S} \right) , \quad (7)$$

where, in our case, A has the value $\approx$ 120.4 A/cm^2 degree^2 [12,13], while the contact surface $S \approx 3.10^{-3}$ cm^{-2}, T= 300°K.

To determine J_S, we have used the logarithmic representation

$$J_S = \lg J / (1 - e^{-\frac{eV}{kT}}) = f(V) ,$$

from (1), there resulting a straight line (Fig.6) at whose intersection with the axis V = 0 the saturation current J_S is to be found (Fig.6).

The experimental results regarding the measurement of the characteristics $J = f(V)$ at the Al-Chl-Cu heterostructure, has led to mean values for J_S ranging in the interval $(1.25-1,30) \cdot 10^{-8}$ A. On introducing the determined values for J_S in (7), we have got the mean values for the Schottky barrier height at the Al/Chl interface, ranging in the interval $V_B=(0.725 - 0.750)$V.

The photoelectric measurements to plot the phoroelectric characteristic $J_L=f(h\nu)$ (J_L representing the illumination current or the photocurrent), has led, as well, to the determination of the Schottky barrier height at the Al/Chl interface. According to Crowell [14], the quantum efficiency, thus the photocurrent at a structure M-S-M, depend on the photons energy according to the relation:

$$\eta \sim (h\nu - h\nu_0)^2, \tag{8}$$

for $h(\nu - \nu_0) > 3kT$.

In relation (8), the energy $h\nu_0$ corresponds to the Schottky barrier height, so that:

$$h\nu_0 = e\, V_B = \phi_B. \tag{9}$$

The photocurrent resulted after the bilateral illumination of the Al-Chl-Cu heterostructure, is also proportional with the square of the energy of the incident photons:

$$J_L \sim (h\nu - e\, V_B)^2. \tag{10}$$

From (8) and (10) there result the dependance $\sqrt{\eta} \sim h(\nu - \nu_0)$ and the dependance of the photocurrent J_L, respectively, function of the energy of the incident radiation, $\sqrt{J_L} \sim (h\nu - \phi_B)$.

On representing graphically the value of the square root from the mean values J_L, experimentally obtained for the illuminating current (photocurrent), function of the energy of the photon flow $h\nu$, with which the bilateral illumination of the Al-Chl-Cu heterostructure has been accomplished, we have obtained a straight line for values of the energy higher than $h\nu_0$ (Fig.7). By extrapolating this straight line up to the intersection with the energy axis h , mean values for the barrier height ϕ_B ranging in the interval (0.95-1.15)eV have been obtained in the intersection point (Fig.7). The study of the spectral characteristics, $U_{CD}=f(\lambda)$, for these types of devices, made evident that the action spectrum of the photoconduction closely followes all the times, the absorption spectrum of microcrystalline chlorophylls, a shift of the red maximum toward the higher wavelength being specific (see Fig.8).

On taking into account the ϕ_B height of the Schottky barrier at the Al/Chl interface, we consider the existence of a quantum tunnelling effect of the carriers through the interface states as well as the posibility of returning to the aluminium AL layer of the electrons which have climbed over the barrier as a result of their interaction with the phonons of the interface structural lattice. Thus, the aspect of the dependency

$$\sqrt{J_L} = f(h\nu) \quad \text{and} \quad U_{CD} = f(\lambda),$$

in the region $h\nu = (0.90 - 1.5)$ eV, (Figs.7 and 8), respectively, make us consider, besides the interaction of the electrons with the field of the absorbed electromagnetic radiation (electron - energy photon $h\nu$ interaction), a photons scattering device on phonons, due to the interaction of the carriers with the lattice oscillations (optical absorption on phonons). Thus, we think that the photoconduction mechanism at the Al/Chl interface is characterized during the optical absorption process both by direct transitions - between states with the same electronic wave vector:

$$|K| = \frac{2\pi}{\lambda_{electron}},$$

$\Delta K = 0$ - and by indirect transitions [16] which take place with the absorption or emission of an energy phonon :

$$\varepsilon_{phon} = \hbar\omega_{phon} = h\nu_{phon}, \text{ (Fig.9)}.$$

Thus, we assume that the direct transitions, characterized by $\Delta K = 0$ from the valence-bond band, to the interface conduction band (Fig.9), determine the sudden increase of the absorption in the regions for which $0.95 \text{ eV} \leq h\nu \leq 1.82$ eV and $2.70 \text{ eV} \leq h\nu \leq 3.02$ eV, (Fig.7-8, Fig.10).

On the other hand, the transitions from the valence zone maximum to the conduction zone minimum at the Al/Chl and Chl/Cu interfaces being supposed as indirect ones (Fig.9), the electrons will be characterized by energetical stages with $K' \neq K$ and interact not only with energy incident radiation $h\nu$ (electron - photon interaction) but also with the interface structural lattice, thus determining the absorption or the emission of a pulse phonon $\hbar\bar{\eta}$,

$$|\bar{\eta}| = \frac{2\pi}{\lambda_{phon}}$$

We consider that these two processes in which these three particles, i.e. photon-electron-phonon, intervene, are characterized by the law of energy conservation [16]:

$$\hbar\omega_e = \varepsilon_f - \varepsilon_i + \varepsilon_{phon}$$
$$\hbar\omega_a = \varepsilon_f - \varepsilon_i - \varepsilon_{phon} \qquad (11)$$

where $\hbar\omega_e$ and $\hbar\omega_a$ represent the energy of the photon absorbed with the emission of the phonon, and the energy of the photon absorbed with the absorption of the phonon, respectively.
In the relations (11), ε_{phon} reprezents the phononic energy while ε_i and ε_f are the initial and the final energy of the

electron, respectively. Consequently, we think that the mechanisms of the optical absorption and of photoconduction at the metal/semiconductor interface from the considered heterostructure are also characterized by the indirect band-band transition in which, in the interaction process between a photon $h\nu = \hbar\omega$ and an electron, an energy phonon ε_{phon} intervenes - see relations (11).

At the same time, we consider that in the case of the indirect transitions, the law of pulse conservation at the electron-phonon interaction has the form:

$$\hbar\bar{K}' = \hbar\bar{K} \pm \hbar\bar{\eta}, \tag{12}$$

where $\hbar\bar{K}$ reprezents the electron pulse, $|\bar{K}| = \frac{2\pi}{\lambda_{electron}}$ and $\hbar\bar{\eta}$ reprezents the phonon pulse, $|\bar{\eta}| = \frac{2\pi}{\lambda_{phonon}}$, $\bar{\eta}$ being the phononic wave vector.

In the following, we assume that the phonon-photon interaction [16] is characterized by the conservation of the pulse and energy:

$$\hbar\bar{q} = \hbar\bar{\eta}$$
$$\hbar\omega = h\nu = \hbar\omega_{phon} \tag{13}$$

where $h\omega_{phon} = \varepsilon_{phon}$.

In the relations (13), $\hbar\bar{q}$ reprezents the photon pulse,

$|\bar{q}| = \frac{2\pi}{\lambda_{photon}}$, and $\hbar\omega$ and $\hbar\omega_{phon} = \varepsilon_{phon}$ are the photon energy and the energy of the phonon with which the photon interacts, respectively.

In Fig.9, we consider the indirect transitions as taking place between virtual states with very short average life, $\tau \approx 10^{-13}$s [16,17], so that we could suppose that the passage of the electrons from the state $K_1=0$ (maximum value of the valence zone at the Al/Chl interface) to the state $K_2=K_{min}$ (from the interface conduction zone), i.e., the transition (2') is accomplished by two mechanisms. Thus, under the action of the absorbed radiation, the electrons pass from the state $K_1=0$ in the valence zone to the state $K'=0$ in the conduction zone (direct transition (1)), a hole remaining on the state $K_1=0$ in the valence zone (Fig.9). This type of direct transition (1) determines the increase of the absorption coefficient α of the photonic radiation in the considered heterostructure (Figs.7, 8.10) for $h\nu > 2.70$ eV.

Taking into account the previously considered model(Fig.9) the electron from the state $K'=0$ has a greater energy than that corresponding to the inferior minimum value with $K_2=K_{min}$ and that is why we think that after a very short time, $\tau \approx 10^{-13}$s [16,17], this will pass in the minimum energy state K_2 absorbing or giving up a phonon - transition (2), (Fig.9).

Another mechanism of the interface photoconduction is that by which, as a result of a photon absorption, the electron is passing through a vertical transition (3), (Fig.9), from alow state in the valence zone $K_1 \neq 0$ to the state $K_2 = K_{min}$ in the conduction zone. The hole from the valence zone makes the transition (4) in the state with $K_1 \neq 0$ near the superior edge of the interface valence zone by the absorption or emission of a photon - relations (11), (Figs.7,8,10).

Consequently, we assume that in the conditions in which $h\nu = \hbar\omega > \phi_B$, where $\phi_B = \mathcal{E}f - \mathcal{E}i$ - relations (11) - the considered indirect transitions will determine a maximum absorption in the region of the high wave lengths as compared with the absorption edge of the metal/semiconductor interface, a fact which is in agreement, experimentally, for $h\nu > 0.95$eV, (Figs.7, 8,10).

The absorption coefficient α corresponding to the considered indirect transition will be made up of two terms corresponding to the absorption α_a and to the emission α_e of the phonon, respectively [16]:

$$\alpha(h\nu) = \alpha_a(h\nu) + \alpha_e(h\nu) \qquad (14)$$

where $\alpha(h\nu) = \alpha(\hbar\omega) = \dfrac{\ln \frac{I_o}{I}}{l}$;

I_o is the intensity of the incident radiation, l is the thickness of the absorption layer at the interface while I represents the intensity of the radiation corresponding to the l thickness.

In the relation (14), the absorption coefficient $\alpha_a(h\nu) = \alpha_a(\hbar\omega)$ corresponding to transition with phonons absorption is proportional to the probability $P(N_{phon})$ of interacting with the phonons, which is a function of the phonons number N_{phon} with the energy $\mathcal{E}_{phon}$ from the interface structural lattice: $P(N_{phon}) \sim N_{phon}$,

$$N_{phon} = \left(e^{\frac{\mathcal{E}_{phon}}{kT}} - 1\right)^{-1},$$

being the Planck distribution function [16]. Under the conditions the tests have been done, for $h\nu > \phi_B - \mathcal{E}_{phon}$, $\phi_B = \mathcal{E}_f - \mathcal{E}_i$, - see relations (11) - , the absorption coefficient $\alpha_a(h\nu)$ corresponding to the phonons absorption [16] has, for this case the form:

$$\alpha_a(h\nu) \approx \frac{(\hbar\omega - \phi_B + \mathcal{E}_{phon})^2}{e^{\frac{\mathcal{E}_{phon}}{kT}} - 1} \qquad (15)$$

The emission probability of the phonon is proportional to $(N_{phon} + 1)$, [16], so that under the condition in which $h\nu > \phi_B + \mathcal{E}_{phon}$, the absorption coefficient $\alpha_e(h\nu)$ for the transition with the phonon emission will have the form:

$$\alpha_e(h\nu) \approx \frac{(\hbar\omega - \phi_B - \mathcal{E}_{phon})^2}{1 - e^{-\frac{\mathcal{E}_{phon}}{kT}}} \qquad (16)$$

As our experiments (at room temperature $T \approx 300^oK$) have been done under the conditions for which $h\nu = \hbar\omega \gg \phi_B$ and, thus, for $h\nu = \hbar\omega > \phi_B + \varepsilon_{phon}$, we consider as being possible both the emission and those with the absorption of the phonons, the absorption coefficient being represented as the sum from the relation (14), $\alpha = \alpha_a + \alpha_e$.

Thus, the dependence $\sqrt{\alpha} = f(h\nu)$, (Fig.10), (the experiments being done at the roome temperature, $T \approx 300^oK$) is characterized by two linear portions corresponding to these two terms from the relation (14): α_e, the absorption coefficient for the phonon emission and α_a, the absorption coefficient for the phonon absorption, respectively. The straight line (a), (Fig.10), corresponding to the dependence $\sqrt{\alpha_e} = f_1(h\nu)$, intersects the abscissa in the point $h\nu = \phi_B + \varepsilon_{phon} \approx 3.02$ eV, while the straight line (b) characterizing the dependence $\sqrt{\alpha_a} = f_2(h\nu)$ intersects the energy axis in the point $h\nu = \phi_B - \varepsilon_{phon} \approx 0.95$ eV, (Fig.10).

Thus, the magnitude of the segment comprised between the intersection points of the straight lines (a) and (b) with the abscissa $h\nu$ (for $T \approx 300^oK$) is $2\varepsilon_{phon} = 2.07$ eV, and the middle point of this values agree with our experimental data at the plotting of both the voltamperic and of the photoelectric characteristics, respectively ($h\nu = \varepsilon_f - \varepsilon_i = \phi_B \approx 1.03$ eV).

The experimental results showed the photovoltaic properties of this kind of mixed photoelement are highly influenced by the duration of the alternative exposition to light and dark, the illumination period playing a special part in the processes occurring at the Al/Chl and Chl/Cu interfaces (Fig.11)

On taking into account the interface lattice structure of the SnO_x/Cd : Na - Cu_xS : Cu - Al - Chl/Cu multijunction, characterized by the existence of some different types of ions (Cu,Mg,Na) we think that the scattering mechanism on phonons during the photonic absorption, is also determined by the existence of the dipole moments. Thus,we shall consider the constituient compounds of the multijunction as a totality of electric dipoles which absorb the photonic energy and we shall assume that the energetic transfer between the photosynthetic pigment molecules is done by the dipolar transition resonance phenomena; the directioning of the energy flow we have experimentally determined, is done from small to high wave lengths in the spectrum - a shift to red, (Fig.8), These reasons being taken into account, we could explain the fact that the strongest photonic interaction experimentally determined, is produced in the region of higher lengths where the photonic radiation frequency is comparable with the dipole oscillations one the process being accomplished with the emission of an optical phonon flow (Figs.8.10).

Consequently, we think that the photonic absorption on the dipolar vibrations of the interface lattice, will determine a decrease in the photovoltaic efficiency, the emission process of the optical photon flow significantly influencing ((under the conditions in which the experiments have been done at $T \approx 300^oK$) the value of the relation time τ_f of the photoanswer (Fig.11).

At the same time, the experimental tests accomplished on the Al-Chl-Cu heterostructure, led to the conclusion that in the study of the photoelectric processes from the Al/Chl and Chl/Cu interfaces, the tunnelling and recombining effects, the contribution of the interface spatial load as well as the influence of the images forces on the height of potential barrier at the metal/semiconductor contact must be taken into account. Thus, when determining the Schottky barrier height at the Al/Chl interface in the presence of an externally applied electric field, we consider that the correction introduced by the images forces (Schottky effect) should be taken into account.

On considering the presence of a Schottky-type barrier at the metal/chlorophyll interface, we have started to study the influence of external electric fields both on the energetic transfer and electron transfer mechanism through the photosynthetic membrane of the pigment molecules, as well as on the electric loading states from the Al/Chl and Chl/Cu interfaces.

Therefore, under the experiments conditions, we could find out a slight increase of the photocurrents values for the samples subjected to external electric fields with intensity values ranging between 10^4 V/cm and 10^5 V/cm; J_{SC}=15-15.5nA/cm^2. This increase is explained both by the quantum tunnelling effect of the electrons through the Al/Chl barrier from Al through the Chl film (the electrons being scattered in the thin Cu layer at the Chl/Cu interface), and by the effect of the images forces at the Al/Chl interface, an effect which can determine a decrease in the Schottky barrier height. The inverse current presented during the tests a slight variation with the applied tension.

On resorting to analogical modelling, and on applying the Crowell and Sze theory [14], the tunnelling phenomena of the carriers (electrons) through the Al/Chl barrier, the recombining phenomenon of the carriers at the Al/Chl interface (due to the surface states) and the effect of the images forces at the metal/semiconductor interface [9,14], have been considered for the determination of the barrier height. Thus, on taking into account the previously discussed facts, we have computed the decreases $\Delta\emptyset$ of the Schottky barrier height at the AL/Chl interface by using the relation:

$$\Delta\emptyset = \left(\frac{e^3 E}{4\pi\varepsilon_0 \varepsilon_s} \right)^{\frac{1}{2}} \quad , [9] . \qquad (17)$$

We have found out a slight decrease of the barrier height, the mean values for $\emptyset$ (under the tests conditions) ranging between $\Delta\emptyset$ = (0.012-0.025)eV (Fig.12).

In spite of the fact that the diminishing of the barrier height is small, we consider that this effect has an important influence on the passage of the current at the Al/Chl interface; thus, the behaviour of the photocurrent when an external electric current is applied, is experimentally explained.

We also consider that in the presence of the applied electric current, the negative loading states at the Al/Chl interface under the influence of the images interface forces, would attract the majority carriers (holes) from the semiconductor, thus reducing the spatial load from the barrier layer and contributing to the decrease of the Schottky barrier at the Al/Chl interface (Fig.12).

We also suppose that the presence of the electric field determines the increase of the tunnelling probability of the electrons by interface states, the wave functions of the metal electrons (AL) in the interface region deeply penetrating the semiconductor and inducing negative loading states in the Al/Chl region. Taking into account that the photosynthetic pigments of the "chlorophyll reaction center" type are organized in two systems [6], P_{690} system(λ=690nm) and P_{700}(λ=700nm), we consider that these are permanently connected under the influence of photons, by an electron carrying chain. Thus, as the result of high hν energy absorption, the P_{700} system emits electrons on an intermediate level, n_3 of unknown electro-negative acceptor (Fig.13).

The P_{700} system regains its lost electron by means of an intermediate compound suggested to be on the form: enzyme + vitamineC energetic level n'), the electron excitation levels of the pigment molecules depending on the wavelenght of the photon that enters the excitation resonance with the pigment molecule "reaction center", similary with the optical pumping phenomenon produced in the LASER effect (Fig.13).

At the same time, during the energy transfer from the chlorophyll molecule P_{690} to the molecule P_{700}, a photon excitation occurs in the electrons from the magnesium atom (Mg^{++} included in pigment structure), this excitation determining two of the electrons of the atom to populate the higher energy levels, thus leaving the chlorophyll molecule, so that the free electrons with a high energy level can be transported (by means of enzymes and vitamins) through the photosynthetic membrane of the pigment molecules. On taking all these facts into account, we assign to chlorophyll molecules a "p"-type semiconduction behaviour.

We also consider that the ionization of the pigment molecule and the water photolysis, $2H_2O \rightleftarrows O_2 + 4H^+ + 4e^-$, probabily determine the maintaining of the energetic transfer of the electronic flow during the excitation period.

We also consider that both the electric field E and the ascorbic acid addition stimulate the protein and enzyme synthesis, resulting in the simplification of the electron transport through the photosynthetic membrane of the chlorophyll molecules, which renders evident a close correlation between the electric field, ascorbic acid, proteic metabolism and photoconduction mechanism in the photosynthetic pigment molecules.

On taking into account the influence of the ascorbic acid on the electron transport through the photosynthetic membrane, we suppose that the Q and X intermediate factors from the P_{690} and P_{700} systems have the form: enzyme + ascorbic acid (Fig.13).

The experimental results obtained so far, have made us come to the conclusion that both the impurities from the I_a and I_b groups (Na,Li,Cu,K), and the photosynthetic pigments can improve the photoeffects at the $Cd-Cu_xS$ junction, by enhancing its photovoltages and photocurrents (U_{CD}=350-450 mV; J_{SC}= 10-10.5 mA/cm^2), [15].

We think that, under the influence of temperature (100-150°C), the Na_2SO_4 and $CuSO_4$ thin films formed during the sputtering process give up the oxygen in the structural lattice of the heterohunction: oxygen diffuses and gets fixed at the CdS/Cu_xS interface, thus determining the generation of a rectifying junction. We also consider that during this process a special part is played by the two physico-chemical parameters, temperature and concentration of the doping solutions, the optimum values that we have found for concentrations being 0.75% for Na_2SO_4 and 1.2% for $CuSO_4$ solutions, respectively, 15.

We consider that during the formation of Cu_xS, a very important role is played by the diffusion, materialized in the ionic exchange that takes place between CdS and $Cu(CH_3-COO)_2$, until the equilibrium is established: $2Cu^{2+} \rightleftarrows Cu^{2+} + Cu^0$:

$$Cu(CH_3-COO)_2 + \overset{NH_2}{\underset{NH_2}{C=O}} + 2H_2O \rightleftarrows CuO + CO_2 + 2NH_3 + CH_3COOH, [5]$$

We consider that in the photovoltaic behaviour of this kind of mixed multijunction, the microcrystalline chlorophyll thin films bring a special contribution. This pigment traps photons and passes them over to different energy levels, by molecular induction or by inductive resonance [1,3,6], the excitation levels of π electrons of the chlorophyll molecules depending on the wavelength of the photon entered the excitation resonance with the pigment molecule. Taking into account the obtained experimental results, we also consider that the electronic transfer between two neighbouring molecules of the photosynthetic pigment is conditioned by the presence both of the electric fields and of the concentration gradients of the dopants solutions; this transfer can be modelled by the mechanism of the quanto-chemical tunnelling in an unielectronic approximation.

We also remark that, for this type of mixed multijunction, the SnO_x negative electrode being semitransparent, it is possible to illuminate the photoelement on its both faces (bilateral illumination), a fact which increases the charge carriers collection gain in the Cu_xS film of the heterostructure.

In Fig.14 we give the volt-ampere characteristic for such a mixed photoelement; here we can see improvement of the maximum values of the open circuit voltage U_{CD} and short circuit current J_{SC}.

We have assigned the slight sudden increase of the absolute value of the photocurrent, on the volt-ampere characteristic, to the influence of the R' series resistance of the photoelement (R' being determined by both the semiconductor resistance and the contact resistance) that produces a sudden decrease of the current through the diode for small voltages: $V < I R'$, according to the equation of the volt-ampere characteristic, where the series R' resistance of the photoelement is also considered:

$$J = J_S \left(e^{\frac{e(V-IR')}{\alpha kT}} - 1 \right) - J_L \qquad (18)$$

where J_S is the saturation current, J_L is the photocurrent, α is a constant of the semiconductor material, ranging between 1 and 2.

By extrapolating the photocurrent corresponding to the point from where the curve descending takes place (Fig.14), the photocurrent for the heterojunction without series resistance is obtained. On considering from the curve the value of the stress and the current in its descending point, we get the approximate value of the series resistence R' - (Fig.14); in our case R' ranges between (0.75-1.52)Ω for the considered samples.

III. FINAL CONCLUSIONS

Our experimental results have made us come to the conclusion that both the photosynthetic pigments and the impurities of the I_a and I_b groups can lead to the improvement of the photoeffects at the CdS-Cu_xS interface, by increasing their photovoltages and photocurrents: U_{CD}=(350-575)mV and J_{SC}=(10.5-15)mA/cm^2.

We think that the photovoltaic characteristics of this mixed multijunction are mainly determined by microcrystalline chlorophyll thin films. This pigment traps photons and passes them over to different energy levels, by molecular induction or by inductive resonance phenomena, the electrons excitation energy levels of the chlorophyll molecule depending on the wavelength of the atom entered the excitation resonance with the pigment molecule. By using the analogical modellation, we have considered this mechanism as being similar to the optical pumping met at the LASER effect.

On taking into account the Schottky barrier at the metal/chlorophyll interface, we have revealed the probability of electronic transfer between two neighbouring pigment molecules by quanto-chemical tunnelling mechanism in the unielectronical approximation.

Thus, on resorting to the analogical modelling and on using the experimental results obtained at the plotting of the photovoltaic and photoelectric characteristics, we have obtained for the Schottky barrier height mean values ranging in the interval (0.95-1.15)eV; this result explains, to a great extent, from an experimental view point, the specific behaviour of the photocurrent and photopotential during our tests. The experimental results obtained at the plotting of the photoelectric characteristics both at the Al-Chl-Cu heterostructure and at the mixed multijunction we have accomplished, have also rendered evident both the presence of a Schottky-type barrier at the Al/Chl interface and the modification of the photocurrent and photopotential behaviour under the influence of the external electric fields, as a result of the decrease of the Schottky barrier height; $\Delta\phi$ =(0.012-0.025)eV in their presence. Under these conditions, we have noticed the fact that, instead of the saturation phenomenon of the inverse current, a slight variation of the current with the applied tension has been produced, the mean values at the Al-Chl-Cu heterostructure ranging between 15 - 15.5 nA/cm^2. We have also remarked that both mean values of the electric field E=7.5 V/cm and the mean concentration of the ascorbic acid of 1.5% stimulate the protein synthesis within the proteic metabolism of the vegetal material, resulting in the intensification of the electronic transfer in the photosynthetic pigment molecule, thus rendering evident a close correlation between the electric field, the ascorbic acid concentration the proteic metabolism and the photoconduction mechanism in the photosynthetic pigment molecule.

The experiments showed, as well, that the interdependence between the optimum values of the physico-chemical parameters, namely, temperature, concentration of the doping solutions and the electric field E, plays a determining part in the physical processes at the CdS/Cu, Al/Chl and Chl/Cu interfaces, considering also the probability of the electronic transfer between the neighbouring pigment molecules by means of the quanto-chemical tunnelling mechanism. We consider that explanation of the mechanism of the energy transfer and electron transport through the photosynthetic membrane, as well as the study of the photoconductive properties of different forms of microcrystalline chlorophyll present a special fundamental and practical interest for the utilization of the photosynthetic pigment films in photovoltaic devices based on anorganic semiconductor materials.

The obtained experimental results have also emphasized the fact that the strongest photonic interaction takes place during the optical absorption process in the region of the highest wave lengths, $h\nu > 0.90$ eV, where probably the photonic radiation frequency is comparable with the dipole oscillations one from the structural lattice; the process is solved by the emission of an optical phonon flow $2\mathcal{E}_{phon} \approx 2.07$eV, which influences, up to a certain degree, the photovoltaic efficiency of this type of multilayer heterojuction.

REFERENCES

1. Tang C.W., Albrecht A.C., Mol.Liq.Crys., 25, 53 (1974)
2. Tang C.W., Albrecht A.C., J.Chem.Phys.., 62, 2139 (1974)
3. Tang C.W. Albrecht A.C., Chem.Phys., 63, 953 (1975)
4. Besson J., Plaques photovoltaiques au sulfure de Cadmium, l'onde electrique, 1 (1975)
5. Zet.Gh., Gherghel Magda, Bul.I.P.Iaşi, tom.XXIX, fasc.1-4, (1983)
6. Zet Gh., Gherghel Magda, Bul.I.P.Iaşi, tom.XXXX, fasc.1-4), (1984)
7. Barghale B.B., Pawar S.H., Indian J.Pure Appl.Phys., 18, 1, (1980)
8. Gherghel Magda, Zet Gh. Ursu D., Vol."Progrese în fizică", 541, Institutul Central de Fizică, Bucureşti, 1988
9. Schottky W., Naturwiss, 26, 843, 1938
10. Tănăsescu F.T. şi colab."Conversia energiei -tehnici neconvenţionale" Editura Tehnică Bucureşti, 1986
11. Melinte Sofia şi colab."On the current voltage characteristics J=f(V) of CdS/Cu_xS solar cells" Lucrările simpozionului naţional de utilizare a energiei solare, Iaşi, 1980
12. Spînulescu I., Pîrvan Radu, "Principiile fizice ale microelectronicii", Editura Stiinţifică şi Enciclopedică, Bucureşti, 1981
13. Spînulescu I., "Fizica tranzistorilor şi principiile microminiaturizării" Editura Didactică şi Pedagogică, Buc., 1973
14. Crowell C.R. et alt.., Phys.Rev., 127, 2006 (1962)
15. Gherghel Magda, Melinte Sofia, "Renewable Energy Sources," vol.1 Solar Energy and Applications, National Research Center Cairo-Egypt and Wayne State University U.S.A., 1990, (179-196)
16. Dima I., Licea I., "Fenomene fotoelectrice în semiconductori şi aplicaţii", Editura Academiei Române, Bucureşti, 1980
17. Bearden A.J., Maklin R.Q., Rev.Biophys., 7 (1976), 131.

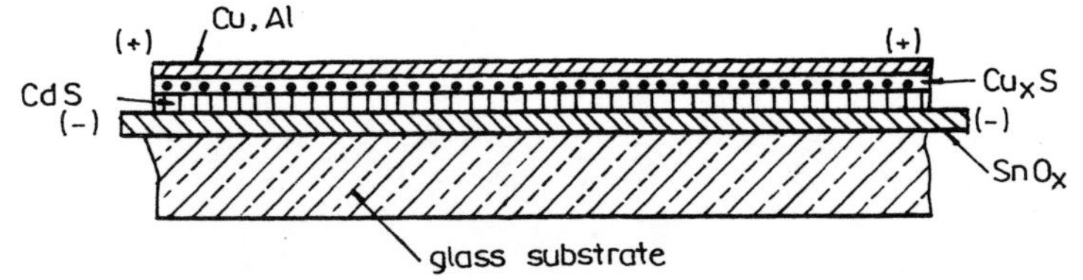

Fig1. The CdS – Cu_xS structure

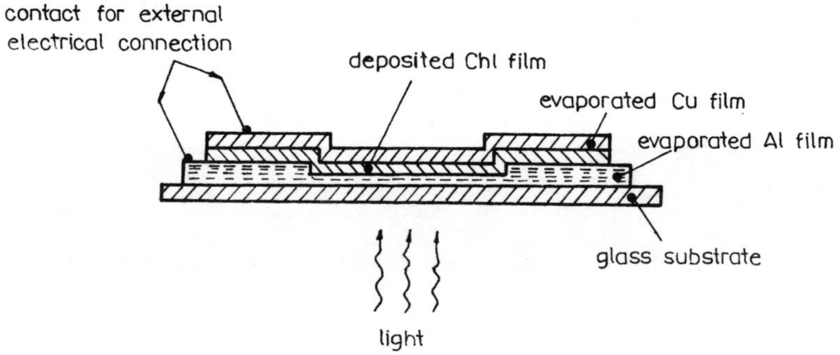

Fig.2 The Al-Chl-Cu structure

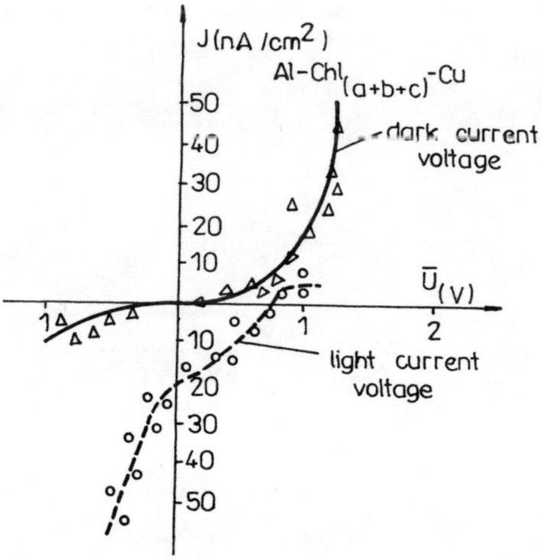

Fig.3 – The J-V characteristics of Al-Chl-Cu

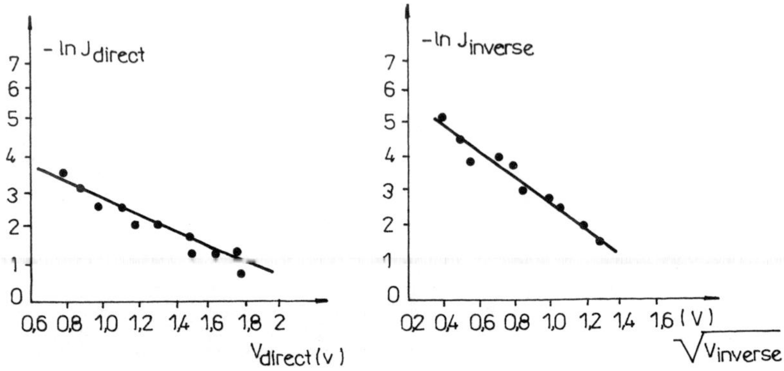

Fig. 4 - Semilogarithmic plot of the dark current voltage characteristics

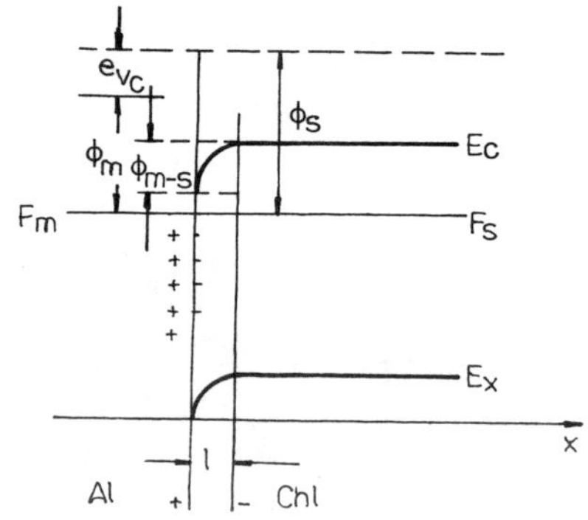

Fig.5 - Energy scheme of Al/Chl interface

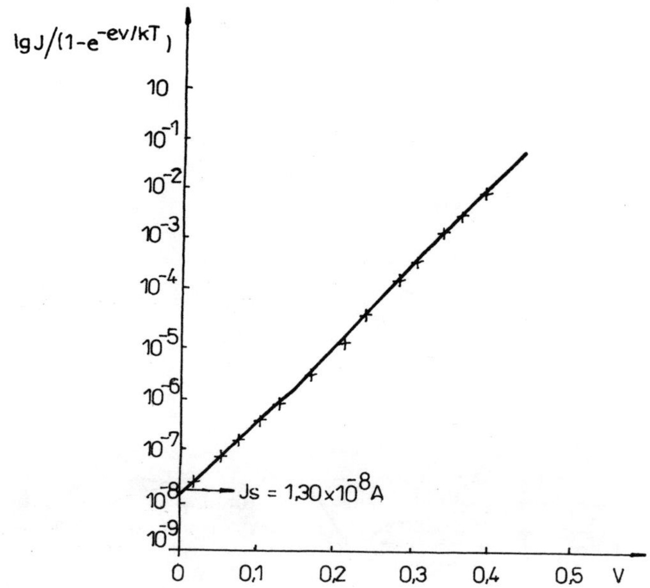

Fig.6 –Semilogarithmic plot of the direct current voltage characteristics

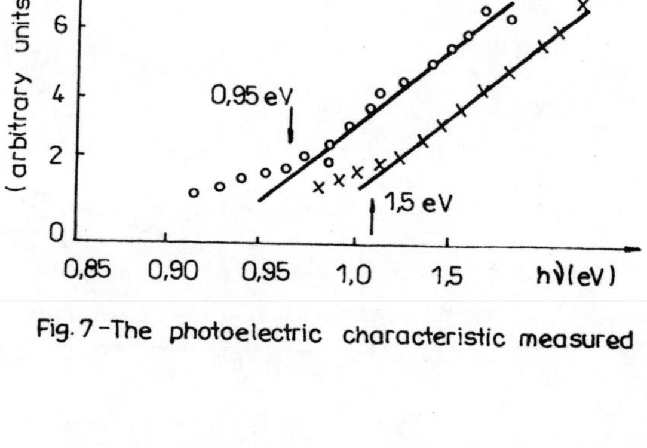

Fig. 7 –The photoelectric characteristic measured

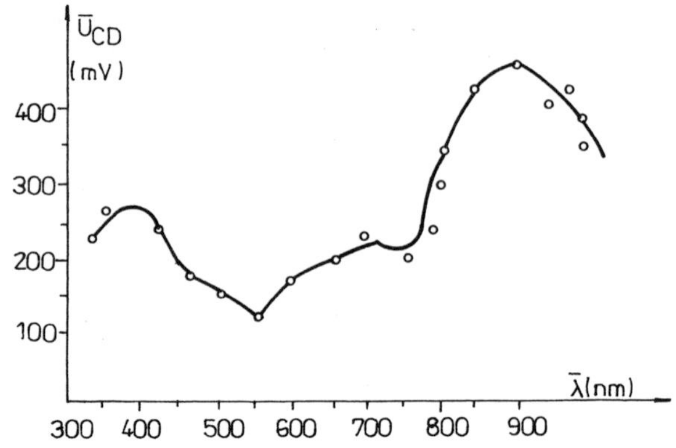

Fig.8 - The $U_{CD} = f(\lambda)$ characteristics

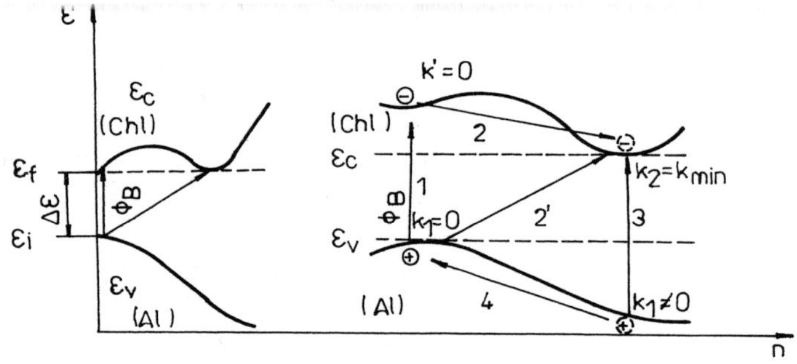

Fig. 9 - Schematic energy band diagram at optical absorption of interface Al /Chl

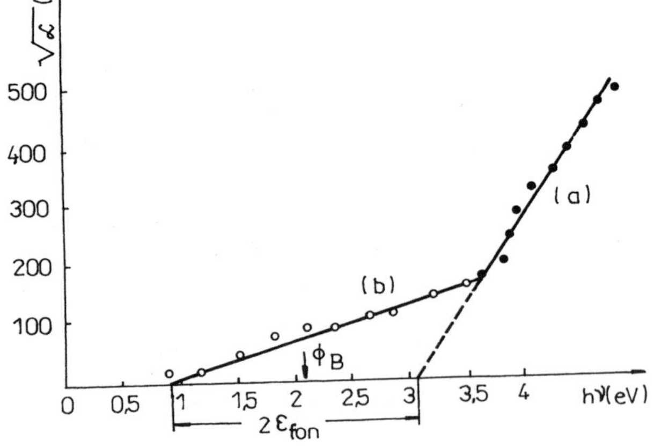

Fig.10 - The $\alpha = f(h\nu)$ characteristics

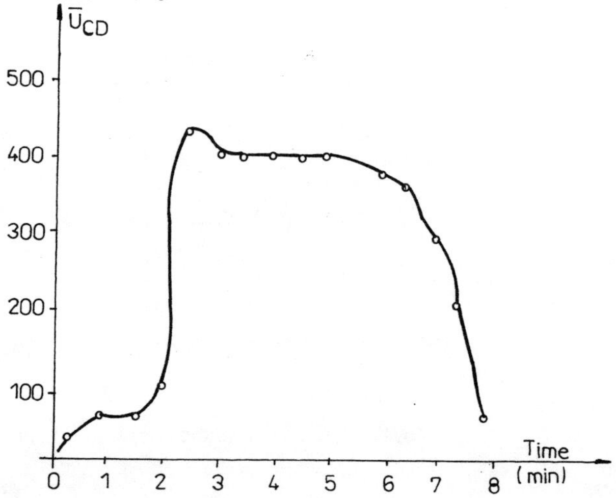

Fig.11 - Variation of photovoltage at room temperature, with illumination time

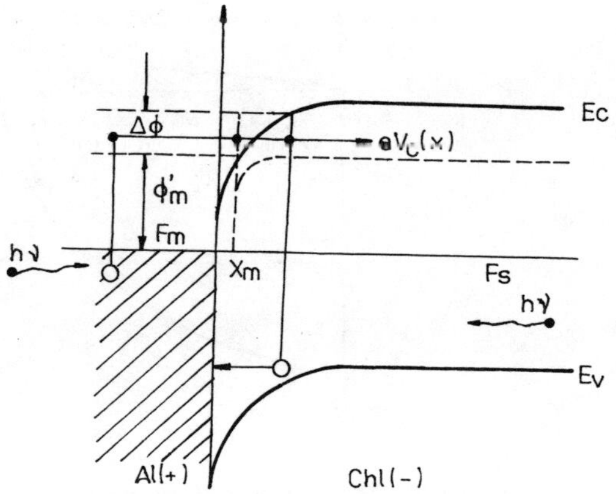

Fig.12 - The barrier layer at Al/Chl interface

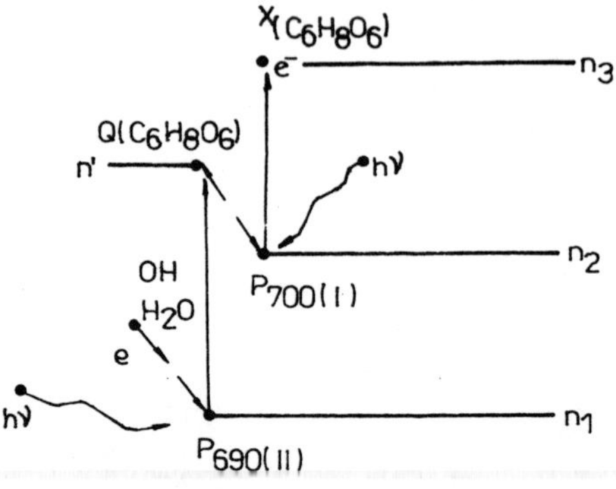

Fig. 13–Scheme of photosynthetic electron transfer

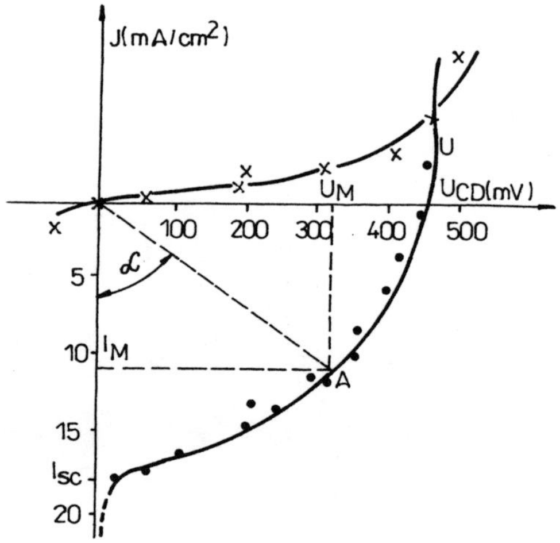

Fig.14–The J-V characteristics for SnOx/Cds:Na–
–CuxS:Cu-Al-Chl/Cu cell

SOLAR FUELS AND CHEMICALS RESEARCH AT THE WEIZMANN INSTITUTE OF SCIENCE

I. Hodara
Solar Research Facilities
The Weizmann Institute of Science
Rehovot, 76100, ISRAEL

ABSTRACT

The Solar Research Facilities Unit of the Weizmann Institute of Science operates two state of the art research tools for concentrating solar radiation. A solar furnace of 20 kw rated power reaching concentrations as high as 10000 suns, and a central receiving tower of 3 mw(th) rated power.

The Solar Fuels and Chemicals Project of the Institute includes the development of a solar chemical heat pipe based on CO_2 reforming of methane, oil-shale gasification and direct solar splitting of water. A research program on biomass gasification has also been recently initiated.

The Solar Research Facilities are described and the latest results obtained in the Solar Fuels and Chemicals Project are presented.

INTRODUCTION

Solar thermal energy and photovoltaics as currently developed are useful mainly for generating electricity when the sun is shining. The markets for solar energy could be greatly expanded through the development of solar thermochemical fuel production. These products could include hydrogen, syngas, methanol and other transportation fuels. Solar fuel production compensates for the intermittent nature of the sun and permits the energy to be used when and where needed.

A Research and Development Program is being carried out at the Weizmann Institute of Science under the sponsorship of the Ministry of Energy and Infrastructure on utilizing concentrated solar energy for power generation, energy storage and fuel production in sunbelt countries.

The research has emphasized the collection of solar heat at higher temperatures than usually used for power production. These high temperatures create the possibility of solar thermochemical fuels production. One of the major development areas has been methane reforming, using solar heat to produce syngas as a means of storing and transporting solar energy in chemical form. The syngas can be used as a fuel or chemical feedstock, or it can be converted to hydrogen or liquid fuels. New areas of research at the Institute include direct thermal splitting of water to produce hydrogen and gasification of oil-shale and of biomass.

THE SOLAR RESEARCH FACILITIES UNIT

The research facilities include two solar radiation concentration devices complementary to each other. A solar furnace of 20 kw rated power reaching concentrations around 10000 suns and a solar tower rated at 3000 kw thermal where the concentration that can be obtained is only a few thousand suns. In this way a wide range of power and concentrations is made available to the research teams.

The Solar Furnace [1]

The main components of the solar furnace are a heliostat, a concentrator, a shutter and an attenuator as shown schematically in Fig. 1.

The heliostat (Arco) is constructed of 64 flat mirror facets creating 100 sq.m. of reflecting area. The position of the sun is computed every second and a command is released from the control computer to two D.C. motors to move accordingly.

The concentrator is a quasi-paraboloid mirror made of 590 spherical mirrors mounted on a spherical structure of 4.56 m. radius. The mirrors are made of 4 mm. thick float glass whose back surface is silver coated. The concentrator diameter is 7.3 m. providing an optimal rim angle of 65 degrees. There are 13 concentric rings, each ring holding a number of mirrors. By fine adjustment of the mirror holders, each mirror is made to reflect the incoming beam exactly towards the target.

A shutter made of a polished aluminium plate 80x80 cm. in size and equipped with extended surfaces for air cooling, provides fast blocking or exposure of the target. The amount of radiation impingent upon the concentrator can be changed by using the building doors as an attenuator.

In this furnace, at 800 watts per sq.m. insolation, 20 kw of power at a concentration level of 10000 suns has been obtained.

The Central Receiver Facility [2]

A schematic diagram of the Central Receiver Facility is shown in Fig. 2. Its major components are a north heliostat field, a concrete tower and a building housing the supporting laboratories, offices and services. The heliostat field can provide up to 3000 kw of concentrated solar light with a few thousand times the sun's intensity. Sixty-four heliostats are arranged in a north field with an opening angle of 95 degrees. Each heliostat has a reflecting surface of 54.3 sq.m. and is spherically curved. Each individual heliostat is equipped with a local control unit that handles the two azimuth and elevation motors and transmits information from two optical encoders and limit switches. Each group of 32 heliostats communicate via fiber optic cables with the group controller which serves as interface between the field controller and the heliostat controller.

The control system is capable of focusing the solar radiation on five possible targets at five different levels independently. The allocation of specific heliostats to each of the active targets is commanded through the field controller. The field controller interfaces with the operational control system (OCS) computer. The OCS is the primary means for coordinating the experiments. Its function is to monitor and supervise the facility during various operating modes. The OCS computer serves as an interface through which five experiment computers can control five groups of heliostats independently. This enables each experimental station to introduce the allocated beams of the heliostats into the target in a controlled manner according to the experimental requirement.

The tower rises 53 m. above ground. Its cross-section is rectangular. Since the Central Receiver Facility serves a wide variety of experiments, maximum flexibility was sought in the design. Five experimental set-ups can be mounted simultaneously without interfering with each other during solar operation. Platforms can be located at the interior of the tower at 3.5 m. intervals. The platforms are placed on steel bridge elements that are easily dismountable and handled by a top crane.

Passive thermal protection for the tower from heliostat beams is provided on the north face by concrete panels that are attached loosely to the tower structure. A 10 cm. air gap is created between the panels and the tower constructive wall. The panels can be replaced easily if damaged.

SOLAR FUELS, CHEMICALS AND HYDROGEN PRODUCTION RESEARCH

The principal objective of this Program is the development of production processes that can benefit from the unique properties of the clean high temperature heat source supplied by concentrated solar radiation. In these processes, hydrogen and syngas are the primary products. Both materials can be used as energy storage and transport media and as feedstocks for the production of many valuable chemicals.

Storage and Transport of Solar Energy [3]

In order for solar energy to capture a large scale of the energy market, it will be necessary to find ways to store it and transport it from the production site to distant locations where it is needed. Israel is a typical sunbelt country whose geography and population distribution pattern has forced the concentration of its industries in the north. More than half of the country is covered by the Southern Desert (the Negev) where the sun shines for over 300 days a year. Solar energy collected and concentrated in the desert must be transported to the north for distances of about 300 km. before being delivered to the user. This can be achieved by using the concept of the closed loop chemical heat pipe.

The basic principle behind the solar chemical heat pipe is demonstrated in Fig. 3. A highly endothermic reversible reaction is carried out at the solar site. The products are cooled to ambient temperature for storage and transport. Heat is released at the consumer site by forcing the reverse exothermic reaction to occur. The original products are recovered and transferred back to the solar site for recycling. The process operates as a closed loop, and therefore it is environmentally clean.

The reaction chosen for the process is the CO_2 reforming of methane:

$$CH_4 + CO_2 \;=\; 2H_2 + 2CO \qquad \Delta H = 250 \text{ KJ/mole}$$

This reaction can also be used in an open mode where methane is reformed to produce synthesis gas using solar energy. Syngas opens the road to the synthesis of other fuels such as hydrogen or methanol and to the production of other chemicals by well known petrochemical processes.

To adapt this reaction to the intermittent nature of solar energy it was necessary to find a catalyst that functions under non-steady conditions [3,4]. After a thorough screening of a large number of catalysts in the laboratory, a Rh on alumina catalyst from Engelhard Corp. was chosen as the best candidate. Starting with the same initial gas mixture, as many as 1100 closed cycles were run in the laboratory, using electric heaters as a heat source [5].

Solar reforming experiments were carried out with a number of receiver/reactor configurations. The first major experiment was done with an integrated receiver/reactor that uses a molten sodium heat pipe to absorb solar energy and transfer it to the reactor tubes*. It performed well and absorbed as much as 8 kw solar power. This is the first time that a sodium reflux receiver/reactor was ever operated under solar conditions [6].

After having tested tubular reactors of various configurations, a new phase of research has begun with the study of direct absorption receivers where the catalyst is directly illuminated through a quartz window. The first direct absorption volumetric receiver as shown in Fig. 4 was built with a sapphire window [7]. Since then, a number of other reactors with quartz windows have been tested, where the Rh catalyst was supported on ceramic foams or ceramic honeycombs. So far, as much as 6.5 kw of solar power was absorbed and reaction rates 10 times faster than those measured with metal tube reactors were obtained.

A closed loop solar reforming/methanation unit was operated including a vertical inconel tube reactor as a solar reformer and a methanator consisting of a number of adiabatic reactors. The procedure followed was carrying out solar re-

---

* The reactor was built at SANDIA Laboratories in Albuquerque, NM, as part of a joint project with Drs. J.D. Fish and R.B. Diver from the U.S. Department of Energy and Prof. J.T. Richardson from the University of Houston, TX.

forming for a number of hours, and compressing the products into the storage cylinders. The stored synthesis gas was then used for the methanation. The products of methanation, mainly methane and carbon dioxide were compressed and stored in cylinders and then used for the next reforming cycle. In all cases the conversions were kept at a level of 80-90 %. The closed loop was operated, so far, with the same gas mixture for over 60 cycles, giving very satisfactory results and without any sign of changes in gas composition or carbon deposition. To the best of our knowledge, this is the first time that such a closed loop was ever operated under real conditions in a solar environment.

The Solar Reforming/Methanation Pilot Plant [8]

Our solar chemical heat pipe development project includes a technological feasibility demonstration task that follows the laboratory studies and the small scale solar experiments described above.

A 350 kw., 3-stage methanator has been built and is operating on site at our central receiver facility. Its purpose is to demonstrate and develop a methanation process for useful recovery of the chemical energy stored in the reformed synthesis gas and to produce methane for recycle into the reformer system.

The methanator is capable of handling low hydrogen syngas mixture (H_2/CO = 1/1) resulting from carbon dioxide reforming of methane, and high hydrogen syngas mixture (H_2/CO = 3/1) resulting from steam reforming of methane.

The heat produced by the methanation reaction is used to generate saturated and dried steam at 20 bars.

The first reformer to be tested in a closed loop configuration with this methanator will be installed on the roof of the solar tower in our central receiver facility. It will have a capacity of 450 kw. It consists of an insulated receiver cavity housing vertical inconel reactor tubes. This type of receiver/reformer is being currently designed.

Oil Shale Gasification [9]

The Israeli oil shale resource, containing enough fossil fuel for about one century of Israel's needs, is the only large fossil fuel resource known within the country. Proven reserves currently are 11 billion tons and the potential resource is likely to be much greater [10].

We are investigating the high temperature gasification of oil shales using solar energy, followed by catalytic conversion of the products into synthesis gas. This process can increase considerably the energy content of the shales.

Solar gasification experiments were carried out at the central receiver facility [9]. Reactor tubes were made either of inconel or transparent quartz. In order to introduce the required flux into the reactors, nonimaging compound parabolic concentrators (CPC) were used*. Typically, with a 3-D CPC irradiating an inconel tube reactor, 20 g of fluidized oil shales were pyrolized at 810°C with a solar power input of 0.8 kw. The degree of decomposition, determined by TGA, was 87% which seems to be quite high when compared to experiments conducted with conventional heating methods. The analysis showed that 80% of the gases were accounted for as CO, CO_2, H_2 and CH_4. The main component was CO. With a 2-D CPC and a quartz tube, an average temperature of 915°C was reached in the bed and the decompositon, measured by TGA was 99%.

Biomass Gasification [11]

The gasification of carbonaceous feedstocks is a process that requires substantial amounts of external energy input at temperatures readily achieved with solar concentrators. The gasification process is carried out in three steps. The first step, pyrolysis, involves heating the solid to a temperature above about 500°C, releasing a volatile fraction of hydrocarbons, H_2, CO, CO_2 and H_2O. The solid left behind (char) is rich in carbon and is gasified by reaction with CO_2 or steam, the second step. The third step involves reforming the gaseous hydrocarbons with CO_2 or steam. The end products are synthesis gas, ash, small amounts of sulfur and nitrogen compounds and usually some residual CO_2 and/or water.

According to data of the Israel Ministry of Energy, the biomass potential of Israel is about 820 000 tons of oil equivalent (TOE) per year. If that much biomass could be processed with solar-thermal energy, the syngas product would have a caloric content of 980 000 TOE. The value of this product would be about USD 180 million. The annual insolation in the Negev desert is about 2500 kw-h per sq.m. per year. Assuming 55% solar field efficiency and 80% receiver efficiency the net available energy equals 1100 kw-h per sq.m. per year. To produce 980 000 TOE requires 1.78 million tons of biomass and a solar energy input of 1679 kcal. per kg. of biomass, equivalent to 3.47×10^9 kw-h/year. If the costs of heliostats could be reduced to USD 75/sq.m., then the solar field direct cost would be USD 236 million, about 1.3 times the magnitude of the product value.

The technology developed could be utilized in California, the Mediterranean basin and wherever large urban industrial sites are located in proximity to a region of high insolation.

----

* The 3-D CPC was kindly lent to us by Prof. Roland Wilson from the University of Chicago, Chicago, IL.

The Weizmann Institute's proposed experimental program comprises three phases:

The first phase consists of exploratory experiments for testing various configurations of receiver/reactors at a scale of solar inputs of up to 15 kw and biomass throughputs of 1 to 2 kg/h. In the second phase, a variety of feed materials will be characterized for their reaction velocities, gas composition and volatiles fraction using a fluidized bed solar receiver. The feed materials will include common municipal, agricultural and industrial wastes, and some non-biomass (residual oil, oil shale, low grade oil) materials. A pilot plant of 50 to 150 kg/h capacity will be designed and built in the third phase. In the design stage of this phase, special attention will be given to ensure a clean off-gas of minimum volume.

THE DIRECT SOLAR WATER SPLITTING EXPERIMENT

The direct solar water splitting experiment for the production of hydrogen is part of the Solar Fuels and Chemicals Research Program of the Weizmann Institute and has been described in detail at a separate lecture in this Conference.

ACKNOWLEDGEMENTS

I should like to gratefully acknowledge the assistance I received from my colleagues, M. Epstein, I. Spiewak and M. Levy, in preparing this presentation. The program described has been funded in part by the Israel Ministry of Energy and Infrastucture whose support is appreciated.

REFERENCES

1. I. Levy and M. Epstein, The Solar Furnace at the Weizmann Institute, W.I.S. Report, October, 1986.

2. M. Epstein, Central Receiver Facility at the Weizmann Institute of Science. Proc. 3rd. International Workshop on Solar Thermal Central Receiver Systems, June 23-27, 1986, Konstanz, Germany. M. Becker, Editor, Vol. 1, pp. 187-197, Springer-Verlag, 1986.

3. D. Fraenkel, R. Levitan and M. Levy, A Solar Thermochemical Pipe based on the CO_2-CH_4(1:1) System. Int. J. Hydrogen Energy, 11, 267 (1986).

4. R. Levitan, H. Rosin and M. Levy, Chemical Reactions in a Solar Furnace. Direct Heating of the Reactor in a Tubular Receiver, Solar Energy, 42 267 (1989).

5. M. Levy, R. Levitan, H. Rosin, G. Adusei and R. Rubin, Storage and Transport of Solar Energy by Thermochemical Pipe, Proc. of the 4th International Symposium on Solar Thermal Energy, Santa Fe, NM. B.P. Gupta and W.H. Traugott, Editors. p. 527-536. Hemisphere Publishing Corporation, New York, 1990.

6. R.B. Diver, J.D. Fish, R. Levitan, M. Levy, H. Rosin and J.T. Richardson, Solar Test of an Integrated Sodium Reflux Heat-Pipe Receiver/Reactor for Thermochemical Energy Transport. Solar Energy, in press.

7. M. Levy, H. Rosin and R. Levitan, Chemical Reactions in a Solar Furnace by Direct Irradiation of Catalyst. J. Solar Energy Engineering, 111, 96 (1989).

8. M. Epstein and I. Spiewak, A 350 Kw. Methanator for Solar Chemical Heat-Pipe. Workshop on Methane Reforming, DLR, Koeln, Germany, June 11-13, 1991.

9. G. Ingel, M. Levy and J. Gordon, Gasification of Oil Shales by Solar Energy. 5th Symposium on Solar High Temperature Technologies, August 27 -31, 1990, Davos, Switzerland.

10. J. Yerushalmi, Development of Synthetic Fuels and Israel's Plan Toward Commercial Utilization of its Oil Shale Resources. 13th Congress of World Energy Conference, Cannes, France, October. 1986.

11. I. Spiewak, A Proposal for Development of a Process for Solar Thermochemical Fuel Production. Weizmann Institute of Science, July, 1991.

Fig. 1. Schematic Diagram of the Furnace

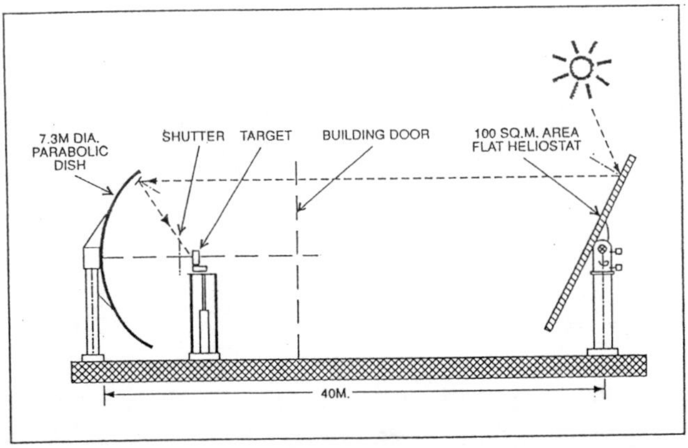

Fig. 2. General Schematic of the Central Receiver Facility

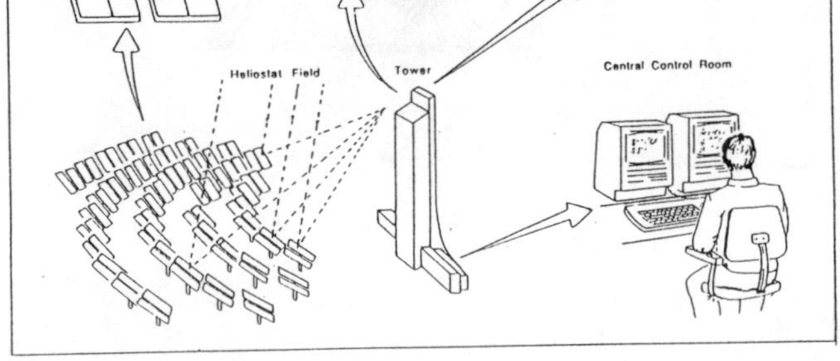

Fig. 3. A Solar Chemical Heat Pipe Based on Methane Reforming and Methanation Reaction

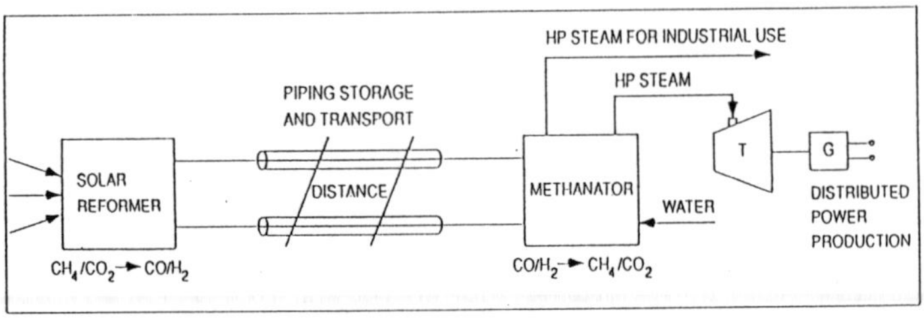

Fig. 4. Direct Absorption Receiver for Small Scale Solar Reforming Experiments

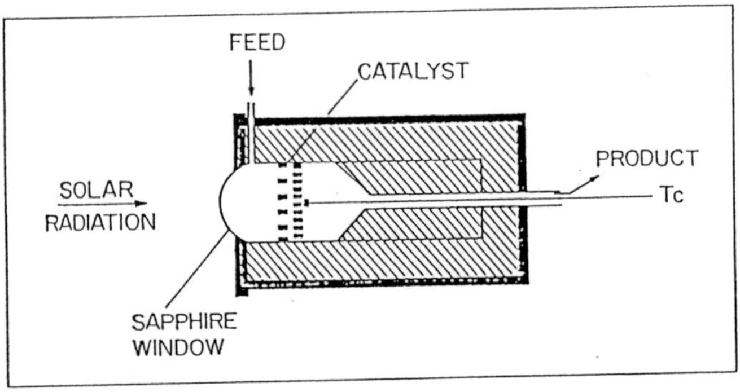

RENEWABLE ELECTRICITY AND HYDROGEN USING SOLAR THERMAL PROCESSES

Roy E. McAlister
President, American Hydrogen Association
219 South Siesta Lane, H-101
Tempe, Arizona 85281

ABSTRACT:

Comparison of several energy conversion technologies are made. Performances of solar thermal technologies including point-focus devices with individual Stirling engine-generators, linear-focus collectors with central turbine expanders, and large-scale heliostat concentrators with central turbine expanders are evaluated. Fuel combustion to elevate the temperature of expansion fluids is reviewed and evaluated. Generation of hydrogen for use as an energy source during low-insolation periods is proposed. Storage of hydrogen in underground reservoirs is recommended. Utilization of hydrogen as a thermochemical energy storage gas is considered.

INTRODUCTION:

In comparisons with other energy conversion options numerous environmental advantages are offered by Solar Thermal Processes (STPs). STPs have no emissions and there is no danger of nuclear accidents. Land requirements are less than photovoltaic, biomass, or hydro-electric systems.

Conversion of solar energy to electricity introduces the question of energy storage. How do you use the sun at night? The answer is hydrogen. For long term storage hydrogen can replenish depleted natural gas and oil reservoirs. By using abundant solar energy to create and store renewable hydrogen we can turn the resources clock back to 1900 when America's natural gas and oil reservoirs were full. By establishing energy abundance with renewable hydrogen, our mental attitude can shift from short-term profits that compromise the environment to investments in long-term prosperity.

STPs can be located in solar rich deserts to send renewable hydrogen and electricity to distant markets. This paper compares selected types of STPs for producing renewable hydrogen and electricity. Renewable electricity can be used on the established grid and to produce hydrogen by electrolysis. At the point of application hydrogen can be used in countless industrial processes, in existing and exciting new engines, or in fuel cells to create prosperity without pollution.

Linear-Focus Concentrators and Heat Engines

Linear-focus concentrators and steam turbine generators provide over 92 percent of the world's electricity that is produced by solar energy. It is produced by a single company. Luz International Ltd. contracts with Southern California Edison to provide electricity generated from solar fired Rankine cycle steam turbines.

Further expansion is underway and when completed in 1994, Luz will have approximately 680 megawatts of capacity for converting solar energy to electricity. Luz's systems are called Solar Electric Generating Stations (SEGS) and are located at several sites in the Mojave Desert of California. Luz supplies enough power to meet the residential electricity needs of approximately 800,000 people. Financing of the SEGS is by limited partnerships with private investors.

Figure 1. shows a process diagram of the generalized SEGS system. Solar collectors heat oil to about 735° F (390°C) which is circulated through heat exchangers to generate superheated steam. The steam is used in a conventional turbine/generator to produce electricity for the utility grid. Design changes have been made in each of the SEGS to increase the power plant efficiency and to improve annual electrical output at utility on-peak hours when electricity rates are the highest. To achieve better return on plant investment natural gas is used for heating the steam to achieve higher thermal efficiencies and to supply supplemental heat at on-peak hours for sales to Southern California Edison.

Rankine cycle thermal efficiency is limited to a fraction of the theoretical Carnot cycle efficiency which is:

$$E = \frac{T_H - T_L}{T_H} \qquad \text{(Equation 1)}$$

Increasing the steam temperature with natural gas heating provides 10 to 20 percent improvement in operating efficiency. In the alternative, SEGS can be operated at night on natural gas. The energy supplied by fossil fuel is limited to 25 percent of the total effective annual solar energy input by the U.S. Federal Energy Regulatory Commission. This limit allows LUZ to qualify for exemption from expensive paperwork requirements. A further benefit is that Federally regulated utilities are required by law to purchase electricity from qualified small suppliers.

SEGS subsystems are arranged in as a solar heated oil loop, a steam generation heat exchanger, an auxiliary natural gas boiler, a steam turbine generator, and a cooling tower. The solar heated oil loop uses parabolic linear trough solar concentrators that track the sun and heat synthetic oil that is circulated through vacuum insulated stainless steel target tubes within glass envelope tubes.

Current SEGS plants are designed to produce full electrical power with steam supplied by solar energy or by natural gas in the gas-fired boiler or by a combination of the two heat sources. In SEGS VII, the boiler has been replaced by a gas-fired Heat Transfer Fluid (HTF) heater which can independently heat the full flow rate of HTF to design temperature. The design goal is to provide maximum power at the lowest fuel cost during the period when utility rates are highest.

From 12 p.m. to 6 p.m. during weekdays of June through September is the on-peak period which coincides well with the ability to operate at full power on free solar energy. Natural gas is used to supplement solar on cloudy days, during selected evening hours, and in instances of solar collector repair and maintenance.

Each Solar Collector Assembly (SCA) includes a parabolic trough solar radiation reflector, a positioning system, and a local control system. The parabolic trough solar collector is a metallic mirror plating on glass which focuses direct radiation to an evacuated stainless steel receiver called the heat collection element (HCE).

SEGS VI contains 800 SCAs, each with a mirror aperture area of 235 square meters. There are 50 loops in the solar field. Each loop has 16 SCAs. Circulation rates of HTF through the SCAs are controlled to produce 304°C to 390°C oil by the available solar energy. The HTF flow is circulated by variable-speed centrifugal pumps in series. Two pumps provide full HTF flow, with a third pump in series for redundancy. Maximum HTF flow is 1.73×10^6 liters per hour.

Superheated steam at 100 bar at nearly the temperature of the HTF is produced for expansion through the power turbine. A portion of the HTF is also diverted to another heat exchanger for reheating of the turbine steam. The power block consists of a reheat steam turbine cycle utilizing a conventional turbine/generator unit and steam-water cycle equipment. Exhaust pressure is approximately 0.07 bar. Condenser cooling water circulates to a mechanical draft cooling tower located near the turbine. Maximum cooling water circulation is about 5 million liters/hour.

Water is taken from local wells or, in the case of the Kramer Junction site, pumped 14.5 km from the California aqueduct system. Incoming water is demineralized and the blowdown from the cooling tower sump is sent to evaporation ponds. Water additions amount to about 450 acre-feet per year for a 30 Mw plant.

Natural gas heat additions are used assure that the SEGS

plants meet their summer capacity targets during periods of cloudy weather. Earlier the natural gas-fired boiler was also used to supply the night time production required to meet the winter time on-peak demand period. Winter night time on-peak rate periods were eliminated and natural gas boiler supplements were devoted to day time peak rate periods.

Natural gas is used first to add heat in the summer on-peak period during cloudy days, or later in the summer when solar operation drops off before the end of the on-peak period. The second highest rate period is the summer mid-peak period. Solar operation supplies the morning mid-peak period, but only a small amount of the afternoon mid-peak production can be supplied by solar operation. Gas heat is used as a supplement during this portion of the day. Any remaining allowance of gas is used to supplement solar operation during the winter mid-peak period.

Malfunction of any single SCA or flow loop reduces the solar field output by the mirror area that has been put out of service. Efficiency of the power plant drops off at part load and solar collection is reduced substantially in the winter. Hybrid operation in the winter increases steam temperature and provides up to 20% improvement in operating efficiency. It is a priority to use gas within the FERC 25% limit in the hybrid mode during mid-peak periods when the most benefit can be gained from operating the plant at full load.

Luz parabolic trough solar collectors utilize a single north-south axis orientation to track the sun. The more common east-west axis orientation gives a slightly higher annual collection capacity and more even annual distribution of electric production. However, the north-south axis biases summer operation when more electricity can be sold at the on-peak rate. The north-south orientation provides for more revenue and more free energy in the generation of electricity. Daylight savings time also helps shift solar production to meet summer on-peak hours.

North-south orientation of the trough collectors cause noon time dip in solar production during winter months. This dip is due to the path of the sun in the winter. At noon the sun is sitting low in the southern sky but the parabolic troughs are facing straight up. This provides the smallest apparent collection area. In the morning and afternoon the incident angle between the collector and the sun is smallest and the highest solar flux is absorbed during these times.

Major problems encountered during SEGS operation since 1984 are Heat Collection Element (HCE) tube breakage and washing of the parabolic reflectors. HCE tube breakage most often occurs because of thermal expansion differences between the outside glass tube and the stainless target tube. Hydraulic shock caused by rapid changes in heat transfer fluid flowrates during solar field morning startup, and nonuniform thermal growth of the HCE receiver during this same startup period causes warpage, breakage, and vacuum leaks. Design revisions have reduced this problem to an acceptable level.

Cumulative mirror surface area contained in SEGS plants amounts to thousands of acres. Cost-effective mirror cleaning is a top priority. New mirrors have reflectivity values of 94 percent. Performance data shows that a drop in average solar field reflectivity of 3 percent results in a drop in solar power output of about 4 percent. The objective of the mirror rinsing program is to maintain an average solar field reflectivity of 88-90 percent. The optimum system at present is use of a high-pressure jet of demineralized water to rinse the mirrors on an approximate two-week cycle.

Small tractors pull lighted trailers carrying a 1900 liter (500 gallon) demineralized water storage tank and two high-pressure pumps. Operators use hand-held spray wands to rinse the trough surfaces. Each plant has a dedicated 3-man mirror rinsing crew.

Mirrors near the power block area are subjected to increased scaling from minerals precipitated out of cooling tower spray. These mirrors are periodically washed with a brushes using a biodegradable cleaning compound to restore reflectivity to acceptable levels.

Cleaning crews have experimented with a self-propelled mirror washing machine developed by Luz Industries Israel. The machine has been in use at the Daggett site. Operated by one individual, the machine washes SCAs using a circular rotating brush followed with a high pressure rinse. Machine-washed mirrors exhibit 2-3 points higher reflectivity on average compared to manually-rinsed mirrors, and machine mirror washing rates are comparable to manual rinsing.

High wind conditions, dust storms, rain, and snow require specialized operational procedures to optimize field performance and protect equipment. The solar fields are structurally designed to withstand a 113 kph (70 mph) wind while in the stowed position at 30 degrees below the eastern horizon. Guidelines are observed to ensure that the fields are adequately protected from high winds. For example, when the average wind speed reaches about 40 kph (25 mph), the fields are configured in a protected mode that involves stowing all perimeter solar collector assemblies. When the average wind speed reaches 56 kph (35 mph), the entire field is stowed.

When rain is predicted (with no threat of hail or dust), the fields are oriented with all solar collector assemblies facing directly up to be washed. Rain provides a clean water rinse. Snowfalls offer even better mirror cleaning.

Developments are planned or in progress to improve cost-effectiveness in both the power block and solar field. The primary areas of development for the solar field are direct steam generation in the collectors to allow increased steam temperatures, durable front-surface glass mirrors, and the use of a small tilt to enhance the incident solar radiation on the receivers.

Point-Focus Solar Thermal Technologies

Point-focus systems are capable of achieving higher temperatures than line focus systems. This provides the potential for increased thermal efficiency in heat engines that use solar energy. Two types of point focus systems, Central Receivers and Parabolic Dishes, have received the most attention.

Current central receiver systems use fields of heliostats to focus in two axes on a single tower-mounted receiver. Solar One is a 10 MW solar thermal central receiver pilot plant for generating electricity. It is located near Barstow, California on the Mojave desert. Solar energy is concentrated by focussing 1,818 heliostats on 160 acres to the top of a 300-foot central power tower where water is heated to steam. Superheated steam at 1,450 psi, 950° F (510°C) either drives the turbine-electric plant or heats an oil transfer fluid, rocks, and sand contained in a thermal storage tank. A conventional turbine/generator produces electricity for distribution by the utility grid. Plans for Solar One were authorized in 1975. Experimental use of Solar One proceeded from 1982 to 1988.

One of the most difficult problems with Solar One was intermittent operation that is caused by interrupted sunlight. Heat recovery from the thermal storage tank produces steam at 450 psi 580° F (304°C) for supplementation of the primary steam. If clouds cause the steam supply to drop below the design minimum the turbine must be shut down.

The Solar One turbine required considerable delay in restarting and considerable loss of energy conversion opportunity. Solar One recorded over 1,500 shutdowns. Restarting Solar One required breaking condenser vacuum and purging with nitrogen gas. Receiver tube failures were attributed to metal fatigue due to the pressure and thermal cycles of each turbine startup and shutdown. Weather outages precluded operation on 40 percent of the days.

This plant is currently undergoing study for major refurbishing, addition of thermal storage, and expansion. Fused salt will store heat and Solar Two will be able to continue to produce electricity during occasional cloud cover and for hours after sunset.

Total Energy From Point-Focus System

The Solar Total Energy Project at Shenandoah, Georgia uses 114 parabolic dish collectors with silicone oil heaters at the focal point of each dish. 750° F (400°C) silicone oil is delivered to a heat exchanger to produce steam for a Rankine cycle turbine generator. In addition to producing 400 KW of electricity this system provides 60 tons (211 kW_t) of air conditioning, and 600 lb/hr (273 kg/hr) of process steam to a knitwear factory.

The Solar Total Energy Project started in 1983. Peak solar-to-total energy conversion efficiency is about 14 percent. Stamped aluminum dish collectors are 23 ft (7-m) in diameter and have aluminized plastic film applied to the surface.

By being characterized as a central turbine system the disadvantages of startup and shutdown similar to the Central Receiver are incurred. This plant is being refurbished and augmented with an auxiliary fossil-fuel heater to avoid shut-down of the turbine generator during cloud cover and to allow continued operation after sunset.

Individual Point-Focus Concentrators

Individual point-focus concentrators as large as 17 meters in diameter track the sun in two axes and focus radiant energy into a receiver. These individual units are called Dish Concentrators. Dish concentrators have been used for more than a century. Often the dish concentrator has been chosen for achieving sufficiently high temperatures to melt ores or to cast metals.

In 1878 a printing press was powered by a dish concentrator that heated an Ericsson engine. Numerous other applications followed. At the turn of the twentieth century Aubry Eneas marketed conical dishes that were 11 meters in diameter. Water was heated to produce steam for driving water pumps.

When a Stirling engine and electricity generator are used in combination with a dish concentrator it is called a Stirling-dish generator system. Advanco Corporation, Cummins, McDonnel Douglas Astronautics Corporation (MDAC) have supplied parabolic point-focus concentrators for use with Stirling engines. See Figure 2.

McDonnel Douglas dish concentrators tracked the sun for more than 175,000 hours or 20 concentrator years during the time from 1982 to 1988. Engine generators on McDonnel dishes have been tested some 12,000 hours of instrumented operation. The Stirling-dish systems have produced more than 100 MWHs of electricity using 25 KW United Stirling AB 4-95 engines. These Stirling-dish generators hold the world record for converting solar radiation to electrical energy. Net energy conversion efficiencies greater than 28 percent have been repeatedly achieved in various locations with the McDonnel Stirling-dish generators.

In addition to holding the record for energy-conversion efficiency, Stirling-dish generators respond more rapidly to solar conversion opportunities than Central Receivers or Linear-Focus systems. Each Stirling-dish tracks the sun and provides rapid solar heating of hydrogen or helium working fluid in the receiver. Stirling-dish systems have provided very high solar utilization characteristics.

Table 1 compares the performance of McDonnell Stirling-dish generators, the Solar One central receiver system, SEGS, and a photovoltaic system called Intersol III. As shown, the Stirling-dish energy conversion provided about two times as much annual electricity (KWH) per square meter of aperture area as the next closest energy conversion system which is the SEGS system. Stirling-dish generators are expected to have about the same first cost as SEGS.

Photovoltaics provided about one-third as much annual electrical energy as the Stirling-dish generators. First costs are much higher for photovoltaics than for Stirling-dish generators. Table 2 shows the costs of some systems that have been built and compares the estimated cost of Stirling-dish generators.

Hydrogen Production and Storage

One of the most common complaints about solar energy is that it is intermittent. Conversion of solar energy to hydrogen as a form of chemical energy storage would allow a future SEGS or Solar Three (Central Receiver System) to operate at night or as clouds pass by generating steam on demand by burning hydrogen. A future Stirling Dish Generator could generate heat on demand by burning hydrogen. Several adequate technologies are available for generating hydrogen including electrolysis of water and thermochemical decomposition of water.

Nature has stored hydrogen, natural gas, and oil for millions of years underground. The petrochemical industry has stored natural gas in salt caverns, aquifers, and rock caverns for more than fifty years. The Electric Power Research Institute showed that there are potential Compressed-Air Energy Storage (CAES) sites in about three-fourths of North America. CAES plants have a 30 year life and store off-peak electricity at an installed cost of about \$425/kW. Storing pressurized hydrogen provides a higher energy storage density than compressed air. Hydrogen Energy Storage (HES) offers even lower installed costs.

An interesting way to store energy and to generate heat on demand is by the following thermochemical reactions:

$$HEAT + CH_4 + CO_2 \text{ ------> } 2\ CO + 2\ H_2 \qquad \text{(Equation 2)}$$

Carbon monoxide and hydrogen are produced by heating methane and carbon dioxide. At the time that heat is needed the reaction is reversed by passing the stored gases over a catalyst:

$$2\ CO + 2\ H_2 \text{ ------> } HEAT + CH_4 + CO_2 \qquad \text{(Equation 3)}$$

Advantages of thermochemical processes such as those suggested above for storing energy include:

1. LONG SHELF LIFE: The chemical species produced by thermal conversion can be stored indefinitely without energy losses. (Thermal storage is always losing energy.)

2. RAPID HEAT PRODUCTION: At the time that heat release is desired, the cold gases pass over a catalyst and release heat.

3. ENERGY TRANSFER TO DISTANT SITE: Heat produced in a solar rich area can be stored and released at distant locations.

The major disadvantage compared to storing pure hydrogen is the toxicity of carbon monoxide.

Conclusions:

Prosperity without pollution can become a reality. Practical ways have been proven to replace the practice of combusting fossil fuels and discarding carbon dioxide to the atmosphere at the rate of 18 billion tons per year. Opportunities to stop atmospheric pollution are offered by point-focus and linear-focus solar collection systems that power heat engines to produce renewable hydrogen and electricity. Renewable hydrogen can be stored in depleted oil and gas reservoirs and delivered by pipelines.

REFERENCES

1. D. Jaffe, S. Friedlander, D. Kearney, "The Luz Solar Electric Generating Systems in California," Proceeding of the ISES Solar World Congress, Hamburg Germany, September 1987.

2. K. Kearney, H. Price, Y. Gilon, D. Jaffe, "Performance of the SEGS Plants, "Proceedings of the Tenth Annual ASME Solar Energy Conference, Denver, Colo., 1988.

3. C. Jensen, H. Price, and D. Kearney, "The SEGS Power Plants: 1988 Performance," 1989 ASME International Solar Energy Conference, San Diego, Calif., April 1989.

4. J.A. Thornton, and J. Lamb, "Evaluation of Cermet Selective Absorber Coatings Deposited by Vacuum Sputtering, SERI/STR-255-3040," Solar Energy Research Institute, March 1987.

5. Y. Harats, and D. Kearney, "Advances in Parabolic Trough Technology in SEGS Plants," presented at the 1989 ASME International Solar Energy Conference, San Diego, Calif., April 1989.

6. Michael Lotker, "Luz In Brief" Luz International Limited, February 1991.

7. William B. Stine, "Progress in Parabolic Dish Technology" SERI/SP-220-3237, June, 1989.

8. "The Ultimate Fuel **Hydrogen on the Horizon**", South Coast Air Quality Management District, El Monte, CA 91731.

9. Charles R. Imbrecht, Chairman, "Energy Technology Status Report, California Energy Commission, June 1990.

10. John C. Schaefer, Project Manager,"Performance of the Vanguard Solar Dish-Stirling Engine Module", Electric Power Research Institute AP-4608, Project 2003-5, July 1986

11. James E. Bennethum, Thomas D. Laymac, (Detroit Diesel Corp) Lennart N. Johansson, Ted M. Godett, (Stirling Thermal Motors Inc.), "Commercial Stirling Engine Development & Applications", SAE International, Future Transportation Technology Conference & Exposition, August 1991.

12. J.E. Bennethum, "Stirling Engine Technology", Detroit Diesel Corporation, October 1990.

13. "Energy Storage: How the New Options Stack Up", Electric Power Research Institute, 1989.

TABLE 1

ANNUAL ENERGY COLLECTION COMPARISON

TECHNOLOGY	ANNUAL ENERGY COLLECTION KWH/SQ.M
STIRLING DISH GENERATOR	565
SEGS	232
SOLAR ONE	180
INTERSOL PV	175

Notes:

1. All systems were measured at comparable locations in the Mojave desert.
2. Data compiled by references 1, 5, 7, 9, and Southern California Edison.

TABLE 2

SOLAR SYSTEM INSTALLED COST COMPARISONS

TECHNOLOGY	TYPE	YEAR	$/KW
INTERSOL PV	PILOT	1982	16,000
SOLAR ONE	PILOT[1]	1982	14,000
PV/CONCENTRATOR	COMMERCIAL	1990	10,000
SEGS	COMMERCIAL	1985	2,500
STIRLING DISH	PRODUCTION RUN[2]	1991	2,400

1. CONSIDERABLE COST REDUCTION WOULD BE EXPECTED IF THE DESIGN WOULD BE TRANSLATED FOR PRODUCTION MANUFACTURING
2. ESTIMATE BASED ON BILL OF MATERIALS USED IN RECORD-HOLDING McDONMNEL DOUGLAS STIRLING-DISH GENERATOR AT PRODUCTION LEVELS OF SEGS SYSTEMS

FIGURE 1.

LUZ SOLAR ELECTRIC GENERATING STATION

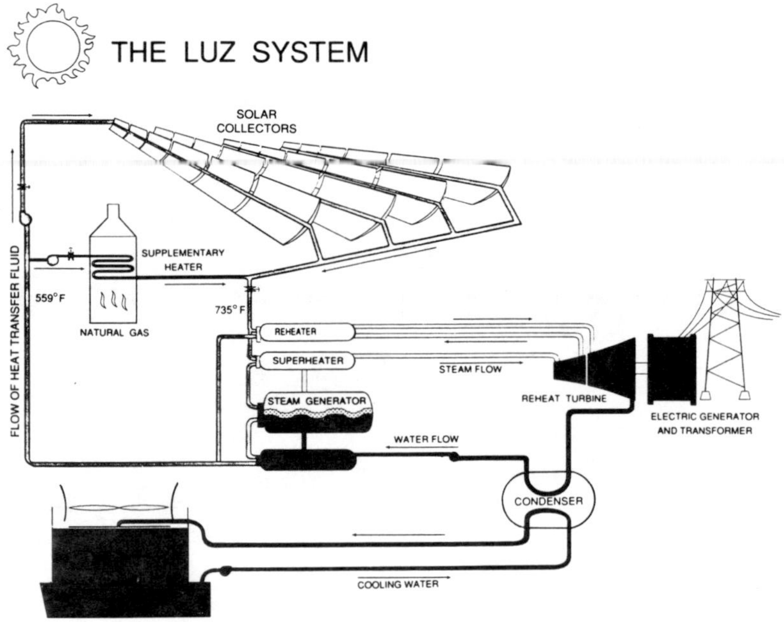

FIGURE 2.

STIRLING DISH SOLAR ELECTRIC GENERATING SYSTEM

A PHOTOVOLTAIC-HYDROGEN-FUEL CELL ENERGY SYSTEM: CONTROL STRATEGY, MONITORING SYSTEM DESIGN, AND PRELIMINARY RESULTS

P. A. Lehman and C. E. Chamberlin
Department of Environmental Resources Engineering
Humboldt State University
Arcata, CA 95521, U.S.A.

ABSTRACT

We report preliminary operational results of a stand-alone renewable energy system using hydrogen as the energy storage medium and a fuel cell as the regeneration technology. The system installed at the Humboldt State University Marine Laboratory consists of a 9.2 KW photovoltaic (PV) array coupled to a high pressure, bipolar alkaline electrolyzer. The array powers the Laboratory's air compressor system whenever possible; excess power is shunted to the electrolyzer for hydrogen and oxygen production. When the array cannot provide sufficient power, stored hydrogen and oxygen are furnished to a proton exchange membrane fuel cell which, smoothly and without interruption, supplies the load.

The control and monitoring systems are nearly complete. For each, we describe the rationale, hardware, preliminary operation, and what issues remain before the systems are complete and bug free. We report on electrolyzer operation and on particular problems which have been solved in adapting a production line unit for PV use. We also report on the system's fuel cell: its performance, problems, and plans to overcome difficulties in future operation.

INTRODUCTION

The Schatz Solar Hydrogen Project began in the fall of 1989. It is a stand-alone photovoltaic energy system that uses hydrogen as the storage medium and a fuel cell as the regeneration technology. Its goal is to demonstrate that hydrogen is practical storage medium for solar energy and that solar hydrogen is a reliable and abundant energy source for our society.

A schematic for the system is shown in Figure 1. It is installed at the Humboldt State University Telonicher Marine Laboratory and the Lab's air compressor system, used to aerate aquaria, is the load. The basic control logic is shown in Figure 2. When PV power is available, it is used directly to supply the load. Any excess power is supplied to the electrolyzer which stores the electricity as hydrogen and oxygen gases. When the array cannot provide electricity, the stored hydrogen and oxygen serve as fuel and oxidant for the fuel cell which provides uninterrupted power. If the storage is depleted, the system returns to utility power.

SYSTEM DESCRIPTION

The PV array consists of 192 Arco M75 modules configured into 12 independent subarrays. Each subarray consists of 16 modules, wired in 8 series pairs for 24 VDC operation. The subarrays are electrically isolated from each other by 60 amp Schottky diodes. The subarrays are individually switchable to either the load or the electrolyzer. If not needed the subarrays can be shorted; this also serves as the safety off configuration.

A 24 VDC, 37 amp-hour nickel cadmium battery is installed in parallel with the load to serve as a buffer and prevent frequent switching of subarrays. Its operation and performance are described below.

The electrolyzer is a Teledyne Energy Altus 20 which has been modified for use with a PV array and to deliver O_2 as well as H_2. It consists of a 12 cell electrolysis module designed to deliver 20 slm of H_2 gas and 10 slm of O_2 gas with an input current of 240 amps at 24 VDC. The gases exit at 700 KPa pressure (with the H_2 pressure slightly greater) and are stored in steel tanks (1000 gal for H_2 and 500 gal for O_2). Storage is approximately 85 KWH at the high heating value for H_2.

The fuel cell is a proton exchange membrane (PEM) cell manufactured by Ergenics Power Systems. It is rated at 1.0 KW, though thus far it has not produced this power output continuously. Since the fuel cell requires an input pressure of at least 250 KPa, only 450 KPa of storage is available for use. The load draws 640 W continuously and if the fuel cell is 55% efficient, the storage provides for 48 hours of operation without replenishment.

The design of the system and the testing of components and subsystems have been reported previously [1,2].

CONTROL SYSTEM

The control system is designed to implement the basic logic described above and to do so automatically. In addition to regular operation, the system monitors various safety interlocks and provides for a shutdown if need arises. The system is configured so that input sensors can be calibrated and software parameters can be varied via on-screen commands. Any or all of the subsystems can be run independently, with manual or automatic control, and a continuous graphic display provides constant updates on system configuration and operating parameters.

Hardware

A Macintosh SE microcomputer (SE) is used for the control system. The SE receives input parameters (both analog and digital) from a MacADIOS II SE data acquisition system via Analog Devices backplanes and Opto-22 digital backplanes. System input parameters are:

- subarray currents (12)
- H_2 and O_2 pressures
- battery current and voltage
- electrolyzer temperature
- H_2 impurity sensor (in O_2 line)
- O_2 impurity sensor (in H_2 line)

- electrolyzer safety interlock
- electrolyzer feedwater pump
- hydrogen gas sensors (2)
- hydrogen hood air movement sensor
- smoke detector
- OK signal from monitoring system
- emergency shutdown switch

The first 6 inputs listed are analog; the remaining 7 are digital. The currents are measured with shunts, the voltages directly, and temperature with a type T thermocouple, each with appropriate signal conditioning and amplification. The digital inputs are delivered to appropriate digital modules which provide a 0 or 5 volt off-or-on signal.

The impurity gas sensors deserve special note. In order to protect against the possibility that a combustible mixture has been produced in the electrolyzer and is being delivered to storage, we have installed in-line gas sensors in both the H_2 and O_2 lines. The H_2 detector is a Figaro TGS 821 tin oxide semiconductor gas sensor. It is specifically sensitive to H_2 gas and when provided with a power source, shows a logarithmic change in resistance with changing H_2 gas concentration. Its sensitivity is such that it can easily indicate whether or not the lower explosive limit of 4.7% H_2 in pure O_2 [3] has been reached. The O_2 detector is a temperature compensated galvanic electrochemical cell which provides an intrinsic voltage when exposed to O_2. Its sensitivity also allows easy detection of the lower explosive limit of 6.1% O_2 in pure H_2 [3]. The system is configured to shut down if either concentration exceeds half of that necessary for combustion.

The SE controls system operation by individually switching relays and solenoids. Output signals are sent by the computer through additional Opto-22 digital backplanes. 120 VAC power for the coils of relays and solenoids is provided by optically isolated switching modules.

Control Algorithm

The control algorithm which allocates subarrays to the load or the electrolyzer begins by determining the subarray input currents and current system configuration. A charging flag indicates whether the battery has been most recently charged or discharged. If most recently charged, the charging flag is off and vice versa.

When the charging flag is on, the program determines the minimum number of subarrays which, when switched to the load, will provide enough PV current to exceed the current drawn by the inverter. It switches these to the load and excess current charges the battery. All other subarrays are connected to the electrolyzer.

When the charging flag is off, the program allocates the maximum number of subarrays which will not cause the PV current to exceed the inverter current. The deficit is provided by the battery. All other subarrays are again connected to the electrolyzer.

There are three circumstances which cause the program to limit the current to the electrolyzer and allocate fewer subarrays than are possible above. They are: 1) The electrolyzer can accept a maximum current of 275 amps under any circumstances. 2) At low currents, the H_2 impurity in the O_2 line is too large so the program prevents operation below 10 amps. 3) At low temperatures, the electrolyzer can accept less than the maximum current. At 30°C and below, currents up to 100 amps are allowable. At 60°C and above, the maximum current of 275 amps is allowable. Between these two temperatures, the current limit increases linearly.

Safety Interlocks

In addition to the gas impurity sensors, the system has 7 more safety interlocks. The electrolyzer has its own programmable controller which monitors operation. If any of the parameters monitored is out of normal range, a signal is produced which, under normal circumstances, would shut down the electrolyzer's DC power supply. The SE monitors that signal and shuts down if it is activated. Power to the feedwater pump is also monitored. This serves as a safety backup to the feedwater level sensor. If the sensor fails in the full position, neither the pump nor a shutdown would be activated. This would cause loss of electrolyte water and electrolyzer module failure. If AC current does not flow to the pump after 50 amp-hours of operation (as it should), a safety shutdown is effected.

Bacharach combustible gas alarms are mounted in the laboratory room and inside the hood which encloses as the hydrogen equipment. If either indicates an alarm condition (set at 40% of the lower explosive limit), the system shuts down. The hood is also fitted with an air movement sensor. If ventilation in the hood fails, a shut down occurs. Finally, if the smoke detector or the emergency push button show a positive signal, a shut down occurs also.

The control system microcomputer and the monitoring system microcomputer also exchange signals to show they are operating properly. If either computer fails or detects an abnormal condition, a shutdown occurs. This protects against the possibility that the control system SE would crash and leave the system in a dangerous configuration.

Operation

The control system has been utilized for several full days of operation. Two of those days are shown in Figures 3 and 4. These graphs show the DC load current, the PV array current, and the electrolyzer current as a function of time. Figure 3 represents a sunny day; Figure 4 represents generally foggy day with filtered sunshine between 1 PM and 3 PM. Several features are worthy of note. First, the general performance of the control system is excellent. It splits the PV current into two components, keeping the load running steadily and furnishing the electrolyzer with any excess power. Second, whenever the electrolyzer current maximum is exceeded as it is on the sunny day, the current is maintained at a safe level by shorting out first 1 and then 2 subarrays.

The behavior of the load current is interesting. At low insolations, the subarray currents are small and the system can control the currents to the load and electrolyzer with good precision. At high insolations (around 1000 W/m^2), one subarray furnishes just the right current for the load (as it was designed to). Under both of these conditions, the battery remains in either a charge or discharge mode and is stable. However, at around 500 W/m^2 insolation, one subarray is too small for the load and two are too large. Under these conditions, the battery goes from charge to discharge mode more quickly, in about 10 minute cycles. The system still operates well and whether this is harmful to the battery or not, we have not yet determined. We plan to test different battery voltage levels in the control software in order to lengthen these cycling times.

MONITORING SYSTEM

The monitoring system is designed to determine the following:

- overall storage efficiency
- array, electrolyzer, fuel cell, and inverter efficiencies
- temperature behavior of the array, fuel cell, and electrolyzer
- effect of environmental conditions on system performance
- real time behavior of the system

In order to accomplish these tasks, the systems displays in real time and records for later analysis the performance of the PV array, the electrolyzer, the H_2 and O_2 gas storage, the fuel cell and other system components. Specifically, the following parameters are measured and recorded:

- temperature: PV modules (3), ambient air, electrolyzer, fuel cell, hydrogen and oxygen storage tanks
- current: electrolyzer, battery, fuel cell, inverter, AC load
- voltage: electrolyzer, battery, fuel cell
- mass flow: hydrogen and oxygen out of the electrolyzer and into the fuel cell
- pressure: hydrogen and oxygen storage tanks
- other: insolation, windspeed and direction

Hardware and Software

The system is made up of an additional Macintosh SE microcomputer that receives data from two data logging systems. They are: 1) A MacADIOS II SE data acquisition system which acquires data from 18 sensors via two Analog Devices backplanes and 2) three stand-alone HydroNET data logging modules which record inputs from 7 sensors in the PV array field and a HydroNET interface module which transmits the data over a 100 foot long line to the SE RS-232c serial port.

The monitoring system software is a compiled application written in Microsoft QuickBASIC and provides for a real time display of sensor values, a display allowing the interactive calibration and testing of sensors, and a recording of inputs on a floppy disk. Observations can be made, displayed and recorded every 20 seconds or averages of multiple readings can be written out and displayed at longer time intervals.

ELECTROLYZER

The electrolyzer I-V curves as a function of temperature have been reported previously [2]. Additionally, the interface with the control system and monitoring of in-line gas purity is described above. Described below are modifications to the gas delivery system and operating results.

Gas Delivery System

Since the electrolyzer is not designed to deliver oxygen nor to deliver dry gases, significant modifications to the gas piping were necessary. The completed gas delivery system is shown in Figure 5.

The oxygen piping begins with a tube-in-tube condenser which is cooled with chilled water at 10°C. This replaced an air cooled condenser furnished with the unit. The gas then flows through a bubbler containing a phosphoric acid buffer and a stainless steel coalescing filter. The purpose of these devices is to remove any trace of KOH electrolyte which would poison the deoxo catalyst. The gas is then heated to 120°C and enters the catalyst bed which oxidizes any H_2 impurity. The H_2 concentration must be kept below 3000 ppm to prevent excessive heating in the fuel cell. A second condenser removes the water formed in the deoxo unit, after which the O_2 is delivered to storage through a backpressure regulator. The drain of the second condenser is not piped to the condensate trap because the small (approximately 15 KPa) pressure drop through the bubbler and deoxo unit would cause KOH electrolyte to be aspirated out of the reservoir.

The hydrogen piping is simpler. The only modifications are the addition of a bubbler to remove KOH and a tube-in-tube heat exchanger to condense out most of the water. Not shown in the figure is a plastic coalescing filter downstream of the backpressure regulator which insures that no trace of KOH will reach the storage tank or fuel cell.

Operation

Electrolyzer operating results are shown for partly sunny afternoon in Figures 6 and 7. In Figure 6, the temperature behavior of the module is shown. Under these conditions, the module required approximately 1 hour to reach its steady state operating temperature of 65°C. At this point, the electrolyzer's cooling fan cycles on and off to maintain this temperature. Two points are notable. First, the heat capacity of the module is high enough so that the temperature is stable even when the current falls below the value high enough for the fan to cycle. More importantly, the short heating time of 1 hour indicates that, under most conditions, the module will heat up quickly enough so that the current to it will not have to be limited by a temperature that is too low.

Figure 7 shows the production rates of H_2 and O_2 and the energy required for H_2 production. It averages 4.70 KWH/m^3 which is a production efficiency of 76%. This value is typical of alkaline electrolyzers though it is less than the efficiency achieved at the HySolar Project in Germany [4].

FUEL CELL

As mentioned above, the fuel cell is designed to provide 1 KW of continuous power. A PEM cell was chosen because of its passive operation, its high efficiency, and its ability to provide power quickly from a standby configuration.

Operation

We have measured the fuel cell V-I curve at various times and the results are shown in Figure 8. As can be seen, performance deteriorated over time and the cell was never able to run the load for more than 1 hour. Probable cause for the poor performance was inadequate water removal. Ergenics is now developing a second generation cell which, it is hoped, will solve this problem.

We were able to test whether or not the cell would pick up the load quickly. When warmed up, the cell easily accomplished this; the motor ran without any interruption.

ACKNOWLEDGEMENTS

The authors gratefully acknowledge generous grant funding from Mr. L.W. Schatz of General Plastics Manufacturing Co., Tacoma, WA, U.S.A. and the assistance of engineering students Tom Herron, Tim Murphy, Gian Pauletto, and Ron Reid.

REFERENCES

1. P.A. Lehman and C.E. Chamberlin, "Design of a Photovoltaic-Hydrogen-Fuel Cell Energy System," *International Journal of Hydrogen Energy*, vol. 16, No. 5, pp. 349-352, 1991

2. P.A. Lehman and C.E. Chamberlin, "A Photovoltaic-Hydrogen-Fuel Cell Energy System: Preliminary Operational Results," *Proceedings of the Tenth European Solar Energy Conference*, 1991, in press

3. Pamphlet G-5, *Hydrogen*, Compressed Gas Association, Arlington, VA, 1974

4. W. Hug, J. Divisek, J. Mergel, W. Seeger, and H. Steeb, "High Efficient Advanced Alkaline Electrolyzer for Solar Operation," *Proceedings of the 8th World Hydrogen Energy Conference*, Pergamon Press, Oxford, 1990

FIGURE 1: SCHEMATIC OF THE SCHATZ SOLAR HYDROGEN PROJECT

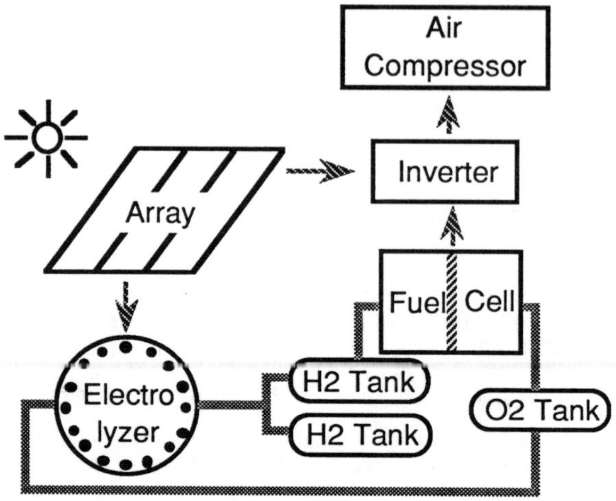

FIGURE 2: BASIC CONTROL LOGIC FOR PV-HYDROGEN-FUEL CELL SYSTEM

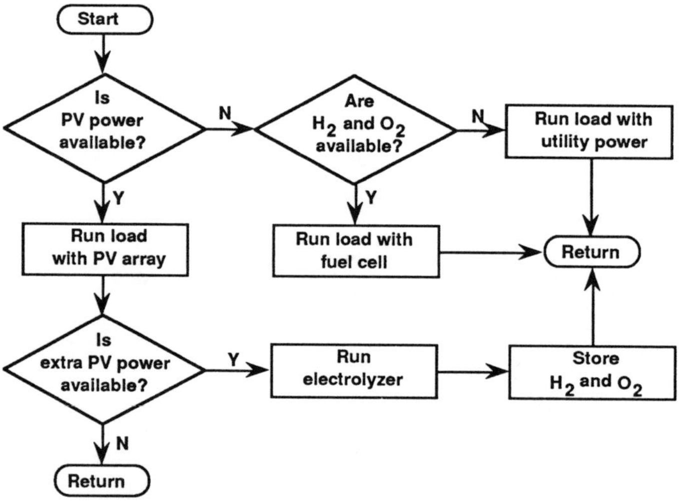

FIGURE 3: CONTROL SYSTEM OPERATION ON A SUNNY DAY

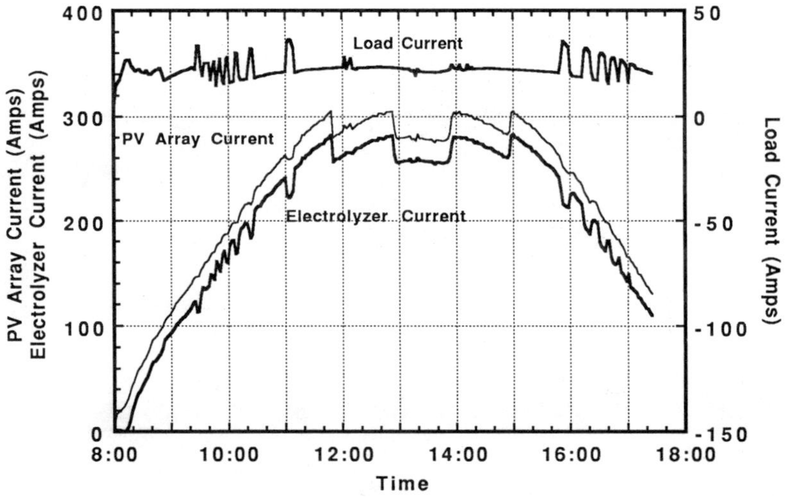

FIGURE 4: CONTROL SYSTEM OPERATION ON A FOGGY DAY

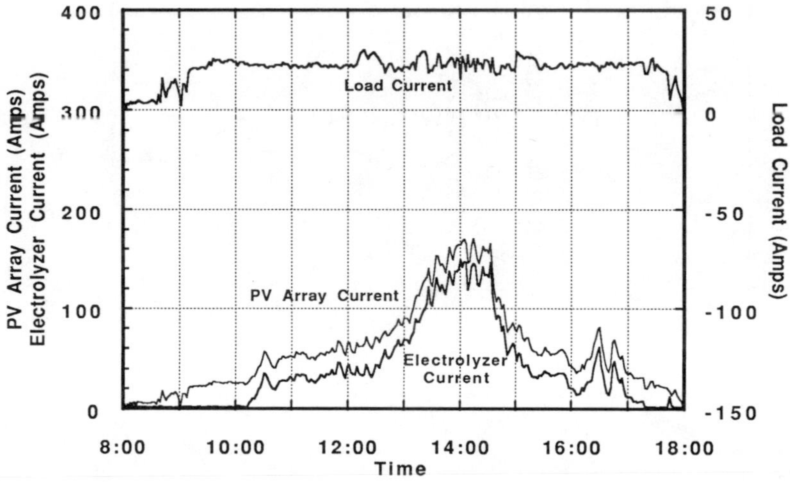

FIGURE 5: GAS DELIVERY SYSTEM FOR HYDROGEN AND OXYGEN

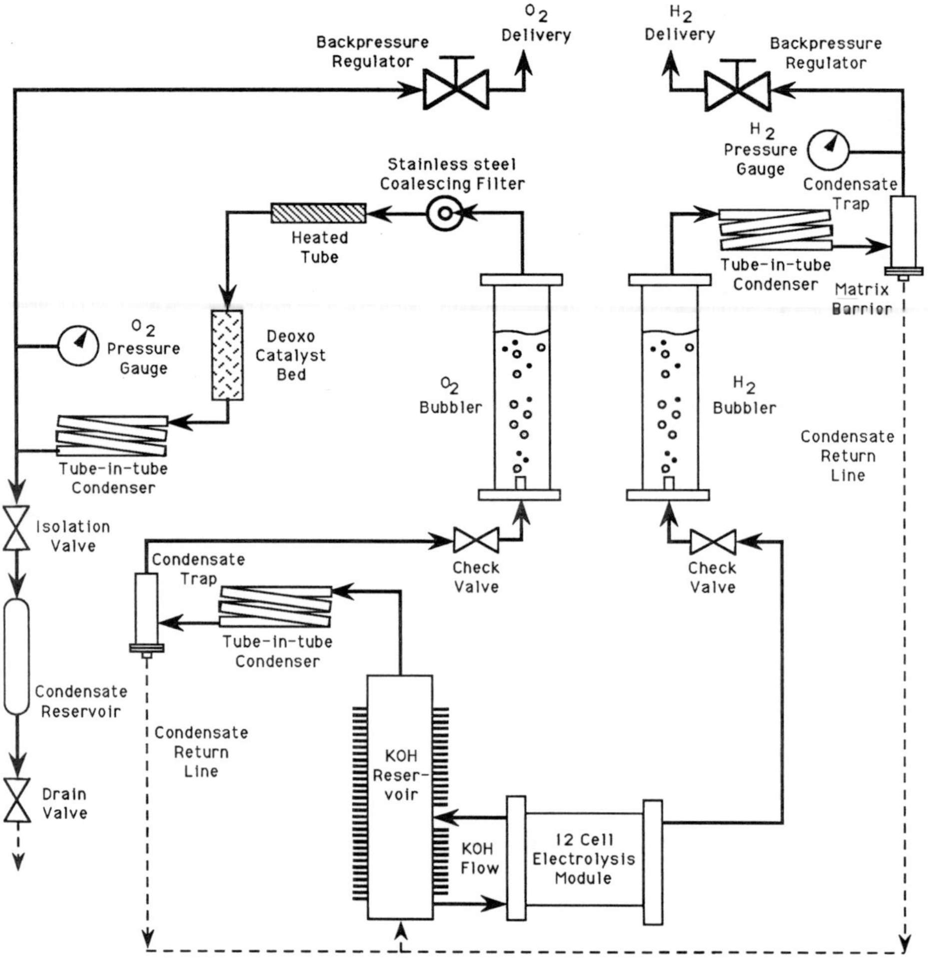

FIGURE 6: ELECTROLYZER CURRENT, TEMPERATURE, AND VOLTAGE EFFICIENCY FOR RUN ON AUGUST 13TH

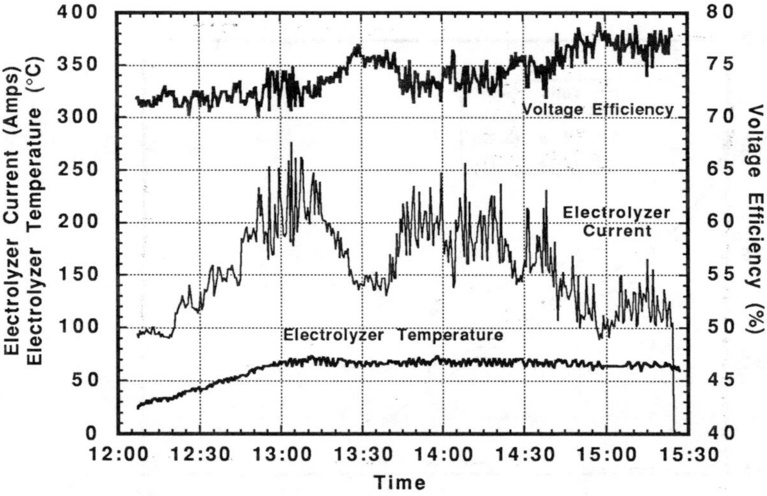

FIGURE 7: HYDROGEN AND OXYGEN PRODUCTION RATES AND HYDROGEN PRODUCTION ENERGY FOR RUN ON AUGUST 13TH

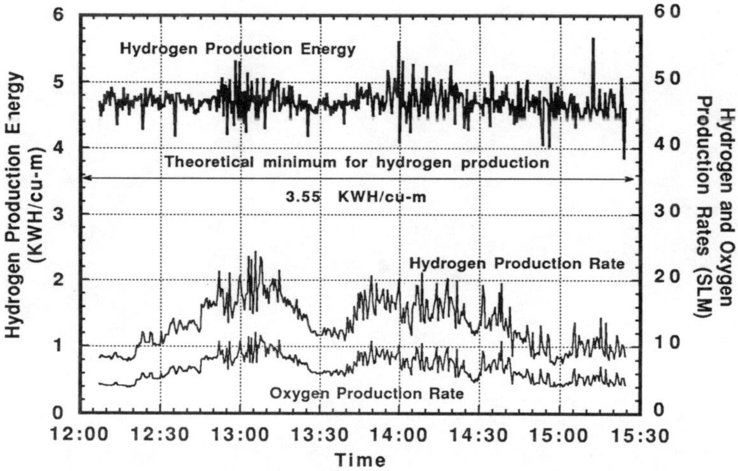

FIGURE 8: FUEL CELL V-I CURVES

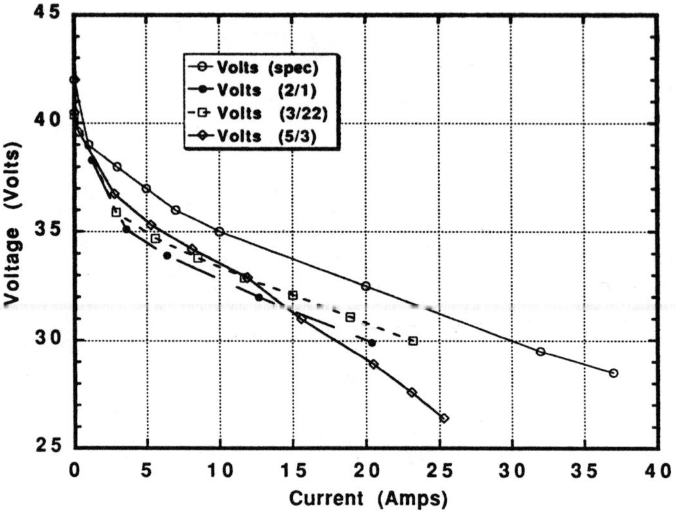

CHAPTER TWO

Hydrogen Production

SMALL PORTABLE HYDROGEN PRODUCER

Helmut Schiller
Hydrogen Service
West Germany

As development of hydrogen technologies continues to progress, applications of these technologies increase. One such application is the creation of efficient power systems in the small to medium power-rate range.

It is our aim, with the development of a small hydrogen producer, to give "ordinary people" a chance to become familiar with this technology at an affordable price. In addition, non-industrial countries will be able to use their renewable energy sources with this power system, as no expensive electric grid or gas pipeline is necessary to produce electricity or gas on site.

The basic module, with the dimension of a suitcase and a weight of 60 kg, stores the cooking gas of a family of four for one week. The hydrogen can also be used in a heater or in a fuel cell for electricity production. In combination with a larger hydrogen storage tank, this unit can power a small house.

The medium-sized units are made to bring more comfort and reduce the energy costs for farms and small villages. In several areas of the European Community, where the cutback of farming subsidies creates big problems, an energy harvest with hydrogen use and sale will bring new prosperity.

There are many different designs to harvest renewable energies. The following pages show some simple and efficient constructions, which are on the market, or will be on the market very soon.

Small Portable Hydrogen Producer

The unit is mounted on a rack which contains a 1kW electrolyzer, a deionization unit, a small battery and a hydride tank for hydrogen storage. The whole unit is 800 mm long, 200 mm wide, and 500 mm high. Combined with a catalytic burner, it is capable of storing energy for a cooking period of one week for a four person family. The total weight, without water, is about 60 kg. This unit is the basic part of a whole small-scale hydrogen system. The purpose of the hydrogen producer is to give "ordinary people" a chance to become familiar with hydrogen technologies at an acceptable price.

The electricity for electrolysis comes from the following renewable sources:

- from seawaves by a special, newly-tested converter
- from wind by simple and efficient wind converters
- from running water by a 1kW crossflow turbine (Dia. 200mm, width 100 mm)
- from special floating water turbines in rivers

An important component of the system is the LYNCH MOTOR. This disc-shaped motor (Dia. 200 mm, 70 mm thick) also functions as a DC alternator. It operates up to 600 amps and 48 Volts. The special arrangement of magnets and wires gives a high efficiency at low speeds, such as 200 revolutions per minute. This motor makes a gearbox unnecessary for the wind converter and the crossflow turbine. In the motor mode, the 13 kg armature reaches similar rates.

In combination with a new fuel cell, which operates like an electrolyzer in reverse, this unit also allows the production of electricity, and can provide more comfort very quickly to many people, making their daily work much easier. It will also reduce the burning of firewood in sensitive areas.

In villages several of the units should be combined with a larger pressure vessel to store the gas for electricity production and cooking. Following is a diagram of the principle construction of the device.

PRINCIPLE CONSTRUCTION OF A PORTABLE HYDROGEN PRODUCER

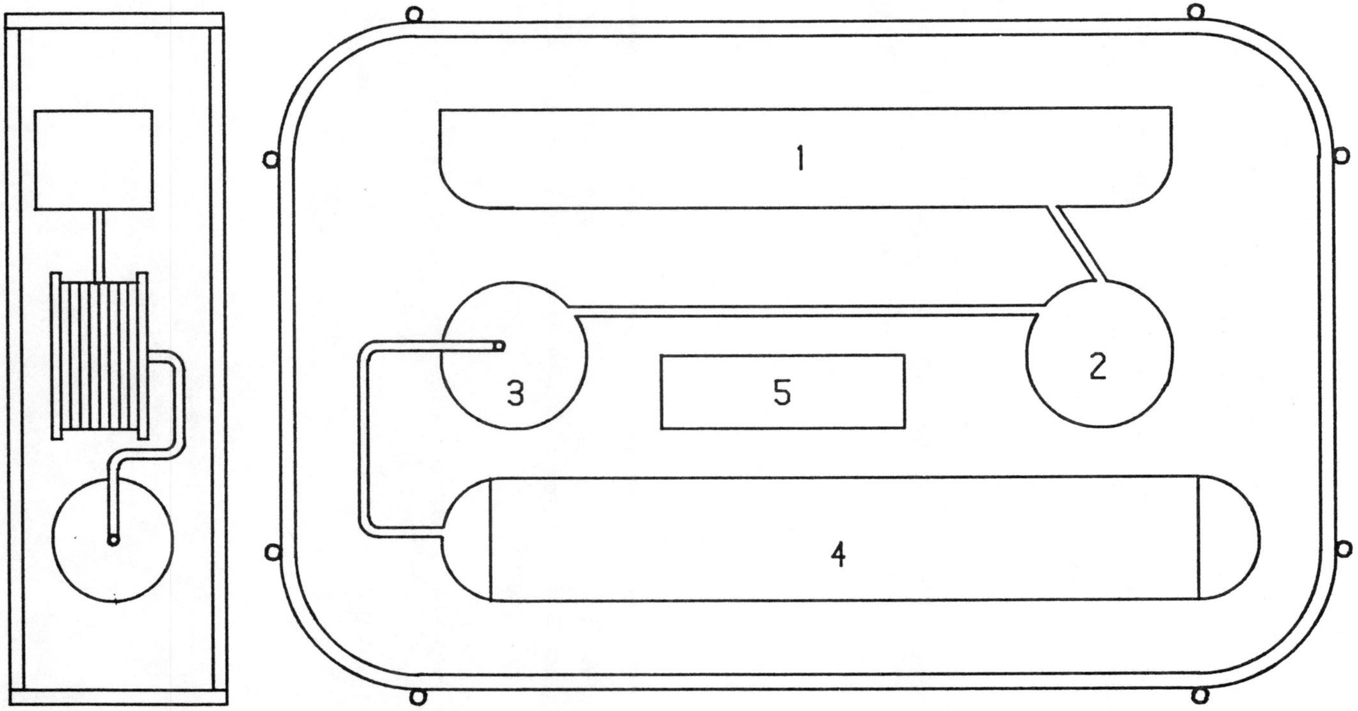

Helmut Schiller
Hydrogen Service
Wasserstoff u. erneuerbare Energien
Beratung, Planung, Ausführung
Scholzenviertel 7 • D 6140 Bensheim-Zell
Tel.(49) 06251-63 213

1. Watertank
2. Deionisation
3. Electrolysecell – Fuelcell
4. Hydridtank
5. Battery and Electronic

The New Solarcells

Solarcells, developed in Switzerland, are based on titanium dioxide. They look like colored glass and provide results similar to those obtained with silicon cells. The needed materials are mass produced at a price of $60 US per square meter. Early next year the first production licence will be granted.

These solarcells will do much to advance hydrogen technology, due to their inexpensive electricity production. The biggest advantage of the solarcells is their good rate of electric power production in cloudy weather.

The electricity consumption of West Germany can be harvested by solarcells which cover 25 percent of all the roofs in the country. The price with silicon cells is about $.86 US per kWh, a price reduction of $.08 US per kWh. With these new technologies, hydrogen will provide electricity storage. A diagram of the operating principle is shown below.

The Lynch Motor

This motor has been tested during the past six years, and will be in production this year. The disc-shaped armature contains 129 magnets and copper wires of 4 x 1 mm cross-section. The low resistance device operates now up to 600 amps and reaches up to 4500 revolutions. The motor gives more than 16 kW of power and an efficiency over 90 percent. The 11 kg device is now being tested in several electric cars in Switzerland and Germany.

This motor also works with similar results in the generation mode, which shows good possibilities for hydrogen production. See the diagram below. Next to the 200 mm diameter unit is a 100 mm unit, and a 600 mm unit for 70 kW is now being developed to power a city bus.

PHOTOVOLTAIC CELL

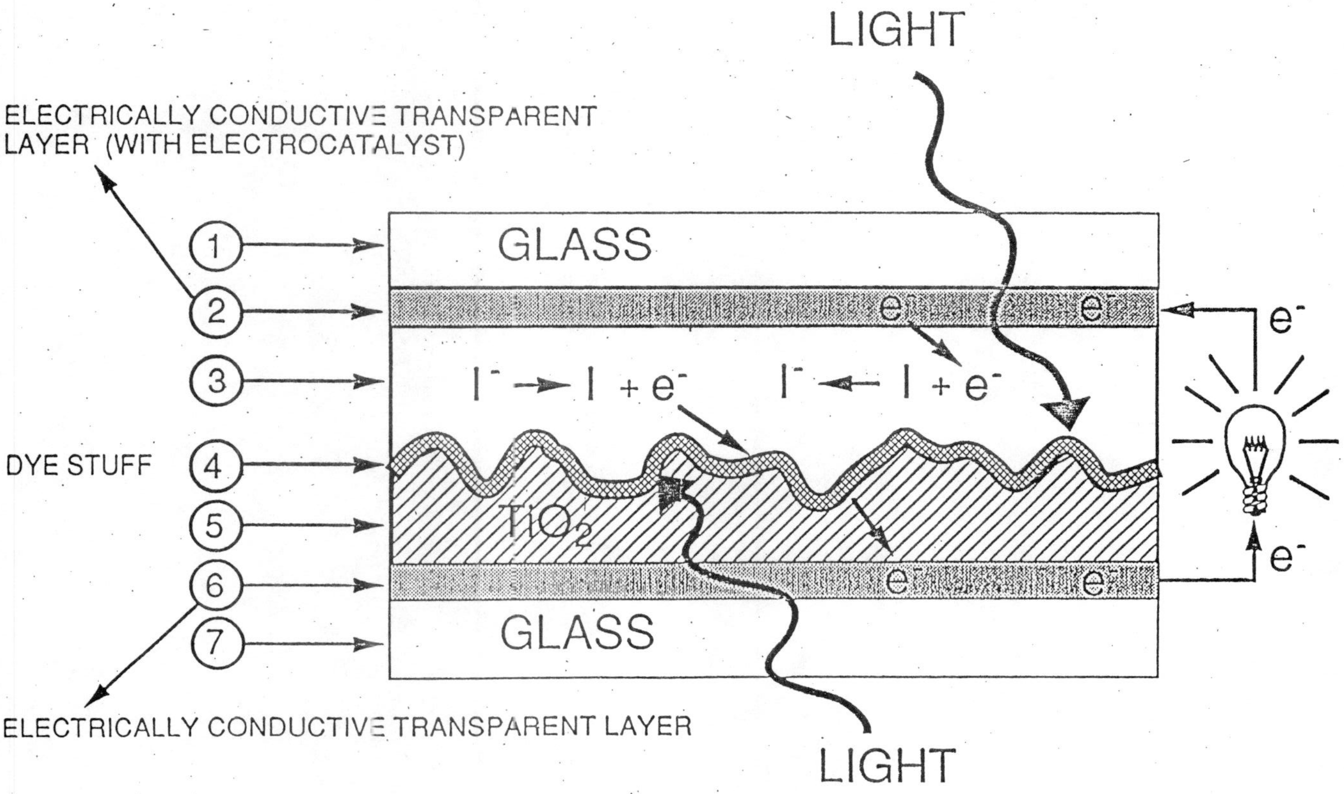

M. Graetzel 1989 (EPFL)

LYNCH MOTOR PERFORMANCE CURVES

C1054
28/12/90

200mm Nominal Size, Nd-Fe-B Magnets, 129-Strip, 48 Volt Supply.

Weight 11 Kg

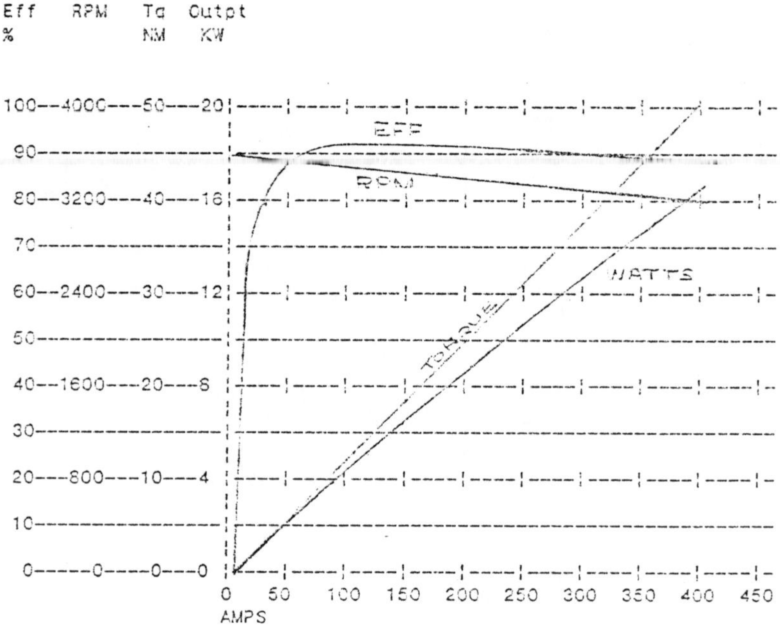

RFd62-12/12/90
(0060)

The Windconverter

The windconverter shown here is made for self-assembling. The calculated costs are the material costs. Construction is a special down-wind application with mechanical self protection. In combination with a new DC-alternator, only a simple belt drive is necessary to increase the revolutions by two to three times. Due to this construction, the weight at the top is dramatically reduced, allowing for a thin mast design. The eleven-meter-diameter windconverter operates on a 12-meter mast and gives an energy output of 20 kW at 10 m/s wind velocity. The whole material cost in Germany is about $12,000 US.

This design is made for the west coast of Ireland, where the average wind velocities are between five and eight m/s. Here the annual energy harvest is between 37,000 kWh and 150,000 kWh. In combination with an ordinary LPG tank and a pressure electrolyzer it is possible to warm a house and drive a car, tractor or boat from the produced hydrogen.

The Waveconverter

The seawave converter shown below is another interesting design. The simple floating construction can be easily built. A device with two 23-meter long rotating drums brings a power output of 50 to 150 kW. On the south coast of Norway this unit will produce 250,000 kWh per year. In the North Atlantic, on the Scottish west coast, the harvest will be more than 1,000,000 kWh. This power plant, built from alloy in Sweden, will cost less than $75,000 US. Also, smaller designs can be useful for islands and isolated locations. Within the next year the first big wave converter will be built, the result of positive results from smaller test converters.

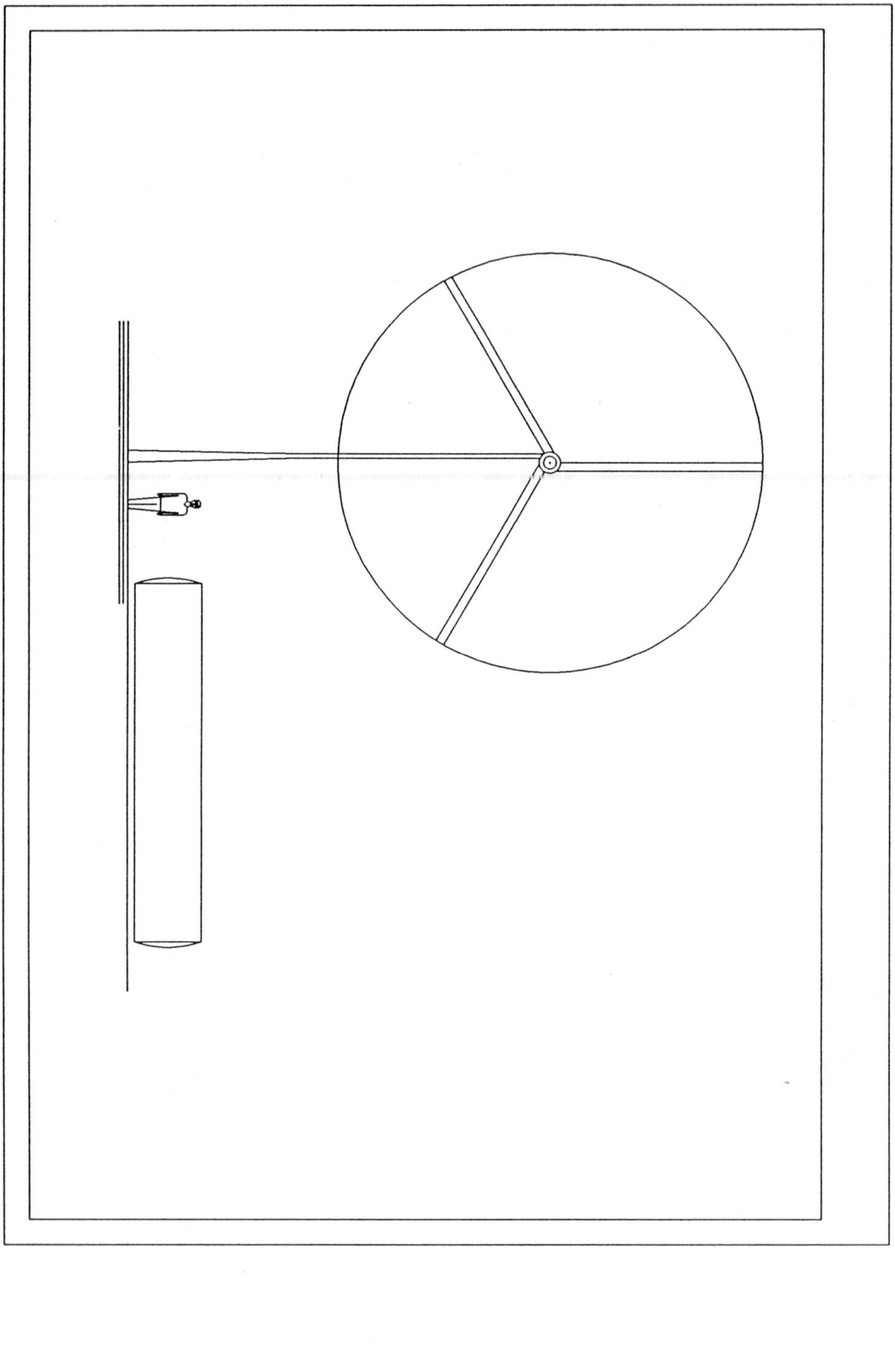

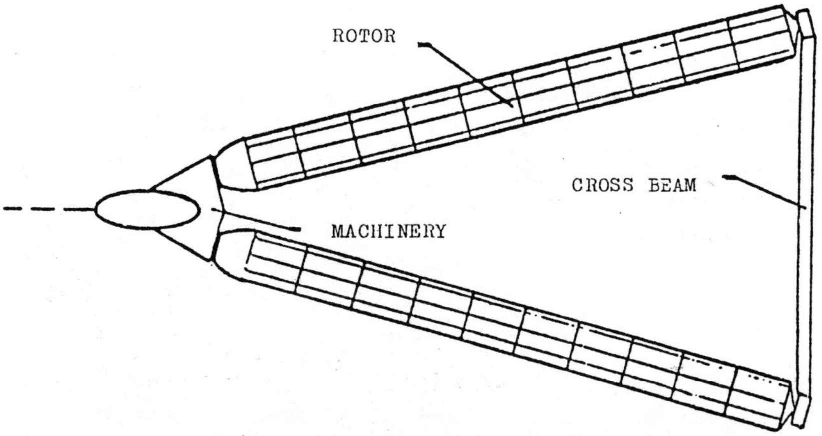

Fig. 1: Diagram of the principal parts of the wave power plant.

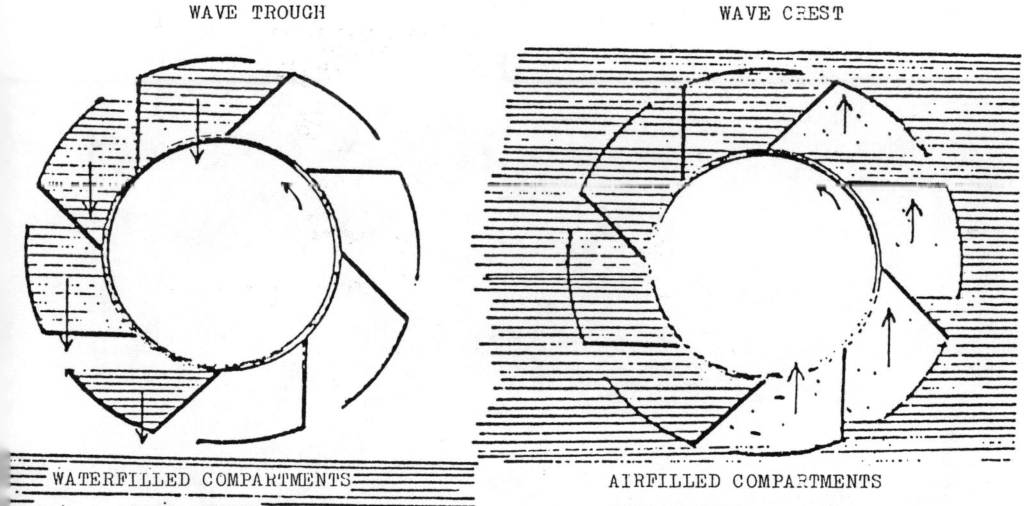

Fig. 2: Functional principle of the wave rotor.

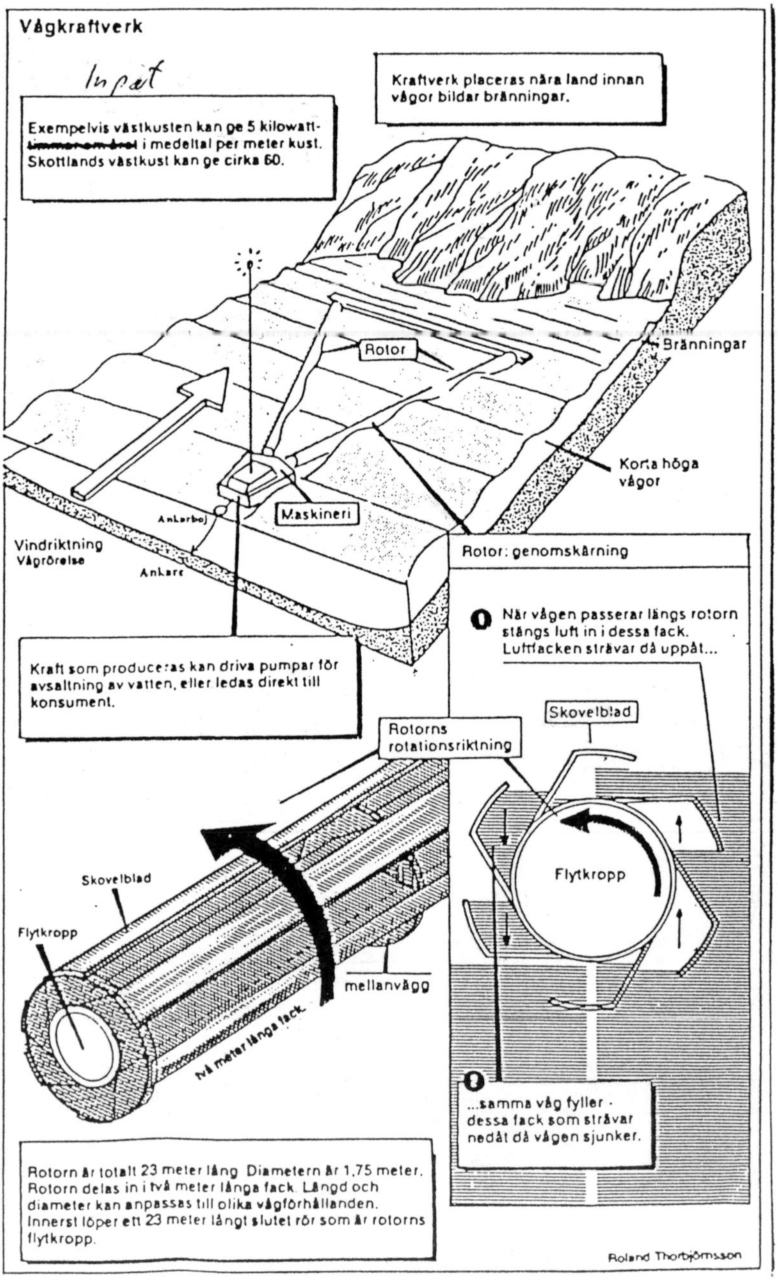
Vågkraftverk
Kraftverk placeras nära land innan vågor bildar bränningar.
Exempelvis västkusten kan ge 5 kilowatt- [redacted] i medeltal per meter kust. Skottlands västkust kan ge cirka 60.
Rotor
Bränningar
Korta höga vågor
Maskineri
Ankarboj
Ankare
Vindriktning
Vågrörelse
Kraft som produceras kan driva pumpar för avsaltning av vatten, eller ledas direkt till konsument.
Rotor: genomskärning
När vågen passerar längs rotorn stängs luft in i dessa fack. Luftfacken strävar då uppåt...
Skovelblad
Rotorns rotationsriktning
Flytkropp
Skovelblad
Flytkropp
mellanvägg
två meter långa fack
...samma våg fyller dessa fack som strävar nedåt då vågen sjunker.
Rotorn är totalt 23 meter lång. Diametern är 1,75 meter. Rotorn delas in i två meter långa fack. Längd och diameter kan anpassas till olika vågförhållanden. Innerst löper ett 23 meter långt slutet rör som är rotorns flytkropp.
Roland Thorbjörnsson

A LABORATORY SIZE SOLAR HYDROGEN GENERATOR

By H. Sofrata[1], Y. Al-Saedi, H. Aba-Oud
KACST, P.O. Box 6086, Riyadh 11442, Saudi Arabia

ABSTRACT

Recently great interest in the Solar Hydrogen Production and Technology is observed worldwide. This is, mainly, due to the fact that energy generation by hydrogen combustion with oxygen or air is pollution free. This energy source does not contaminate the environment. Pure water vapor, which is friendly element is the natural cycle, is the main output.

A small solar electrolyzer has been tested under solar operation. The system has been operating utilizing an activated cathode. Special very thin KREBSKOSMO P.S. diaphragm have been tested.

Using activated cathode a maximum hydrogen flow of 18 normalized liter/hour and oxygen flow of 9 normalized liter/hour have been collected. At a current of 40.5 ampere 2.5 volts. The hydrogen purity has been measured as 99.67 vol.%.

The system has been powered by 6 photovoltaic panels, 38 watt peak. These photovoltaic panels are originally designed for a nominal output of 12 volts. All panels have been adapted to produce a nominal 3 volt output.

The experimental results are displayed to give a model for the mentioned operation goals.

INTRODUCTION

Due to the delay of hardware delivery of some elements, preliminary readings and test runs only have been reported. The rest of the experimental work is going on now and will be published as soon as the data is available.

The electrolytic splitting of water to H_2 and O_2 gas has been considered as a good candidate for fuels expected to replace fossil fuels. The exhaust gases for such fuel is only pure water vapor, which is returned to the natural water cycle. CO_2, which is responsible for greenhouse effect and world warming, is not produced in contrary of coal, oil, petrol or natural gas burning. This new fuel will not contribute to the accumulation of carbon dioxide. Solar energy is an unlimited supply of direct current electric power and has been utilized in several projects world wide. High efficiency photovoltaic cells and long lasting low power consuming electrolyzers are one of the main objectives of such project. [1-7].

---

[1] Author to whom all correspondence must be addressed.

In this paper, a laboratory size solar powered hydrogen generator has been considered.

The system has been operating utilizing an activated cathode. Special very thin KREBSKOSMO P.S. diaphragm have been tested.

Using activated cathode, a maximum hydrogen flow of 18 normalized liter/hour and oxygen flow of 9 normalized liter/hour have been collected. At a current of 40.5 ampere 2.5 volts. The hydrogen purity has been measured as 99.67 vol.%.

The system has been powered by 6 photovoltaic panels, 38 watt peak. These photovoltaic panels are originally designed for a nominal output of 12 volts. All panels have been adapted to produce a nominal 3 volt output.

The experimental results are displayed to give a model for the mentioned operation goals.

TEST RIG

The test rig consists of three main components, namely: the photovoltaic panels set, the storage batteries, controlling switches, the electrolyzer and measuring instruments. In the following, a simple description for each component is given. Figure 1.

PHOTOVOLTAIC PANELS SET

The photovoltaic panels set consists of nine locally manufactured specially designed modules (Figure 2). Each module consists of parallel connected four sets of series connected nine cells rows. The nominal peak power of each module is 50 watts, 4 volts and 30 amperes at the peak point. In Figure 3, a-b, an I-V curve of a typical module is given, the peak power is 50 watts at Vp of 4.31 volts and Ip of 11.75 amperes. The open circuit volt is 5.38 volts and Ip of 70 amperes. (See Appendix I for more details.)

In order to minimize the power losses, blocking diodes have been connected only when the system is running in battery mode; otherwise, the system is connected directly to the electrolyzer.

CURRENT CONTROL PANEL

The requirement of this experiment work is to control the current flow from zero to the maximum power. To carry this task, two methods have been introduced to give an easy, simple and cheap, full control on the current flow; namely:

1. On/Off switch control for each individual solar panel and an overall Master Switch for emergency cases. (See Figure 1)

2. In the solar mode, the orientation to the sun may be also used as a controlling parameter. Putting the solar modules out of sun direction, the flowing current may be controlled. The maximum output will be, of course, when the solar panels are tracking the sun. In order to this task, the solar panels have been installed on a six rollers fram. Simple hand moving will put the solar panel in the proper place. A simple targeting pole is used to help proper

sun orientation.

STORAGE BATTERIES

Four low voltage, two volt each, high capacity, 1100 ampere-hour, liquid acid deep charge batteries have been used. The batteries have been considered for overnight operation phase.

ELECTROLYZER START-UP

1. The electrolyzer is assembled with activated cathode, anode, and very thin KREBSKOSMO P.S. diaphragm. Take care during terminal connection; by tightening the cable nuts, electrodes should not be turned.

2. 600 ml of electrolyte (KOH) has been filled at each side from the top of the gas separator and keep approximately 10 minutes as it is. During this time, the diaphragm slowly absorbed electrolyte and all air bubbles raised to the gas separator. The stagnant bubbles have been moved by knocking with finger on the flexible PVC-tube. The electrolyte filling caps at oxygen side have been closed.

3. Distilled water has been filled into the middle of gas separator bringing the level up to the level of electrolyte.

4. The pole connection from solar panels has been checked again to the electrolyzer.

5. The current has been switched on and adjusted as low as possible by panel by panel and by sun direction. Only little deflection at ampere meter will be enough. However, keep for two to three minutes. Slowly start gas production. The cell voltage has been observed near about 2.0 to 2.8 volts.

6. Gradually and slowly has been increased the current up to 20 ampere; all air has been allowed to purge from gas separators.

7. The two flexible PVC-tubes O_2 and H_2 have been connected in the middle of gas separator with a slight slope towards separator (to avoid carryover of condensate). Now, hydrogen gas will pass through the flexible tube to vent. The other flexible tube for oxygen has been dipped in the water column in pressure adjusting seal and the oxygen gas pressure is adjusted so that the water levels in gas separator for oxygen and hydrogen are equalized.

   In this way, the electrolyzer can run without hydrogen gas analyzer and also at very low load.

GAS ANALYZER

Hydrogen gas analyzer has been taken into operation after adjusting and setting with the test gas. (Appendix II gives more details.)

ADJUSTMENT OF GAS ANALYZER WITH HYDROGEN TEST GAS

Following steps have been carried out:

1. Open cock valve towards moisture remover column.

2. Open safety over pressure cock valve and dip flexible PVC tube into water seal.

3. Fill column with blue Silicagel (approximately 150-200 ml). Attention: In the bottom of the column, there should be one PVC-strainer.

4. Flexible PVC-tube from hydrogen flowmeter to analyzer to be connected and outlet from analyzer to atmosphere (It should always be free to atmosphere.)

5. Hydrogen gas cylinder for test gas, purity should be 99.99%.

6. Connect one pressure reducing valve with two manometers.

7. Open the cylinder main valve and reduce the gas pressure in outlet manometer to approximately 0.1 bar.

8. Connect the given flexible PVC-tube to the test gas. Now, the H_2 gas will pass through cock valve to safety seal pot; gas bubbles are visible.

9. After a few seconds, slowly open the needle valve for H_2 flowmeter and adjust the flow rate between 10 to 20 nl/h.

10. Analyzer will show full deflection (>20 mA). In a few minutes, it will start falling. Wait till the value is constant.

11. With a small screw driver, set the analyzer 0 mA and wait for a few minutes watching the setting value. If it is constant, close the hydrogen cylinder main valve. Watch flowmeter; it will drop to zero.

MEASURING OF PRODUCED HYDROGEN

1. Now, change the product gas direction. Gas will go through condensate catch pot and the flowmeter will indicate the gas flow.

2. If the H_2 gas flow is very low, increase the electrolyzer load.

3. Analyzer will again indicate full deflection, but after a few minutes it will slowly come down. This will take approximately 10-15 minutes. Wait till the value (mA) is constant. Gas purity can be read from figure of gas purity (Figure 6).

4. Adjust the O_2 gas pressure by dipping the pipe in pressure adjusting seal pot to keep the water level in gas separator for O_2 and H_2 equal.

ELECTROLYZER AT HIGH LOAD

In this case, anolyte and catholyte flow are very fast. There may be a big electrolyte level difference in the gas separator, low level mostly being on the anolyte side.

To prevent this level difference, put squeezing clamp on the flexible tube at anolyte inlet and reduce the flow by tightening the screw. The level will come up; adjust the squeezing clamp so that the levels of anolyte and catholyte are equal.

BLUE SILICAGEL

When the originally blue color of the Silicagel turns to light violet, this indicates near exhaustion. Remove the column from the clamp, open the top rubber cock and fill all the Silicagel into one flat beaker glass. Regenerate the Silicagel at approximately 100•C in an oven for two to three hours. The color will turn back to blue. After cooling to approximately 40-50•C, the Silicagel can be filled back into column. For a few minutes, the gas analyzer will indicate full deflection. This is due to the air (O_2) contained in the column.

EXPERIMENTAL RESULTS

The zero hydrogen has been opened and the analyzer warming up is started. After one hour of operation, as recommended by the manufacturer, the test runs have been started.

The operating current is started by letting only one module on. The hydrogen purity has been read on the purity current ammeter. As shown in the current-purity Figure 5, better hydrogen purity is achieved as the operation current exceeds 28 ampere. Increasing the current, the stability running status has been observed to be asymptotic achieved at a current of 2.25 ampere. From the hydrogen analyzer current-vis-purity Figure 6, we have hydrogen purity of 99.7%.

Loading the analyzer with more current by introducing more solar modules the current increase has been achieved with an increase of the voltage applied on the cell. In Figure 7, the system performance is displayed.

SAFETY PRECAUTIONS

General

Hydrogen is an inflammable gas. Mixtures of hydrogen and oxygen or of hydrogen and air form explosive oxyhydrogen, which can be ignited already by sparks.

Oxygen must not come into contact with oils or fats or any other organic compounds to avoid explosive fires. Further hazards can arise from working with D.C.-voltages and handling of caustics (electrolyte).

Any use of fire or open lights and smoking are strictly prohibited in all rooms where hydrogen and oxygen are generated and/or processed. It is further prohibited to store any combustible or self-inflammable materials in these rooms. Fire extinguisher must be kept nearby.

REPAIR DURING OPERATION

It is not allowed to carry out repair or other works where sparks or fire can occur at the electrolyzer or in rooms with possible hydrogen/oxygen atmosphere. Such works can only be done after shut-down of the system.

SHORT-CIRCUITS BY FOREIGN MATTER

Tools, wires or other metal parts must not be kept lying near the electrolyzer as they might cause short circuits, sparks, or fire accidentally by careless handling.

LEAKAGES, CRYSTAL FORMATION

The electrolyzer must be kept dry and clean at all times and be free of leakages and crystals. The latter have to be removed during operation intervals with small wooden rods and by rinsing with water. the cell block and the connecting parts shall also not be encrusted with crystals as these could cause electroytical corrosion and short circuits with formation of sparks.

VENTILATION OF THE ROOM

Good ventilation of the room should be provided for so that any hydrogen which may possibly escape from the plant can quickly be vented from the room. Ventilation facility should be installed at the highest spot of the room and must never be closed.

PROTECTION AGAINST CAUSTIC INJURIES

While preparing the electrolyte, goggles and protective rubber gloves have to be worn. Caustic potash which is used as electrolyte eats into the skin and causes eye injuries. Therefore, a bucket with 3-5% acetic acid, a bottle of 3% boric acid, and vaseline should be kept at hand.

ATTENTION

Demineralized water is NOT DRINKING WATER; it causes heavy intestinal trouble!

DANGER SIGNS

The following danger signs have to be fixed in a position where they will be clearly noticed, as well as at the entrance doors:

1. SMOKING PROHIBITED.

2. ENTRANCE ONLY ALLOWED TO OPERATING PERSONNEL.

APPENDIX I

PV MODULES

Specially designed and manufactured photovoltaic modules have been used in order to match the hydrogen production analyzer. Each module consists of four rows; each has nine cells connected in parallel. The I-V characteristic curve of one of the modules is shown in Figure 3-a and Figure 3-b. In Figure 3-a, the I-V trace is a Spire flash one (SFT), however, the Figure 3-b is produced using a portable tracer (PST), which operates in the field under the solar exposer.

The two Figures reflect the Isc and Voc of the module. The insolation by the SFT is 1000 watt per square meter w/m^2 and the cell temperature has been maintained at 25C°. On the other hand, using the PST in the field, the solar insolation has been measured and found to be 958 w/m^2 and the cell temperature has been evaluated and found to be 44C°.

The peak power is only 22.83 w, that is 45% of the manufacturer given ideal peak power; 50.58 w. The PST short circuit current I_{sc}, however, is 11.72 ampere as compared by 13.06 ampere as given by the SFT. Also, the SFT open circuit volt V_{oc} is 5.38 volt and the PST is 4.73 volt.

Since the maximum portable tracer current is 50 ampere in Figure 4, an I-V curve for only three by three modules is traced by the portable tracer.

APPENDIX II

GAS ANALYZER

A gas analyzer with sealed reference chamber based on the thermal conductivity method has been used. (WLF - Siemens, W. Germany) Figure 8.

DESCRIPTION

APPLICATION

The measurement of the concentration of a gaseous component, e.g., H_2 in binary or quasi-binary gaseous mixtures.

The zero can be selected at any value between 0 and 100%, thereby permitting measurements with the range of interest only.

The analyzer is supplied in a cabinet, together with the devices for gas flow and preparation.

DESIGN

Version for field mounting sheet-steel housing with lockable door. Electrical connection to terminals via two cable glands.

MODE OF OPERATION

The Wlf gas analyzer uses the thermal conductivity of gases to measure their concentration.

The warming up of an electrically-heated wire is dependent on the thermal conductivity of the gas surrounding it. Four such wires are connected to form a Wheatstone bridge. Two are exposed to the sample gas, the other two to the reference gas. A constant direct current heats the wires, in air by about 80K, in hydrogen by about 30K above the temperature of the chamber blocks.

Because of the different thermal conductivities of the sample and reference gases, the wires are warned up by different amounts. The wires therefore assume different temperature and resistance values, thereby unbalancing the wheatstone bridge and causing a current to flow in the bridge diagonal. This current is a measure of the concentration of the component in the case of small changes in the difference in the thermal conductivity and with quasi-binary gas mixtures.

SET-UP WITHOUT DIAPHRAGM PUMP PRESSURE (FIGURE 9):

1. Shut off the sample gas line (close needle valve (1)).
2. Connect a water-filled U tube 1.5 meter long (7) to the gas outlet nipple.
3. Turn cock (4) into position.
4. Open needle valve (1) and close it again when the U tube (7) indicates a pressure of about 0.1 bar.

5. The set-up is leakproof if the pressure in the U tube drops by less than 1% in 20 minutes; otherwise, eliminate leak and repeat test.

6. Remove U tube and reestablish operating condition.

ADJUSTING THE MECHANICAL ZERO:

1. Switch off the main voltage.

2. Adjust the built-in indicator to zero using the mechanical zero adjuster.

3. Adjust any external indicators or recorders to zero.

ADJUSTING THE ELECTRICAL ZERO:

To do this, the zero gas must be applied to the measuring chambers. The composition of the zero gas is indicated on a label on the chamber block and depends on the filling of the reference chambers, e.g., dry air (1) or pure gas, such as helium (He), argon (Ar), methane (CH_4), nitrogen (N_2), ammonia (NH_3) or a gas mixture such as 50% methane (CH_4) and 50% carbon dioxide (CO_2) in the case of digester gas measurements.

The gas analyzer must have attained the operating temperature.

1. Connect a zero cylinder through a reducing valve as in Figure 6 to 8.

2. Close the needle valve (1) and turn the cock (4) into position.

3. Open valve of zero gas cylinder and pass gas into analyzer at a flow rate of 0.3 l/min.

4. Turn the adjuster for the electrical zero so that the correct value for the concentration of the zero gas is indicated (initial scale or full-scale valve).

REFERENCES

[1] Divisek, J., Mergel, J., Schmitz, H., "Improvement of water Electrolysis in Alkaline Media at Intermediate Temperatures", Int. J. of Hydrogen Energy, Vol. 7, No. 9, pp 695-701, 1981.

[2] Carpetis, C., "An Assessment of Electrolytic Hydrogen Production by Means of Photovoltaic Energy Conversion", Int. J. of Hydrogen Energy, Vol. 9, No. 12, pp 989-991, 1984.

[3] Winter, C. J., Nitch, J., "Hydrogen Energy - A Sustainable Development - Towards a World Energy Supply system for Future Decades", Int. J. of HydrogenEnergy, Vol. 14, No. 11, pp 785-796, 1989.

[4] Dutta, S., Block, D., Port, R., "Economic Assessment of Advanced Electrolytic Hydrogen Production", Int. J. of Hydrogen Energy, Vol. 15, No. 6, pp 387-395, 1990.

[5] Winter, J., Nitsch, J., "Wasserstoff als Energietraeger", Springer Verlag, Berlin, W. Germany.

[6] Sidz, J., "Modern Concepts of Electrochemistry", No. 18, p 304, 1986.

[7] Krebskosmo, "Private Communication", Berlin, W. Germany, 1991.

Fig 1 System Flow Chart.

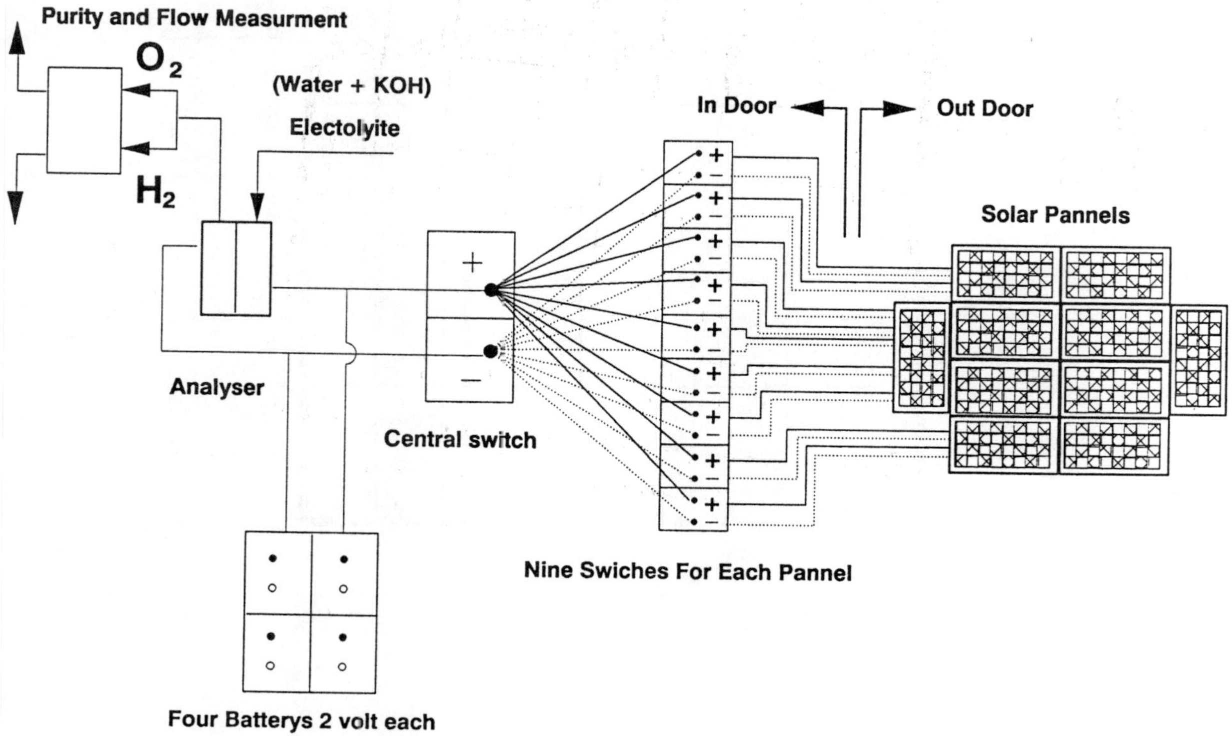

Fig 2 Photovoltaic Special Pannel

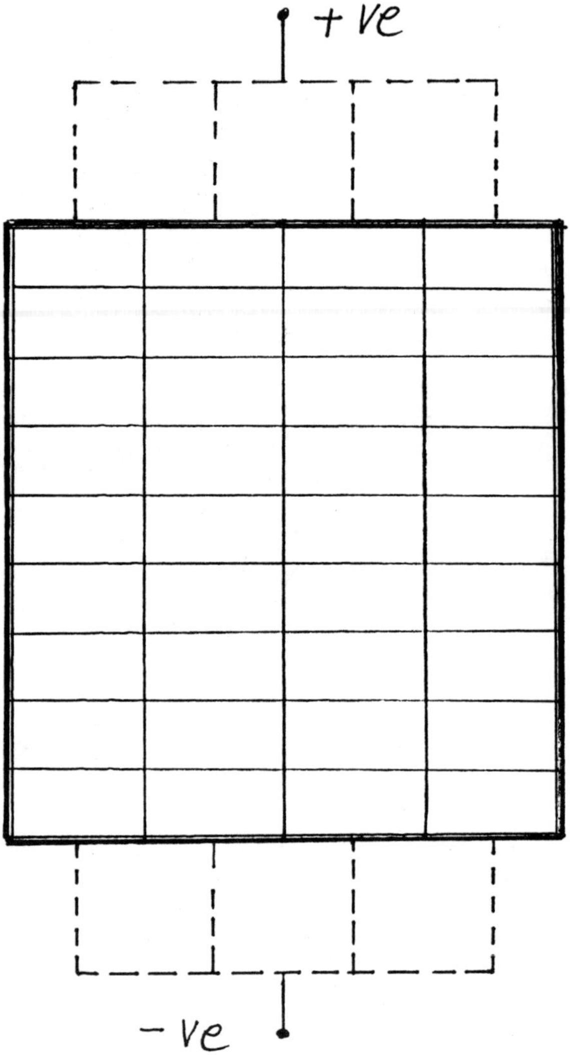

Fig 3 I-V curve for a typical module as traced by
a- Spire Flash Tracer (SFT)
b- Portable Solar Tracer (PST)

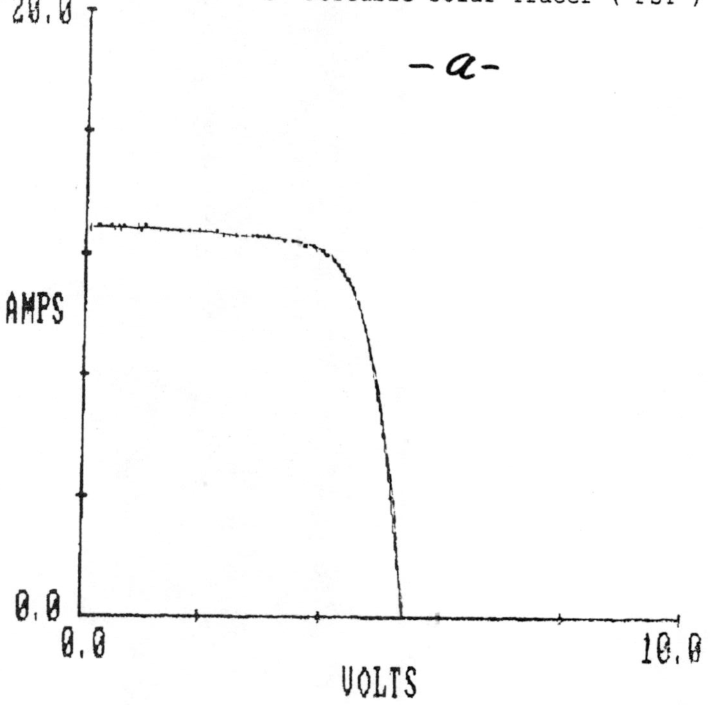

Serial# 10259
15-07-1991, 08:39:33

Sun = 100 mW/cm2
Temp = 25 degC, 29
Voc = 5.38 V
Isc = 13.06 A
Pmax = 50.58 W
Vmp = 4.31 V
Imp = 11.75 A
FF = .72
Rs = 0.05 Ohms
Vld = 4.25 V
Ild = 11.84 A
Pld = 50.32 W
Ef,c = 13.6 %

Fig 3 -b-

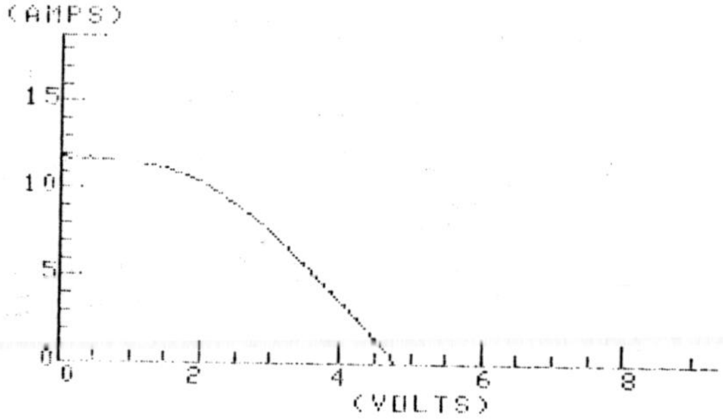

DATE 07-30-91 TIME 11:17:15

SITE	SV	
SUB-SYSTEM	H2 PROD	
MODULE	NO.6 S/N10259	
ARRAY TEMP	44	C
IRRADIANCE	FIX TILTED TG/.958	KWM
ISC	11.72	A
VOC	4.725	V
P PEAK	22.83	W
I AT PEAK	8.424	A
V AT PEAK	2.710	V
FILL FACTOR	41.22	%
DATE	07-30-91	
TIME	11:16:08	

Fig 4 I-V curve for three modules connected in parallel as traced by the PST

-a-

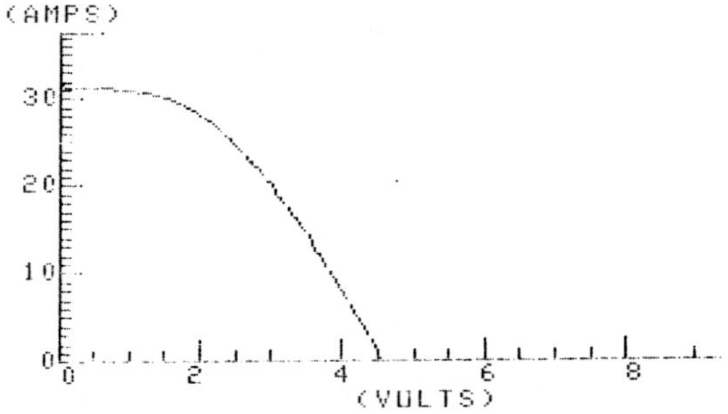

DATE 08-07-91 TIME 11:16:22

SITE	SV
SUB-SYSTEM	H2 PROD
MODULE	PV MODS1,2,3.
ARRAY TEMP	49 C
IRRADIANCE	TMI/.953 KWM
ISC	31.35 A
VOC	4.505 V
P PEAK	61.55 W
I AT PEAK	22.70 A
V AT PEAK	2.710 V
FILL FACTOR	43.57 %
DATE	08-07-91
TIME	11:15:09

Fig 4 -b-

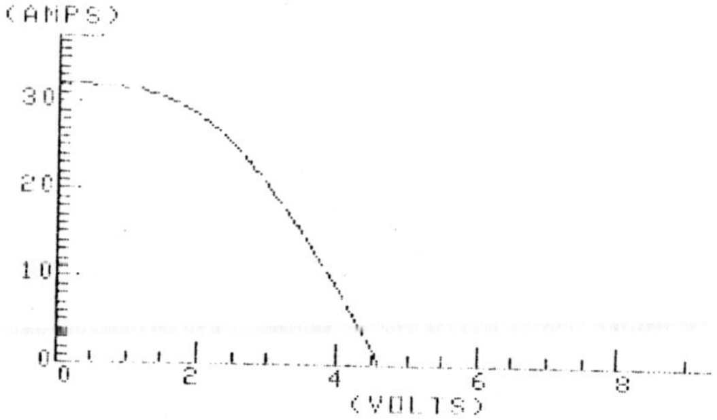

DATE 08-07-91 TIME 11:22:05

SITE	SV
SUB-SYSTEM	H2 PROD
MODULE	PV MODS4,5,6.
ARRAY TEMP	49 C
IRRADIANCE	TNI/.953 KWM
ISC	32.08 A
VOC	4.541 V
P PEAK	62.74 W
I AT PEAK	23.14 A
V AT PEAK	2.710 V
FILL FACTOR	43.05 %
DATE	08-07-91
TIME	11:20:56

Fig 4 -C-

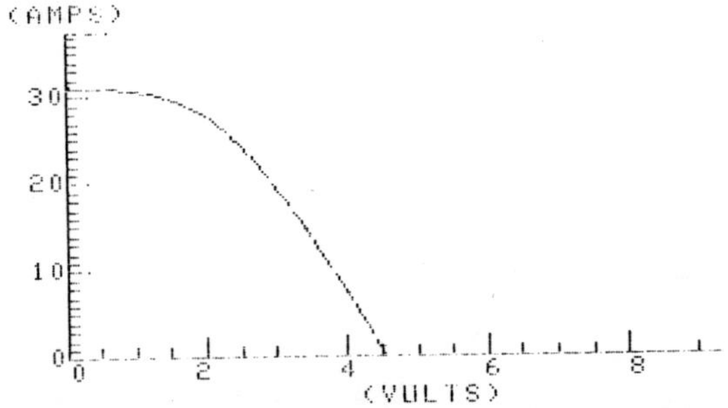

DATE 08-07-91 TIME 11:27:11

SITE	SV	
SUB SYSTEM	H2 PROD	
MODULE	PV NOS7,8,9.	
ARRAY TEMP	49	C
IRRADIANCE	TNI/.953	KWM
ISC	31.06	A
VOC	4.505	V
P PEAK	60.35	W
I AT PEAK	23.88	A
V AT PEAK	2.527	V
FILL FACTOR	43.13	%
DATE	08-07-91	
TIME	11:26:04	

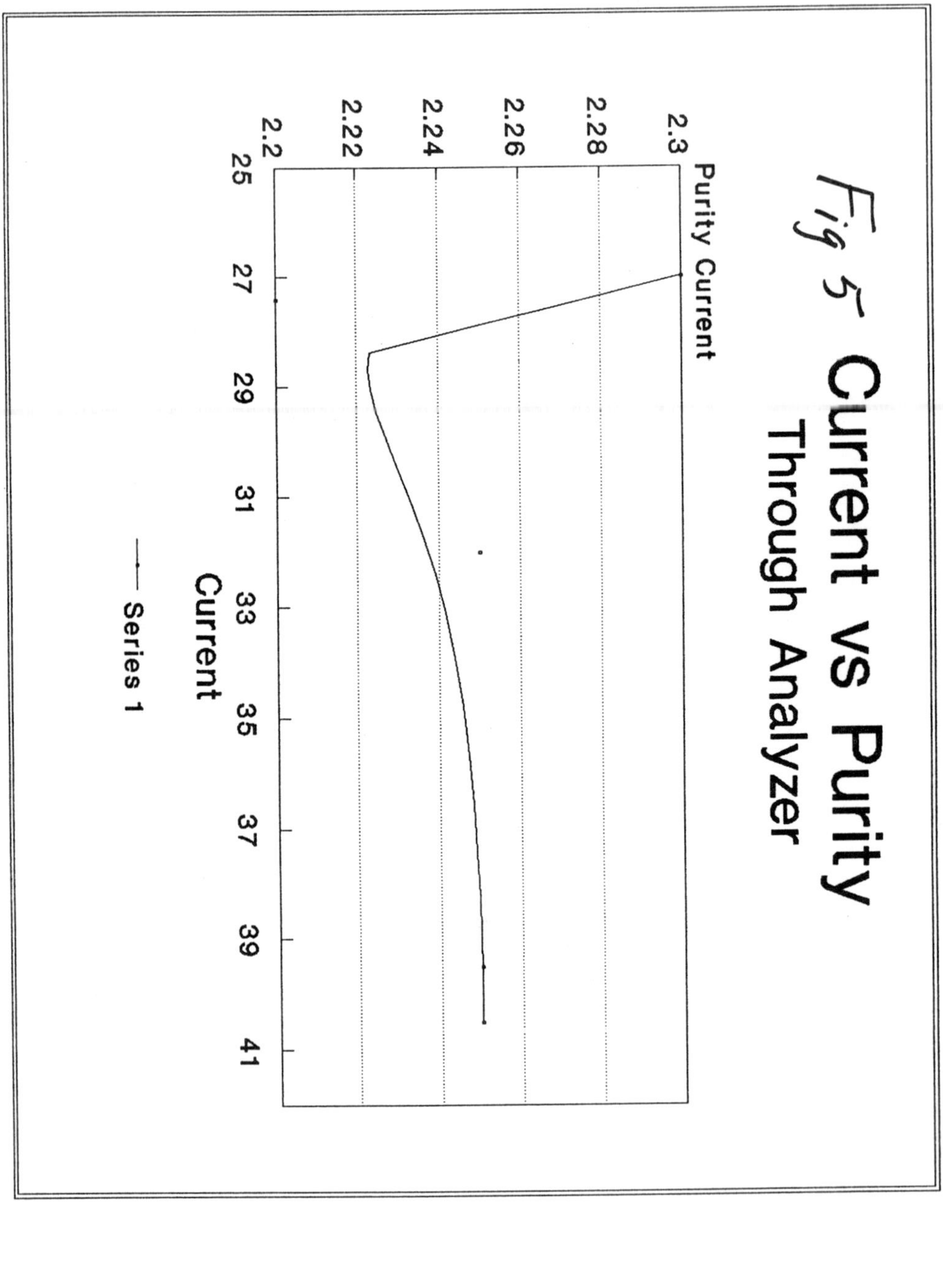
Fig 5
Current vs Purity
Through Analyzer
Purity Current
2.3
2.28
2.26
2.24
2.22
2.2
25
27
29
31
33
35
37
39
41
Current
Series 1

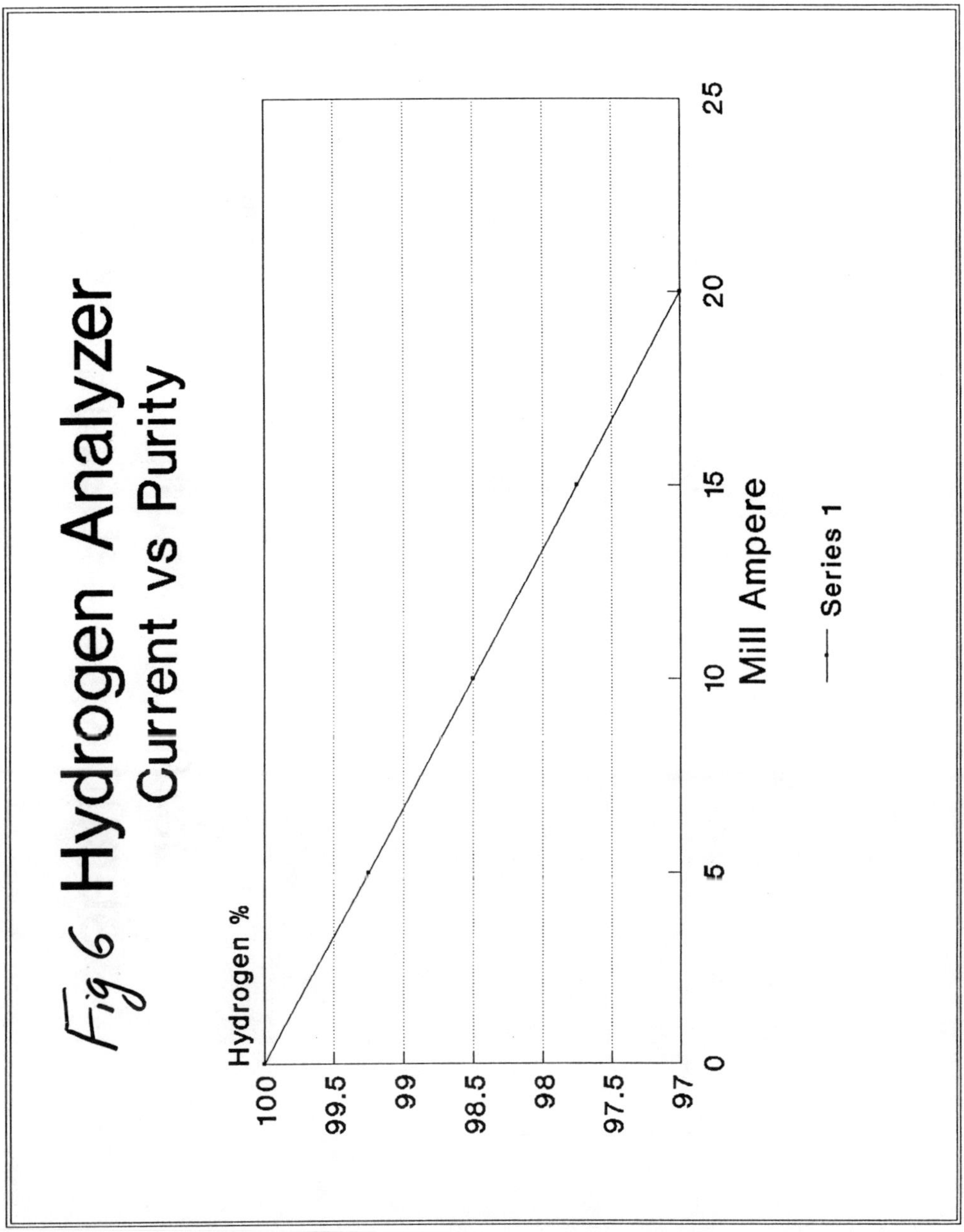
Fig 6 Hydrogen Analyzer
Current vs Purity
Hydrogen %
100
99.5
99
98.5
98
97.5
97
0
5
10
15
20
25
Mill Ampere
Series 1

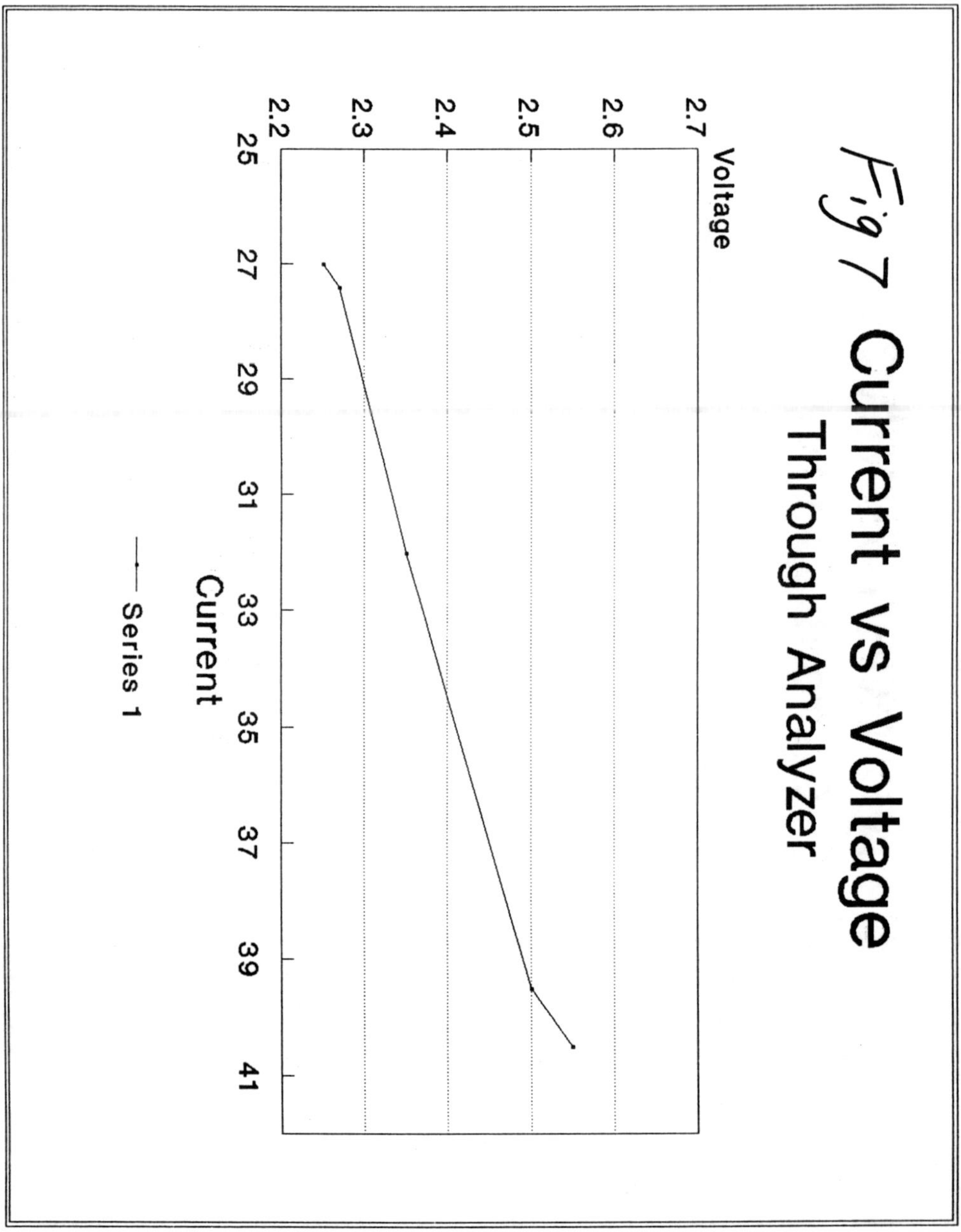
Fig 7
Current vs Voltage
Through Analyzer
Voltage
2.7
2.6
2.5
2.4
2.3
2.2
25
27
29
31
33
35
37
39
41
Current
Series 1

Fig.8 Analyzer with sealed reference chamber, field mounting, dimensions in mm

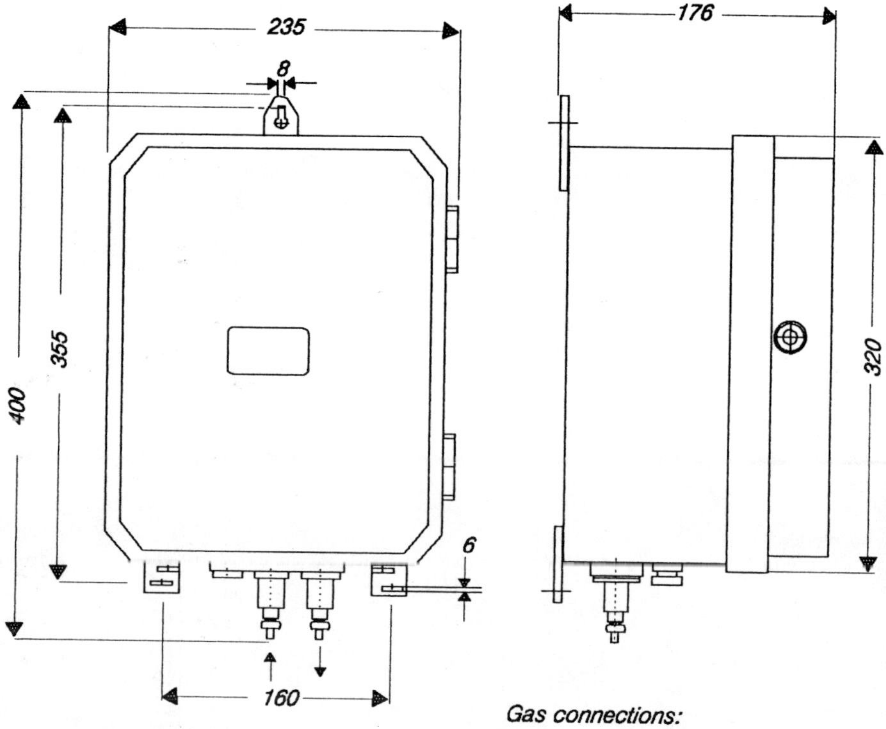

Gas connections:

Tube nipples with 6 mm o.d.

Fig. 9 Measuring set-up without diaphragm pump (pressure)

1 Needle valve
2 Filter bottle
3 Membrane filter
4 Three - Way cock
5 Capillary
6 Gas analyzer
7 U tube for leak test
8 Gas cylinder with reducing valve

Mk sample chamber
Vk Reference chamber
M Sample gas

Cock positions

cock	4	1
Leak test Electrical zero		Open/Closed
test		Closed
Operation		Open

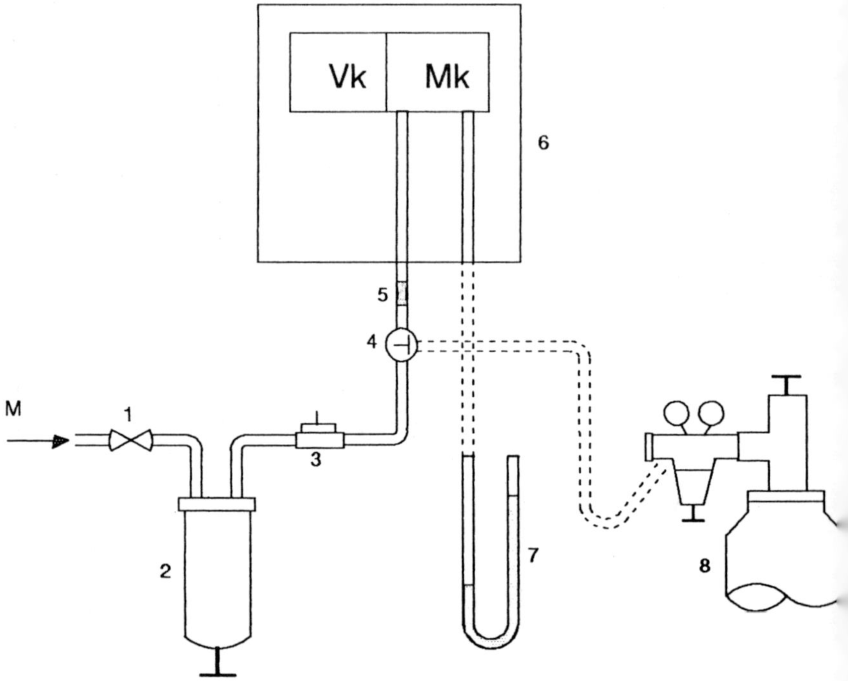

FEASIBILITY STUDIES OF PHOTOASSISTED HYDROGEN GENERATION ON MEMBRANE SUPPORTED CHLOROPHYLL-PLATINUM

S.Nema, N.Gontia, S.Khare & S.K.Nema
Macromolecular Research Centre,
R.D.University,
Jabalpur, INDIA-482 001

ABSTRACT

The feasibility experiments for Hydrogen generation by employing Chlorophyll a-Platinum and Chloroplast-Platinum in a manner very close to solid polymer electrolyte (SPE) technique are presented. Perfluoro sulphonated ionomer membrane is used as SPE. A fine colloidal dispersion of Platinum has been interspersed in the matrix of Nafion-117 as SPE membrane. Chlorophyll-a and chloroplast were coated on SPE by solvent evaporation process. The structural and morphological change in chlorophyll before and after photoirradiation in visible region have been observed. The Diffused Reflectance Spectrum of the chlorophyll containing perfluoro ionomer membrane was similar to absorption spectrum of chlorophyll. The results of spectroscopic (Infra red and UV-VIS) studies have been presented. The wide variation in amount of Hydrogen produced by photoirradiation and irreproducibility has been attributed to lower quantity of Hydrogen due to very small quantities of Platinum and chlorophyll and trapping of Hydrogen in hydrated membrane.

INTRODUCTION

The plausibility of photoassisted Hydrogen by employing Chl-a-H_2O aggregate coated on a transition metal electrode system has received worldwide attention. Tributsch and Calvin [1] used a semiconductor electrode on which chlorophyll was deposited. Photogalvanic effects in metal-chlorophyll-metal sandwich cells have been reported by Tang and Albrecht [2]. Hence, the main approach behind these researches is to search a more compatible photochemical conversion of solar energy storage systems. Specific efforts towards water photolysis system is one of the most attractive and promising direction of Hydrogen economy. Photolysis of water to Hydrogen fuel could be envisaged by the use of artificially designed systems that duplicate functions of the natural process. The various systems include colloidal system, semiconductor electrode and photosensitizer-semiconductor electrode system etc.

The photoelectrolysis at a Chl-a-water aggregate coated Platinum electrode as a photocathode has been studied by Fong et al, they showed Hydrogen generation using Platinized polycrystalline chlorophyll-a electrode by water splitting reaction [3,4].

$$2H_2O \xrightarrow[\text{Chl-a/Pt}]{\text{visible light}} 2H_2 + O_2 \quad ---(1)$$

The photobiological processes that produce Hydrogen can be designed by the separation of chloroplast from plants and exclusion of the enzymes involved in CO_2 fixation.

Photoproduction of Hydrogen has also been reported from isolated chloroplasts in the presence of heterogeneous Platinum-colloids in the presence of bipyridinium charge relays acting as an electron carrier [5]. This communication deals with the photoreaction of Chl-a/Pt arranged in SPE mode. The present studies are the part of project entitled "Photoassisted Water Splitting Using Functional Polymers" (PAWAS), wherein various semiconductor materials, transition metal complexes and chloroplast pigments have been incorporated in few polymers and in perfluoro ionomer membrane to explore the possibility of utilizing solar energy conversion systems. The polymer based photoenergy conversion system imparts structural integrity and make available large surface area for the photoreaction.

Our efforts are mainly focussed for the systems based on direct photodecomposition of water and to develop photo-excitable electrodes modeled on the PS I and PS II involved in the primary photosynthetic process of green plants. The chloroplast which has been extracted from in-vivo structure is incorporated in Nafion membrane i.e. in-vitro system. The efforts to place chlorophyll in membrane matrix is apparently intended to mimic the environment and functional characteristics associated with the leaf matrix. The photolysis of water is the central event of photosynthesis and there is every reason to suppose that chlorophyll is involved in this process. Thus, Witt and coworkers [6] found that a long wavelength chl-a(II^-) 435-682 nm is involved in the photolysis of water. In the proposed system it is supposed that the Platinum particulates catalyses the redox reaction of chlorophyll. The chlorophyll take adequate energy from visible light to yield excited chlorophyll*. The chlorophyll* release an electron to a proton in the electrolyte solution and generate Hydrogen. The resulting cation radical (chl^+) accepts an electron from OH^- ion to regenerate chlorophyll. Besides establishing feasibility of Hydrogen generation comparative study of chlorophyll activity before and after white light illumination was made possible through FT-IR and UV-VIS spectroscopic studies.

EXPERIMENTAL

Chlorophyll was isolated from fresh spinach leaves by usual procedures. All the reagents viz acetone, petroleum ether (60^oC-80^oC) and methanol used in the extraction and separation of chlorophyll were 99.99% pure and of AR Grade. 1% solution of Hydrogen Hexa Chloroplatinate (IV) hydrate (H_2 Pt Cl_6.$6H_2O$) used to get a fine dispersion of Platinum on SPE membrane was obtained from Aldrich Chemical Company, USA. Nafion-117 used as SPE membrane was obtained from du Pont, USA. The membrane had an equivalent weight of 1100 corresponding to an ion-exchange capacity of 0.83 m eq/g.

Preparation of Chlorophyll-a

50 gm of dried and powdered spinach leaves were digested with 125 ml of acetone containing 15-20% water for 10 hrs. From the filtrate thus obtained the solvent was removed by distillation and chloroplast (CPT) was vacuum dried. Further separation of chlorophyll-a (Chl-a) and chlorophyll-b (Chl-b) was done by applying distribution technique. About 2.5 gm of dried chloroplast was dissolved in 100 ml of Petroleum ether to which 100 ml of 90% aqueous methanol was added. Chl-a goes to upper light petroleum ether layer along with certain epiphasic carotenoids and Chl-b goes to bottom aqueous methanol layer with few hypophasic carotenoids & xanthophylls. Finally, amorphous Chl-a was obtained by the removal of petroleum ether. The amorphous material was dissolved in limited quantity of 100% absolute alcohol and crystalline Chl-a was obtained by the evaporation of alcohol.

Pretreatment of Nafion Membrane

The Nafion membrane was pretreated by immersing in boiling HNO_3-H_2O (1:1) for 30 min to remove impurities and then boiling it in deionized water for an hour to remove nitrates and introduce reproducible amount of water into the membrane.

Dispersion of Platinum on SPE Membrane

A very fine micro dispersion of Platinum on Nafion membrane has been done by the procedures developed in our laboratories. 0.05 ml of 1% methanolic chloroplatinic acid was spread homogeneously on one side of the pretreated Nafion membrane of 4 cm^2 size and solvent was evaporated at 80^oC. This membrane was heated upto a temperature of 200^oC for 3 hrs under Nitrogen atmosphere to yield Platinum loading of approximately 0.05 mg Pt/cm^2 or 1.0×10^{-6}M Pt over four cm^2 membrane.

Preparation of Chlorophyll-Platinum Supported Membranes

The Nafion membranes having Platinum-dispersion were coated with methanolic chlorophyll-a and chloroplast solution on the same side of Platinum by solvent evaporation process. 0.089 mg Chl-a was spread on the membrane which was expected to yield 1×10^{-7}M Chl-a/cm^2 membrane. Similar SPE membranes with CPT and Chl-a were prepared by keeping the molar concentration 10^{-6}M to 10^{-5}M/cm^2. The resultant membranes were dried at room temperature under vacuum.

Characterization

The conversion of H_2PtCl_6 into Platinum particulates has been ensured by the pH experiments and measurements of ion-exchange capacity, this technique is developed in our laboratory [7]. The contents of chlorophyll-a, chloroplast and Platinum in $\mu g/cm^2$ over Nafion membrane are given in Table 1. The Diffused Reflectance Spectra of chlorophyll incorporated Nafion membrane was performed on Perkin-Elmer's UV-VIS Spectrophotometer Model Lambda-6. In the light induced Hydrogen generation experiments the amount of Hydrogen generated with time was recorded on

Applied Photophysics Biopotentiostat. The Chl-a/Pt and CPT/Pt SPE membranes were immersed in the cell of Biopotentiostat containing double distilled water. This water was deaerated by purging with Argon gas for 20 min. The SPE membrane was illuminated by 900 W Photoirradiator. The system was continuously operated under visible light for 2 hrs 30 min in each of the cases as reported in Table 1. The chlorophyll coated on Nafion membrane was extracted by CCL_4 and the FT-IR spectra of this chlorophyll was taken by using Perkin-Elmer's FT-IR Spectrometer Model 1720X.

RESULTS AND DISCUSSION

The urge for studying photocharge generation in the natural components of chlorophyll has prompted us to study it with other photosynthetic pigments present in the chloroplast. Therefore, the experiments with chloroplast and chlorophyll-a without much purification have been performed. Fong et al. [8] found that hydrated microcrystalline chlorophyll-a is superior to the amorphous aggregate in light induced charge separation. Thus, we have designed our system on SPE on the basis of the studies conducted by Fong et al. wherein the photoelectrolysis has been reported at Chl-a-water aggregate coated platinum electrode. Therefore, it became imperative that the nature of chlorophyll-water interaction should be established with regard to our system. In the proposed system the chlorophyll has been incorporated on the Nafion membrane with water. The IR Spectrum shows the absorption band at 1715 cm^{-1} and 1745 cm^{-1} which could be assigned for the ester (C=O) group of hydrated chlorophyll-a as reported by Katz et al. [9]. There is a ester peak splitting which on comprison with assignment of Katz for chlorophylls indicate the hydrated nature of chlorophyll [9]. The disappearance of the absorption peak for free keto (C=O) group at 1700 cm^{-1} indicates that the chlorophyll-a has chlorophyll-chlorophyll interaction involving Keto carbonyl-water interaction, wherein water is coordinated to the magnesium; C=O---HO(H)---Mg. Peaks obtained at 1515 cm^{-1}, 1565 cm^{-1}, 1630 cm^{-1} and 1670 cm^{-1} show the presence of chlorophyll-b and other photosynthetic materials along with chlorophyll-a.

The perfluoro ionomer membrane (Nafion) has been chosen as solid polymer electrolyte due to its excellent chemical and thermal stability. Even after heating at 200° C under Nitrogen atmosphere, the ion-exchange capacity of the membrane is retained [10] though its morphology changes from clustered to homogeneous one. In our experiments Ion-exchange capability for Platinum containing SPE has been found to be 0.0415 meq/cm^2. Platinum particulate plays important role to catalyse the water splitting reaction of chlorophyll. They can be considered as micro-electrodes on the surface of Nafion as well as within its matrix.

In the photoassisted water splitting experiments the SPE membranes with CPT-01, CPT-02 and CPT-03 did not respond to white light illumination hence no Hydrogen generation was observed. On subjecting SPE membranes CHL-01, CHL-02 and CHL-03 to irradiation, no Hydrogen generation was observed for first half an hour in each of the cases. Afterwards slow change

in potential indicated the feeble generation of Hydrogen. The maximum Hydrogen generation observed was 2.8×10^{-3} cc, 2.1×10^{-3} cc and 1.4×10^{-3} cc H_2 per 40 ml of water at different time duration in CHL-01, CHL-02 and CHL-03 respectively. Further no more Hydrogen was evolved as is evident in Figure 1. No Hydrogen was generated when Nafion and Pt modified Nafion was photoirradiated under identical conditicn.

In addition to photolytic reactions as shown by eq. (1), Photoreduction of CO_2 into formic acid by Platinized (Chl-a-$2H_2O$) has been reported [11]

$$2CO_2 + 2H_2O \xrightarrow[\text{Chl-a}]{\text{visible light}} 2\,HCOOH + O_2 \quad \text{----(2)}$$

The Ar purging to drive CO_2 and O_2 present in the system has become essential. The Hydrogen generation in the photocatalytic system containing chlorophll-Platinum may have the generalized mechanism as shown below

$$[Chl] \xrightarrow{h\nu} [Chl]^* \xrightarrow{Pt} [Chl]^+ + e^- \quad \text{----(3)}$$

$$e^- + H^+ \longrightarrow H \quad \text{----(4)}$$

$$H + H \longrightarrow H_2 \quad \text{----(5)}$$

$$[Chl]^+ + e^- \text{ (from } OH^-) \longrightarrow [Chl] + \tfrac{1}{2}O_2 \quad \text{----(6)}$$

The amount of H_2 generated is quite low which is attributed to a very small surface area of the membrane exposed to photoirradiation. For our future experiments a photochemical cell with larger surface area available for photoirradiation has been designed and fabricated, which can accomodate large surface area membrane with suitable photoirradiating system. The schematic diagram of this cell is shown in figure 3. Larger quantity of H_2 may be generated due to exposure of large surface area and the precise estimation of H_2 may become feasible as it will be generated in nonaqueous zone. Further experiments are underway.

Since the stoppage of Hydrogen generation was observed after 3 hrs, the structural changes in the chlorophyll has been noted by studying the spectrophotometric analysis of the samples in the blue & red region. The ratio of the absorption in blue (422 nm) to in red (665 nm) has shown the variation from 2.70 to 3.45 this indicate the deterioration of chlorophyll structure. This observation however, is at variance from what was reported by Fong [11] wherein it has been suggested that the deterioration occurs at Chl-a-Pt interface rather than that of chlorophyll. The Diffused Reflectance Spectrum of the membrane soaked with chlorophyll solution shows maximum absorption at 700 nm. The λ max indicate that chlorophyll incorporated on the Nafion membrane has photoactivity very close to chlorophyll solution (absorption maximum is closer to solid one). The results of photo-splitting of water experiments with Chlorophyll-Platinum over the surface of the membrane in view of the SPE mode operation are reported. The experiments are underway wherein chloroplast/chlorophyll-Platinum will also be incorporated throughout the membrane using the technique being developed in our laboratory [12].

Evidently, chlorophyll photoactivity is poisoned by oxygen in both in vivo [13) and in vitro [14] light reactions. It is credible that carotenoids protect triplet excited chlorophyll from oxidation [15]. The immense role played by β-carotene is in the stabilization of chlorophyll from oxidation and it can efficiently store the photoenergy. It also prevents chlorophyll degradaion in similar manner as in vivo [16]. It may be concluded that when light falls on the electrode the Chlorophll-Platinum constitute a micro-electrode onto (or within) Nafion matrix.

CONCLUSION

The feasibility of photocatalytic Hydrogen generation has been established by photoirradiation of SPE supported Chl-a/Pt. The reproducibility of the experiments with chloroplast-Pt containing membranes could not be attained and the investigations with improved experimental coditions are planned. We deem to get better production of Hydrogen by utilizing a more purified sample of chlorophyll-a in SPE electrode.

ACKNOWLEDGEMENTS

The authors are grateful to Dr.V.R.Gowariker and Dr.K.V.C.Rao erstwhile members of IDRG, VSSC, Trivandrum, India for constant inspiration and helpful discussions. The kind support for spectroscopic analysis by Dr.M.Sharon, IIT,Bombay and Dr.Hitesh Malviya, Bioscience Department, R.D. University, Jabalpur is gratefully acknowledged. We profusely thank our colleagues M/S S.D.Namdeo and Rajendra Dubey of MRC for secretarial assistance.

REFERENCES

1. H.Tributsch and M.Calvin, Photochem. Photobiol.,(14), 95, 1971.
2. C.W.Tang and A.C.Albrecht, J.Chem. Phys., (63), 953, 1975.
3. F.K.Fong and L.Galloway, J.Am. Chem. Soc., (100), 3594, 1978.
4. L.Galloway, D.R.Fruge, G.M.Haley, A.B.Coddington and F.K.Fong, J.Am. Chem.Soc.,(101) , 229, 1979.
5. P.Cuendet and M.Gratzel, Photobiochem. Photobiophys., (2), 93, 1981.
6. H.T.Witt, Naturwissenschaften, (55), 219, 1968.
7. "The Estimation of conversion of Electrocatalytic grade Platinum by thermolytic dissociation of chloroplatinic acid on surface of polymer films using pH measurements", Macromolecular Research Centre, R.D. University, Jabalpur, 1991 Technical Report No.MRC:TECH-REP:GEN-2:91.

8. F.K.Fong and N.Winograd, J.Am. Chem. Soc., (98), 2287, 1976.

9. Karlheinz Ballschmiter and J.J.Katz, J.Am. Chem. Soc.,(91:10), 2661, 1969.

10. J.Surowiec and R.Bogoczek, J.Thermal Analysis, (33), 1097, 1988.

11. D.R.Fruge, G.D.Fong and F.K.Fong, J.Am. Chem. Soc., (101:13), 3694, 1979.

12. "Incorporation of chlorophyll in Nafion membrane",Macromolecular Research Centre, R.D.University, Jabalpur, 1991 Technical Report No. MRC:TECH-REP:GEN-3:91.

13. W.Junge in "Photosynthesis I". A.Trebst and M.Avron, Eds. Springer-Verlag, Heidelberg, Germany, 70, 1977.

14. L.Galloway, D.R.Fruge, L.M.Fetterman and F.K.Fong in "The Chemical Conversion and Storage of Solar Energy", R.Bruce King, R.R.Hautala and C.R.Kutal, Eds., Humana Press, Clifton, N.J. 1979.

15. M.Griffiths, W.R.Sistrom, G.Cohen-Bazire and R.Y.Stannier, Nature (London) (176), 1211, 1955.

16. Masuo Aizawa, Morio Hirano and Shuichi Suzuki Electrochimica Acta (23), 1185, 1978.

TABLE 1 : CONCENTRATION OF CHLOROPHYLL-a, CHLOROPLAST AND PLATINUM OVER NAFION MEMBRANE

SAMPLE DESIGNATION	Chl/CHLOROPLAST (g/cm^2)	PLATINUM CONCENTRATION (g/cm^2)
CPT-01	8920.0	50.0
CPT-02	892.0	50.0
CPT-03	89.2	50.0
CHL-01	8920.0	50.0
CHL-02	892.0	50.0
CHL-03	89.2	50.0

FIGURE 1 : HYDROGEN GENERATION AS A FUNCTION OF TIME

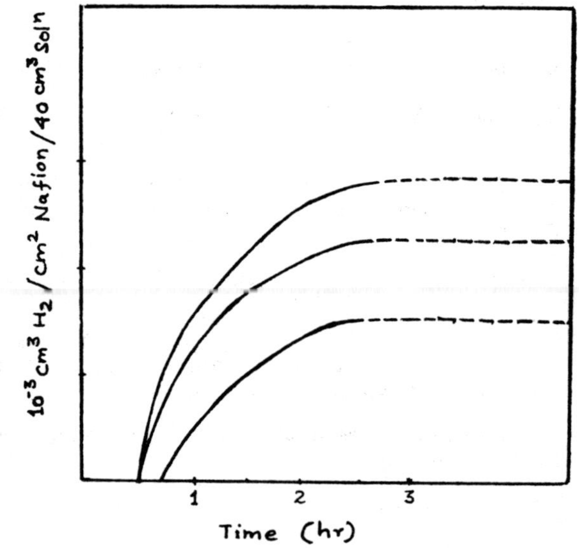

FIGURE 2 : PLOT OF ABSORPTION INTENSITY Vs WAVELENGTH

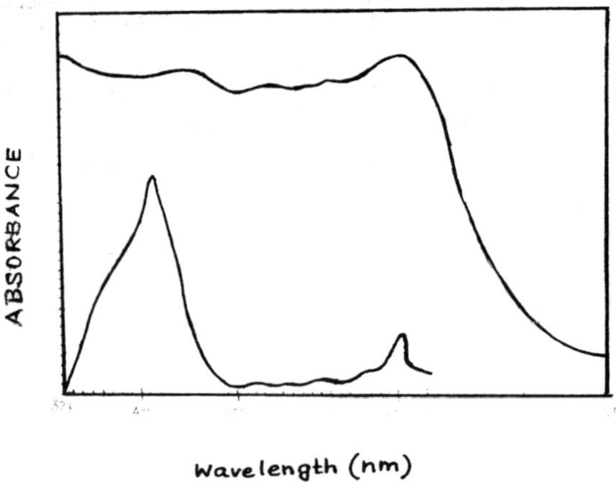

FIGURE 3 : SCHEMATIC DIAGRAM OF PHOTOELECTROCHEMICAL CELL

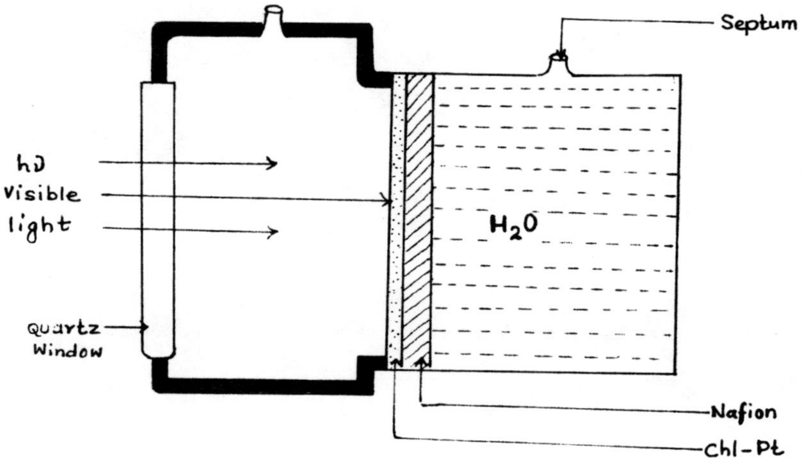

EVALUATION OF IME-SPE HYBRID MEMBRANES FOR HYDROGEN PRODUCTION

S.K.Tiwari, M.K.Gupta, V.Mitra* & S.K.Nema
Macromolecular Research Centre,
Rani Durgavati University,
Jabalpur, INDIA-482 001

*Department of Non-Conventional Energy Sources
New Delhi, INDIA

ABSTRACT

The use of membranes for the production of high purity hydrogen by water electrolysis has drawn considerable scientific attention in recent years. The Solid Polymer Electrolyte (SPE) and Inorganic Membrane Electrolyte (IME) have emerged as potential technologies for water electrolysis in future as they offer specific advantages with regards to efficiency, cost effectiveness and ease of scale up. In the present studies, we have envisaged IME-SPE hybrid membrane system wherein thermally and chemically stable inorganic ion exchangers are incorporated in perfluoro ionomeric membranes. Incorporation of polyantimonic acid (PAM), hydrates of zirconium oxide (ZrO_2) and zirconium phosphate (ZrP) have been incorporated by technique developed in our laboratories. The physico-chemical and mechanical properties of hybrid membranes have been determined and further the electrochemical properties evaluation are underway. The results of this preliminary studies has, however, indicated that IME-SPE hybrid membranes may find applications in Energy Conversion Systems and Nuclear waste recovery management.

INTRODUCTION

Ever increasing global energy demands with depleting fossil fuel sources has made the mankind desparate to find an alternate source of energy for future requirements. Future fuel should be long lasting with no pollution hazards. Hydrogen can be an alternative source of energy as it is non pollutant (on combustion gives water) and could be produced by water electrolysis at competitive costs. The water electrolysis has thus assumed far greater significance in view of the role of Hydrogen in future energy scenario and as well as feed stock to numerous industries. Solid Polymer Electrolyte (SPE) and Inorganic Membrane Electrolyte (IME) technologies for H_2 production by electrolytic method have emerged as potential candidates for low and medium temperature water electrolysis.

Solid Polymer Electrolytes (SPE) are generally perfluoro ionomer membranes. Nafion, a perfluoro ionomer membrane has been the centre of attraction for last few decades because of its (i) exceptional chemical stability (ii) unique morphology in which exchange sites, counter ions and water phases are separated from fluorocarbon backbone. Though, Nafion is generally employed in deionised water electrolysis, it has excellent thermal, chemical and mechanical stability in alkaline media. Neutralization of sulfonic acid group increases the glass transition temperature from 110 to 220°C [1,2]. However, its use is limited to low alkali concentrations (~ 10% NaOH) at high terperatures because of its optimum conductivity [1]. Fischer reported that Nafion is attacked severely in 50% KOH above 150°C [3].

Inorganic Membrane Electrolytes (IME) were developed [4,5,6] to replace asbestos diaphragms in order to increase the operating temperatures from 70-90°C to 120-150°C in alkaline medium. IME membranes consisted of organic binders such as polysulfones (PS), polyvinylidine fluoride (PVDF) and polytetrafluoroethylene (PTFE) with inorganic ion exchangers such as polyantimonic acid (PAM), zirconium oxide (ZrO_2), titanium oxide (TiO_2) and their phosphates. Excepting phosphates, which tend to hydrolyze, PAM and ZrO_2 are resistant to alkali at high temperatures [7,8].

Recently, Morimoto et al.[9] has developed a new concept of AZEC system of alkaline water electrolysis. This employes AZEC membranes developed by surface modification of Flemion DX (a perfluorocarboxyl membrane) with porous non-electrode thin layers making the surface more hydrophilic. It helps eliminating IR drop (ohmic losses) across the membrane caused by adhesion of gas bubbles. Consequently, it is possible to operate the electrolyzer in alkaline medium with zero or minimum gap, enabling both anode and cathode to be loosely held in contact with membrane surface.

Taking the cognizance of the development hitherto undertaken and with a view to evolve a new system which could be used in alkaline medium. The novel membrane being developed shall have increased surface hydrophilicity, ionic conductivity at low water content and ion exchange capacity. We have incorporated inorganic ion exchangers (IIE) into Nafion. Nafion was choosen because of its high ionic conductivity, low gas permeability and excellent chemical, thermal and mechanical stability. Inorganic ion exchangers onto its surface may render the surface hydrophilic and the IIE present onto/into the Nafion matrix will enhance its total ionic conductivities.

The present communication is essentially the results of a part of studies being conducted at our laboratory to realize IME-SPE hybrid membrane for project "Solid Electrolyte Electrolyzer Development (SEED)" sponsored by Department of Non-Conventional Energy Sources, Govt. of India.

EXPERIMENTAL

Materials

Membranes of Nafion-117 (E.I. du Pont, USA) having equivalent weight (EW) of 1100, thickness of 0.185 mm and ion exchange capacity (IEC) of 0.83 meq/gm were used. Pretreatment of purchased membranes was done by immersing it in boiling HNO_3:H_2O (1:1) for 30 min and than boiling in deionised water for 1 hr.

Antimony (V) chloride and Zirconyl oxychloride were obtained from Fluka, USA, and were used as such. Other chemicals were phosphoric acid, methanol, and hydrochloric acid, all A.R.Grade from Merck (India), were used without further purification.

IME-SPE Hybrid Membrane Preparation

The IME-SPE hybrid membranes were prepared by incorporation of inorganic ion exchangers (IIE) in Nafion. The procedure involves the precipitation of these exchangers taking advantage of ion exchange (or Donnan's absorption) and then repeated exchange and precipitation (REP) technique.

Polyantimonic acid (PAM) was incorporated by immersing methanol swelled Nafion into a 2% solution of Antimony (V) chloride in methanol for 24 hrs followed by immersing the resultant membrane in 0.5N HCl solution at $60^{\circ}C$ for another 24 hrs. The resultant product is designated as NAF PAM-101. Additional PAM was incorporated by repeating the same process and each incorporation is termed as number of loading and the samples are designated as NAF PAM-102, NAF PAM-103 and so on.

Zirconium phosphate (ZrP) was incorporated by refluxing exchanged/soaked films in 2% Zirconyl oxychloride-methanol solution in 10% phosphoric acid for 2 hrs. This sample is designated as NAF ZRP-201. Additional ZrP was incorporated by repeated loading and samples are designated as NAF ZRP-202, NAF ZRP-203 and so on.

Zirconium oxide (ZrO_2) was incorporated by soaking dry acid Nafion films with a 2% solution of Zirconium oxychloride in methanol for 24 hrs. Exchanged/soaked films were boiled in slightly alkaline water for 1 hr washed and dried at $110^{\circ}C$ for 4 hrs. The resultant product is designated as NAF ZRO-301. Additional ZrO_2 was incorporated by repeated loadings and the samples are designated as NAF ZRO-302, NAF ZRO-303 and so on.

Characterization

All the samples prepared were characterised using Thermogravimetric Analyzer (Perkin-Elmer), Systronics μ pH meter, Instron's Universal Testing machine and a cell to measure burst pressure.

Water Content

Water content of wet membranes was measured employing a novel technique using TGA [10]. Samples (7-10 mg) were kept at $30^{\circ}C$ under nitrogen purge and temperature was raised to $220^{\circ}C$ with heating rate of $10^{\circ}C$/min. The temperature is kept at $220^{\circ}C$ for 1 hr and the difference in weight recorded as water content. The water content in all hybrid membranes is given in Table 1.

Ion Exchange Capacity (IEC)

Ion Exchange Capacity was determined by the technique developed in our laboratory [11]. Membranes in H^+ form were kept in 0.1N NaCl solutions of different pH for 24 hrs. The pH of resulting solutions was measured and the total concentration of H^+ ions released was calculated. Polyantimonic acid (PAM) shows high exchange capacity for Ag^+ ions [12]. The NAF PAM membranes were dipped in $AgNO_3:HNO_3$ (1:1) solution for 24 hrs and than boiled in water to remove excess $AgNO_3$. The exchange of Ag^+ ions was determined by TGA method. Datas are given in Table 2.

Mechanical Properties

The tensile strength, elongation and tensile modulus was measured for different samples in accordance with ASTM/D/882 using Instron's Universal Testing machine with cross head speed of 50 mm/min. Burst strength was determined using a two compartment cell. A pressure of 5 kg/cm^2 was applied to 20 cm^2 membrane samples. Results of mechanical properties are represented by Table 3.

RESULTS AND DISCUSSION

Water contents of Nafion and hybrid membranes are shown in Table 1. It is evident from the table that all the hybrid membranes have lower water contents than that in Nafion. These results are in accordance with the assumption that the presence of inorganic particulates inside the ionic clusters (which holds the maximum water) may reduce its water content. In light of the inherently bound water with IIE the actual water content in clusters of Nafion shall be lower than what has been determined in the case of hybrid membranes. It is reported that PAM, ZrP and ZrO_2 has 25, 13 and 27 percent associated water respectively which is lost on heating above 600°C [13,14,15]. The results are also reported for the percent water associated with IIE and ionic phase of membrane in Table 1.

As discussed previously [16] and experimental section of this paper the retention of IIE in the membrane primarily depends upon the IEC of membrane (self diffusivity), IEC exhibited by individual IIEs towards the respective ions and diffusion of alcoholic solutions of respective electrolyte precursor into the matrix of Nafion. These three pimary processes may be responsible for each IIE loading in the hybrid membranes. In view of these multiple parameters responsible for IIE loading in the membranes, it is rather difficult to theoretically predict the quantity of IIE which can be incorporated into the Nafion. The quantities of IIE retained in the membrane at different loadings and as determined by TGA techniques are given in Table 1. It is evident that IIE content is increasing with successive loading. PAM content in NAF PAM hybrid, however, has not increased unlike other IIEs. This may be attributed to low diffusivity of pentavalent Antimony ions into Nafion and may be due to rather high reactivity of Antimony (V) chloride with water. Expectedly, the ZrP content in NAF ZRP hybrid was maximum in comparision to other IIEs. ZrO_2 in NAF ZRO hybrid was also reasonably high. Considering the fact that associated water is present with each hydrated IIE, the values of IIE content (hydrated one) will assume much higher values and this has been shown in column 5 of Table 1.

Ion exchange capacities (IEC) of different hybrid membranes at various pH values are presented in Table 2. The results show the decrease in IEC (meq/gm) of hybrid in order of NAF PAM > NAF ZRP > NAF ZRO. Polyantimonic acid (PAM) is a cation exchanger and its IEC depends on its method of preparation [17]. When the IEC of NAF PAM was taken at lower pH values the pH of the solution due to exchange with Na^+ ions reaches nearly 2 which is very close to pK_a value of Nafion (i.e. 1.98) [18]. This may have reduced the exchange of Na^+ ions by Nafion because of onset of equilibrium for ion exchange between Na^+ and H^+. However, at higher pH values (11.32 pH) it shows appreciable increase in IEC (~ 13%). The ion exchanging nature of PAM of NAF PAM was rechecked based on its high IEC for Ag^+ ions [19]. Ag^+ exchanged NAF PAM shows the residue left equivalent to Ag^+ ion exchange with PAM. Under the identical conditions no residue was left in unmodified film of Nafion. This has established that the PAM incorporated in NAF PAM has an IEC of ~5 meq/gm for Ag^+ ions as reported earlier [19]. Zirconium phosphate (ZrP) behaves as cation exchanger [20] and the low ion exchange capacity could be attributed to pH dependence of the ion exchanger. The maximum IEC could be obtained at about 12 pH. The low IEC of ZrP may be attributed to the reason that the pH value within clusters could not have reached desired level because of the poor diffusivity of electrolyte in Nafion. Zirconium oxide (ZrO_2) behaves as cation and anion exchanger depending on the pH conditions [21]. It is evident from Table 2 that the IEC of NAF ZRO is lower at low pH. It could be attributed to anion exchange behaviour of ZrO_2 of NAF ZRO which becomes dominent with lowering of pH. However, even at higher pH values there is no appreciable change in IEC which is attributed to the pH within the clusters (where IIE is located) could not reach the desired value. i.e. 11 pH [21). This is primarily due to the poor diffusion of electrolyte as well as the drop of pH outside the membrane to a level of 2.

Mechanical properties of samples are represented by Table 3. It is evident from the table that even after drastic experimental conditions for the incorporation of IIEs, the mechanical strength is not affected to a greater extent. In case of NAF PAM hybrid the tensile strength has been affected to some extent. The tensile strength of NAF ZRP hybrid remained almost same as of Nafion. Interestingly, the NAF ZRO hybrid shows higher tensile strengths. Modulus of NAF PAM shows a drop of 50%. While the Modulus of NAF ZRP remained comparable to Nafion, the Modulus of NAF ZRO has gone up considerably. These results, however, suggest that hybrid membrane can be employed in electrolyzers operating at high pressure. Good mechanical properties exhibited by membranes have also been demonstrated as these membranes possess reasonable burst strength. All the samples withstood a differential pressure upto 5 kg/cm^2.

Hybrid membranes with different loadings have been subjected to thermal stability evaluation using TGA method and these have been reported in our previous publication [22]. TGA studies reveals a two step decomposition and first decomposition occuring at 250°C and above. In this publication we have undertaken thermal analysis (DSC) which reveals that the glass transition temperatures of wet acid form of hybrid membranes are around 160°C in comparision to 120°C of wet Nafion. However, the Tg's of sodium salt of wet hybrid membranes and as well as wet Nafion remains comparable (i.e. 160°C). In view of enhanced Tg of hybrid in H^+ form of these could be employed for deionised water electrolysis at much higher temperature than reported for Nafion membrane in SPE technologies.

The hybrid membranes developed by the incorporation of different IIEs in these studies have demonstrated physicochemical characteristics better than unmodified Nafion. In view of this the hybrid membrane may find applications not only in water electrolysis but also in brine electrolysis, water photolysis, gas separation, desalination of water and the nuclear waste recovery.

CONCLUSION

The method of incorporating polyantimonic acid (PAM), hydrated zirconium oxide (ZrO_2) and zirconium phosphate (ZrP) onto/into perfluoro ionomer membrane have been described. The physicochemical and mechanical characteristics hitherto resultant membranes known as IME-SPE hybrid membranes have been determined. The electrochemical characteristics of hybrid membranes are being evaluated and the hybrid membranes shall be used in "state of the art" electrolyzer to determine their utility in the system. It may be concluded that in light of desirable characteristics exhibited by hybrid membranes they may find applications in Energy Conversion Systems, brine electrolysis and recovery from nuclear waste etc.

ACKNOWLEDGEMENT

The authors expresses sincere thanks to Dr.U.D.N. Bajpai, R.D. University, Jabalpur, for helpful discussion and Dr.K.N. Ninan and Mrs. Ambikadevi, VSSC, Trivandrum, for their help extended in mechanical characterization. The secretarial assistance from M/S S.D. Namdeo and Rajendra Dubey is gratefully acknowledged.

REFERENCES

1. R.S.Yeo, J.McBreen, G.Kissel, F.Kulesa and S.Srinivasan. J.Appl. Electrochem. BNL Report 27136.

2. S.C.Yeo and A.Eisenberg. J.Appl. Poly. Sci. 875,(21),1977.

3. J.Fischer, H.Hoffman, C.Luft and H.Wendt. EEC report EU6085, Brussels, p.227, oct. 1978.

4. H.Vandenborre, R.Leysen and L.H.Baetsle Int. J. Hydrogen Energy. 165,(5),1980.

5. A.Dravnieks and J.J.Bregman. Chem. Engg. News. 42,(39),1961.

6. G.Alberti. Atti. Accad. NaZ. Lincei, Rend Class Sci. Fis. Mat. Nat., 3427,(31),1961.

7. V.Vesely and V.Pekarek. Talanta. 219,(19),1972.

8. C.B.Amphlett. Inorganic Ion Exchangers. Elsevier, Amsterdam, 1964

9. T.Morimoto, M.Yoshitake, S.Morikawa and Y.Oda. Int. J.Hydrogen Energy. 501,(11),1986.

10. Macromolecular Research Centre, R.D.University, Jabalpur. Technical Report No. MRC:Tech-REP:PAWAS-1:91.

11. Macromolecular Research Centre, R.D.University, Jabalpur. Technical Report No. MRC:TECH-Rep:PAWAS-3:91.

12. L.H.Baetsle and D.Huys. J.Inorg.Nucl. Chem. 639,(30),1983.

13. M.Abe and T.Ito. Bull. Chem. Soc. Japan. 2366,(41),1968.

14. A.Clearfield and J.A.Stynes. J.Inorg.Nucl. Chem. 117(26),1964.

15. G.H.Nancollas and R.Paterson J.Inorg. Nucl. Chem. 259,(22),1961.

16. S.K.Tiwari and S.K.Nema proceedings of 8th IMS Conference held at N.Delhi, Jan. 28-29,1991.

17. M.Abe and T.Ito. Bull. Chem. Soc. Japan. 333,(41),1968.

18. Z.Twardoswki, H.L.Yeager and B.O'Dell. J.Electrochem. Soc. 328,129(2),1982.

19. L.H.Baestle and D.Huys. J.Inorg. Nucl. Chem. 639,(30),1983.

20. C.B.Amphlett. L.A.McDonald and M.J.Redman. J.Inorg. Nucl. Chem. 220, (6), 1958.

21. Idem, ibid., 236,(6),1958.

22. S.K.Tiwari and S.K.Nema. communicated to Eur.Poly.J.

TABLE 1: WATER AND IIE CONTENTS OF HYBRID MEMBRANES

	1	2	3	4	5
SAMPLE DESIGNATION	WATER CONTENT (%)	IIE CONTENT (DRY %)	WATER WITH MEMBRANE(%) (Col.1-Col.4)	WATER* WITH IIE (%)	ACTUAL IIE CONTENT (Col.2-Col.4)
NAFION-001	23.0	--	23.0	--	--
NAF PAM-101	23.0	0.11	22.9	0.03	0.14
NAF PAM-103	--	1.66	--	0.55	2.21
NAF PAM-105	22.6	4.15	21.2	1.38	5.54
NAF ZRP-201	22.8	8.78	21.5	1.31	10.09
NAF ZRP-203	--	13.74	--	2.05	15.79
NAF ZRP-205	22.0	34.84	16.7	5.25	40.09
NAF ZRO-301	17.2	8.34	14.1	3.08	11.42
NAF ZRO-303	--	11.51	--	4.26	15.77
NAF ZRO-305	15.5	16.92	9.2	6.26	26.92

*Calculated on the basis of water associated with various IIEs as reported in literature [13,14,15].

TABLE 2 : ION EXCHANGE CAPACITY OF HYBRID MEMBRANE

SAMPLE DESIGNATION	ION EXCHANGE CAPACITY (IEC)		
	AT pH	VAULE (meq/gm)	%age VARIATION*
NAFION-101	--	0.83	--
NAF PAM-101	2.98	0.80	-3.61
NAF PAM-101	5.06	0.81	-2.41
NAF PAM-101	11.32	0.94	+13.25
NAF ZRP-201	2.98	0.61	-26.50
NAF ZRP-201	5.06	0.68	-18.07
NAF ZRP-201	11.32	0.82	-1.20
NAF ZRO-301	2.98	0.21	-74.70
NAF ZRO-301	5.06	0.33	-60.24
NAF ZRO-301	11.32	0.37	-55.42

* % variation is with respect of NAFION-001

TABLE 3 : MECHANICAL PROPERTIES OF HYBRID MEMBRANES

SAMPLE DESIGNATION	TENSILE STRENGTH (Kg/cm^2)	ELONGATION (%)	MODULUS (Kg/cm^2)
NAFION-001	163	137	1670
Naf Pam-101	101	66	660
NAF PAM-102	106	97	880
NAF PAM-103	97	79	880
NAF PAM-104	96	83	890
NAF PAM-105	100	90	830
NAF ZRP-201	135	75	1390
NAF ZRP-202	127	92	990
NAF ZRP-203	139	95	1480
NAF ZRP-204	158	92	1880
NAF ZRP-205	140	83	1709
NAF ZRO-301	198	91	1280
NAF ZRO-302	193	75	3030
NAF ZRO-303	172	75	2080
NAF ZRO-304	212	43	4170
NAF ZRO-305	220	40	3950

DIRECT SOLAR WATER SPLITTING EXPERIMENT

I. Hodara, A. Kogan and A. Gorodnev
Solar Research Facilities
The Weizmann Institute of Science
Rehovot, 76100, ISRAEL

ABSTRACT

The feasibility of hydrogen production by direct solar thermal splitting of water is investigated. Water can be dissociated only at temperatures exceeding 2000 K. This temperature is obtained by concentrating the solar radiation using a system of mirrors consisting of a 56 sq.m. area field heliostat and a 63 cm. diameter parabolic mirror. The hottest parts of the reactor are built from zirconia ceramics prepared in this laboratory. Emphasis is put on the separation of hydrogen from the product stream while the gas mixture is still hot, preventing thus its recombination with oxygen. The separation method chosen is effusion through a porous ceramic membrane.

INTRODUCTION

When the environmental effects and the fuel requirements of existing hydrogen production processes are considered, using the sun as an alternative energy source becomes a very attractive way of producing hydrogen from water. Solar hydrogen production is based entirely on energy derived from the sun and does not depend on burning fossil fuels. It is a non polluting process and therefore, environmentally acceptable.

There are a number of ways to produce hydrogen from water using solar energy. The only technology that has reached the demonstration stage at a scale of hundreds of kilowatts is photovoltaic electrolysis [1,2]. A second way is by thermochemical cycles developed in the past 15 years in various laboratories [3-5]. These processes were originally designed for storing excess heat from fission reactors [6].

Direct solar splitting of water is another alternative method of producing hydrogen using solar energy. Although, basically, direct splitting is a straightforward reaction, its realization depends on the availability of solutions to a number of technical problems: a means of producing highly concentrated solar radiation by which the reactor can be heated homogeneously to very high temperatures; ceramic materials having suitable thermal properties and chemical stability in an environment containing steam, hydrogen and oxygen; and a technique of separating hydrogen from excess steam and its dissociation products while they are still hot.

HYDROGEN BY THERMAL WATER SPLITTING

Fletcher and his collaborators at the University of Minnesota [7,8] stressed the thermodynamic advantages of a one-step process with heat input at as high a temperature as possible. The theoretical and practical aspects of the reaction were studied, among others, by Lede et. al. [9,10] and by Ihara [11]. The main emphasis of these works were on the thermodynamics and the demonstration of the feasibility of the process. On the other hand, the cardinal problem of the separation of the products prior to their recombination was entirely neglected. Quenching methods were proposed by various authors, but quenching of course is a very inefficient way of producing hydrogen, and moreover does not solve the problem of product separation. By quenching the dissociation products, most of the thermal energy contained in the remaining 80 percent undissociated water becomes unretrievable, rendering the process uneconomical.

Two trends characterize the thermal dissociation of water vapour. It is favoured by high temperatures and by low pressures. Detailed thermodynamic calculations of equilibrium concentrations are given by various authors [7,12] and computer codes easy to work with for performing these calculations are available *.

The calculations show that at 2500 K and 0.1 atm. pressure, about 20 percent of water vapour is dissociated resulting in an equilibrium composition of 8.4 mole percent of hydrogen and 3.2 percent oxygen. If the dissociation products are allowed to cool to 2000 K they recombine almost completely to obey the new thermodynamic equilibrium conditions.

CONCENTRATION OF SOLAR RADIATION

In order to collect efficiently the solar radiation in a cavity reactor operating at 2500 K it is necessary to concentrate the radiation by a factor of at least 10000. Our solar research facilities include two separate installations that enable us to reach these high concentrations. A solar furnace of 20 kw. nominal power consisting of a 100 sq.m. area flat heliostat combined with a 7.5 m. diameter parabolic concentrator produces, at its focal point, a fire ball of about 8 cm. diameter. We have measured concentrations exceeding 10000 suns at the focal plane of this furnace. A second facility producing highly concentrated radiation is our central receiving tower of 3 mw thermal rated power. Here, the field heliostats are slightly curved and each heliostat can concentrate the incoming solar radiation only by about 50 times. When working with a single heliostat, a secondary parabolic concentrator is used to obtain the desired concentration level. Figure 1 shows a schematic diagram of the optical arrangement we have used in the present investigation. A 63 cm. diameter

----

* For performing our own calculations, we used the PC Version of the NASA Program CET-85 written by M. Pepho, T. Kappauf and E. Whitby from the Department of Engineering of the University of Minnesota. For the original version of the program, see Ref. [13].

parabolic concentrator (Optical Radiation Corp., Azusa, Ca.) installed at the roof of the central receiver tower faces a field heliostat of 56 sq.m. total reflecting area. A zirconia target placed at the focal plane of the concentrator melts immediately when exposed to concentrated radiation, producing a molten cavity of about 24 mm. diameter as shown in Fig. 2. This experiment shows that a temperature around 3000 K has been reached, equivalent to the melting point of zirconia.

The temperature distribution that could be expected in our reactor when exposed to concentrated radiation was evaluated by a Monte-Carlo exchange analysis using a program written by A. Steinfeld* and modified by A. Segal from this laboratory. For these calculations a reactor model shown in Fig. 3 was used. It consists of a cylindrical cavity with a crucible shaped membrane installed on its back wall. The inner surfaces of the reactor were divided into any desired number of hypothetical rings and an estimate of the temperature distribution on the walls was obtained by calculating the temperature of each ring. For a first approximation, the analysis was performed with a bundle containing only 2000 rays. The cavity was assumed to be perfectly insulated and the effects of medium participation as well as the presence of a heat sink were not considered at this stage. Therefore, with the present code, only the upper limit of the temperatures that can be expected is calculated. The variables input used for these calculations was, with reference to Fig 3, as follows:

Reactor radius, R1, M	0.05
Crucible radius, R2, M	0.03
Shadow radius, R3, M	0.05
Concentrator radius (projected), R4, M	0.32
Reactor length, L1, M	0.20
Crucible length, L2, M	0.10
Twice the Focal Length, P, M	0.35

The absorptivity both of reactor and crucible walls was assumed equal to 0.5 and the flux of concentrated radiation entering the cavity for each aperture diameter was calculated using experimental calorimetric data. In Fig. 4 are shown the upper limits of the temperatures reached at the face of the membrane at various aperture diameters. The numbers on the abscissa correspond to concentric ring numbers on the surface of the membrane on the plane facing directly the aperture, with numbers 1 and 10 corresponding to the center and outer rings respectively. These preliminary calculations show that with the present experimental arrangement the reactor can be heated up to the desired range of temperatures.

The flux distribution on and around the focal plane was measured by water calorimetery. The amount of heat absorbed by the calorimeter was calculated from the measured difference in temperature of the inlet and outlet streams of cooling water and the known water flow rate. Moving the calorimeter on the optical axis of the parabolic concentrator, the flux pattern shown in Fig. 5 was

---

* Present address: Paul Scherrer Institute, CH-5232, Villigen, Switzerland.

obtained. Two sets of calorimetric data corresponding to two aperture diameters are shown. The peak values of absorbed power are 3.2 and 7.1 KW respectively for 3 and 5 cm aperture diameters.

The accurate measurement of high temperatures is not straightforward. The use of thermocouples is naturally excluded because of the relatively low melting points of the metal couples. Thermocouples can be used, however, at a safe distance a few centimeters away from the focal plane as the temperature gradient along the optical axis is very sharp. The thermocouple tip must, however, be properly shielded to avoid radiation induced errors. We have attempted to measure the temperature distribution on the optical axis of the concentrator on both sides of the focal plane as follows: A type K, inconel sheathed thermocouple junction was implanted on the back of a small zirconia disc. The disc was placed at one end of a water cooled cylindrical tube whose axis coincides with the optical axis of the secondary concentrator. The temperature was measured at different points along the optical axis moving on both sides away from the focal plane. Fig. 6 shows the results obtained. Even with such a simple arrangement, one can get a feeling of the temperature gradients to be expected around the focal plane of the concentrator.

Infra-red pyrometry is the generally accepted technique for measuring high temperatures, although the interpretation of the results requires an accurate knowledge of the emmissive properties of the target. Data on the emmissivity of zirconia is available from the literature [14], but the spread of experimental values dictates critical use of the data.

THE REACTOR

To construct a reactor operating around 2500 K in contact with steam and its dissociation products, special materials are required. At these temperatures, the choice of candidate materials is restricted to a few natural or man-made materials. Among them we consider only oxide type ceramics as they have high melting points and are the least likely to react with steam or its dissociation products. Further requirements from the materials are mechanical and thermal stability and fabricability. We have decided to use calcia-stabilized zirconia for building our reactor as we beleive that this material fulfills all these requirements better than others.

Figure 7 is a schematic representation of the reactor. It consists of a cylindrical 10 cm inner diameter and 20 cm long stabilized zirconia housing. One end of the housing is closed by a circular zirconia disc with an aperture of 3 cm diameter. A porous wall Degussit zirconia crucible (Friedrichsfeld, Germany) is installed at the opposite end of the housing backed by a water cooled metal flange. The ceramic reactor is enclosed in a vacuum tight, fused silica bell-jar (The General Electric Co., Germany). The space between the zirconia receiver and the surrounding silica bell-jar is filled with insulating material made of zirconia fiber reinforced with zirconia cloth (Zircar Products Inc., Florida, N.Y.). The insulation thickness was determined from the maximum temperature the quartz wall was permitted to reach. The design value of the wall temperature is around 800 ° C, well below the maximum working temperature of fused silica.

All the zirconia parts of the reactor, except the crucible, were fabricated in this laboratory from commercially available zirconia powder stabilized with 5% calcium oxide (Cotronics Corp., Brooklyn. N.Y.). The end product, although brittle to a certain extent, can be drilled and machined if handled carefully. The material has a good resistance to thermal shock and can be heated up to 2200 °C without showing any macroscopic or microscopic structural damage. Figure 8 is a photograph of the reactor parts showing the ceramic components fabricated in this laboratory.

Superheated steam at a constant preset flow is introduced into the reactor cavity. Steam flow regulation is achieved with the aid of a simple throttled nozzle arrangement by control of the upstream temperature and pressure [15]. For a solar energy input of 5 kw., the net amount of energy available to run the reaction after substracting the radiation and conduction losses, is about 2.5 kw. This quantity of energy enables feeding the reactor with steam at a rate of about 20 g. per minute. The additional operating conditions are 2500 K at 0.05 atm. pressure.

When the membrane wall temperature exceeds 2000 K steam adjacent to the membrane begins to dissociate. Two streams of gas are extracted in parallel from the ractor cavity. One stream is extracted by diffusion across the membrane's porous wall. Under suitable pressure and temperature conditions this stream is enriched in hydrogen. The second stream by-passes the membrane wall and is depleted of hydrogen.

COMPOSITE REFRACTORY MEMBRANES

In order to achieve efficient gas separation by gas diffusion it is necessary to maintain a Knudsen flow regime across the porous membrane. The molecular mean free path in the gas must be greater than the average pore diameter. From kinetic theory we have estimated that at 1 atm. pressure and 2500 K the mean free path of the molecule is 0.6 μm. This value dictates the use of membranes having pore diameters less than 1 μm. If membranes with larger pore sizes are used, the operating pressure must be reduced accordingly. In this work we are testing commercially available zirconia crucibles as membranes. They have an open porosity of 20-30 % and an average pore size of 5 μm. The cavity pressure corresponding to a membrane pore size of 5 μm. is 0.05 atm.

Since zirconia membranes with pore size smaller than 1 μm. were not available in the market, we decided to develop our own technique of producing membranes of required pore size and thickness. We are working now on the development of a sol-gel method of coating coarse zirconia substrates with a thin film of zirconia. The technique is based on the same principle as for preparing semipermeable membranes used in reverse osmosis. Fig. 9 shows a schematic cross-section of the type of coating required.

In a preliminary experiment, we prepared, starting from zirconium propoxide, a thin coating of zirconia on a fused silica surface by the sol-gel technique. The coating had a thickness between 100 to 150 nanometers and a pore size between 40 and 100 nanometers. An SEM micrograph of such a film after heat treatment at 700 deg C. is shown in Fig. 10.

ACKNOWLEDGMENTS

The authors wish to gratefully acknowledge the financial support received from the Heineman Foundation for Research, Education, Charitable and Scientific Purposes, Inc., Rochester, N.Y. We also wish to express our sincere thanks to Prof. Edward Fletcher for his valuable advice and for making available to us the PC Version of CET-85 computer code written by his students.

REFERENCES

1. J.M. Ogden and R.H. Williams, Solar Hydrogen, Moving beyond fossil fuels, World Resources Institute, October 1989.

2. C.-J. Winter, Hydrogen Energy Progress VIII, Proc. 8th World Hydrogen Energy Conf., Honolulu and Waikoloa, Hawaii, 22-27 July 1990, Pergamon, 1990, Vol 1, p. 3-47.

3. G.H. Parker, Solar Thermal Hydrogen Production Processes, Final Report, Westinghouse Electric Corporation, Report No. DOE/ET/20608-1, 1982.

4. J.H. Norman, G. Besenbruch and L. Brown, Solar Production of Hydrogen Using the Sulfur-Iodine Thermochemical Water Splitting Cycle, General Atomic Co. Report No. GA-A16493, 1982.

5. S.E. Foh, J.D. Schreiber, J.B. Pangborn and T.P. Whaley, Thermochemical Hydrogen Production, Gas Research Institute, Report No. GRI-80/0039, 1980.

6. G.E. Beghi, Int. J. Hydrogen Energy, 11, 761 (1986).

7. E.A. Fletcher and R.L. Moen, Science, 197, 1050 (1977).

8. E.A. Fletcher and R.C. Yu, Energy, 4, 373 (1979).

9. J. Lede et. al., Int. J. Hydrogen Energy, 7, 939 (1982).

10. J. Lede et. al., Int. J. Hydrogen Energy, 12, 3 (1987).

11. S. Ihara, Int. J. Hydrogen Energy, 5, 527 (1980).

12. S. Ihara, in Solar-Hydrogen Energy Systems, T. Ohta, Ed. Pergamon, 1979, pp.59-80.

13. S. Gordon, and B.J. McBride, F.J. Zeleznik, Computer Program for Calculation of Complex Chemical Equilibrium Compositions and Applications, Supplement 1-Transport Properties, NASA TM 86885, October 1984.

14. Y.S. Touloukian and D.P. DeWitt, Thermal Radiation Properties: Non-Metallic Solids, in Thermophysical Properties of Matter, Y.S Touloukian, Series Editor, Vol. 8, pp. 526-7, Plenum, New York, 1972.

15. A. Kogan, I. Hodara and M. Vana, Int. J. Hydrogen Energy, 16 (1991), in press.

FIGURE 1: HELIOSTAT-SECONDARY CONCENTRATOR CONFIGURATION

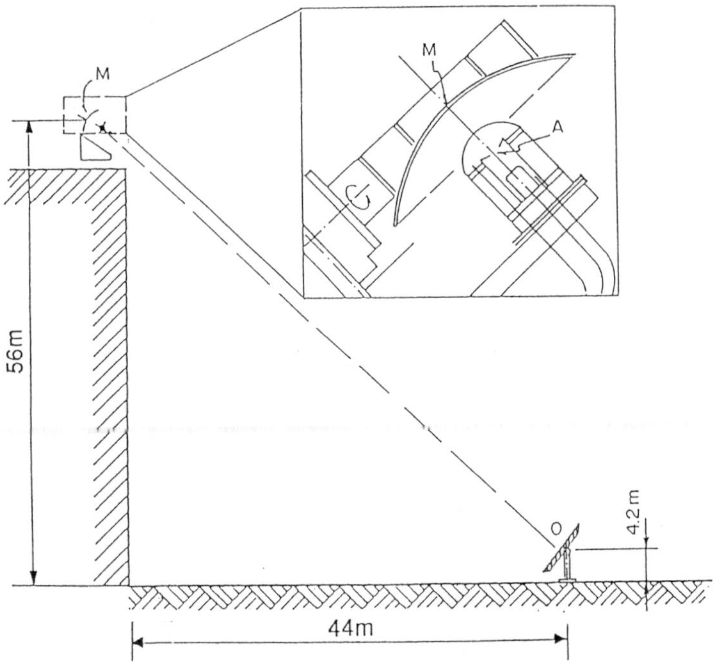

FIGURE 2: ZIRCONIA DISC EXPOSED AT THE FOCAL PLANE OF THE SECONDARY CONCENTRATOR

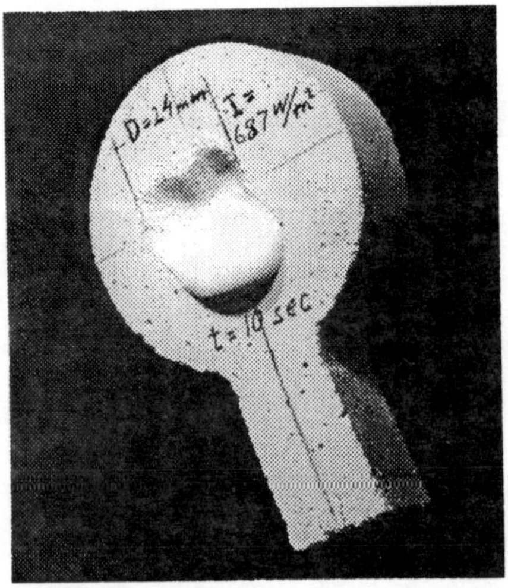

FIGURE 3: REACTOR CONFIGURATION FOR MONTE-CARLO RADIATION EXCHANGE ANALYSIS

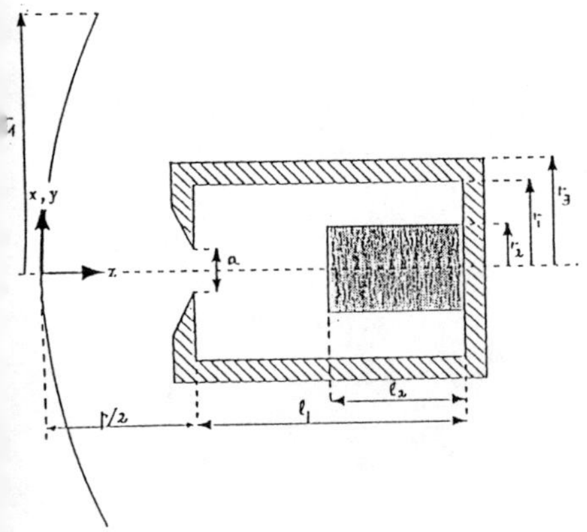

FIGURE 4: MONTE-CARLO COMPUTATION RESULTS OF MEMBRANE FACE TEMPERATURE AT VARIOUS APERTURE DIAMETERS (A) LI = 0.200M, L2 = 0.100M

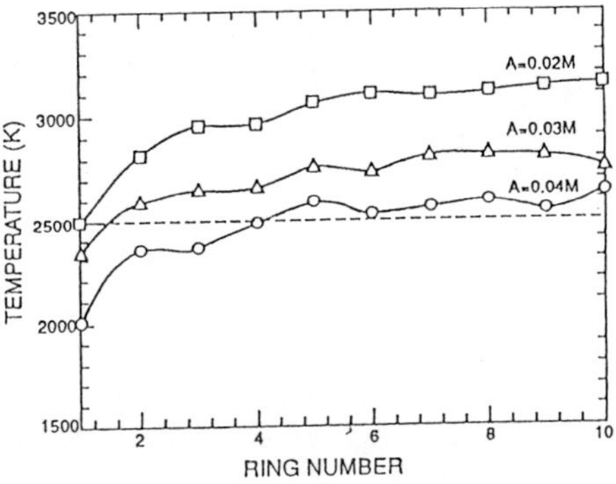

FIGURE 5: CALORIMETRIC MEASUREMENT OF ABSORBED POWER AT 750 W/M^2 INSOLATION

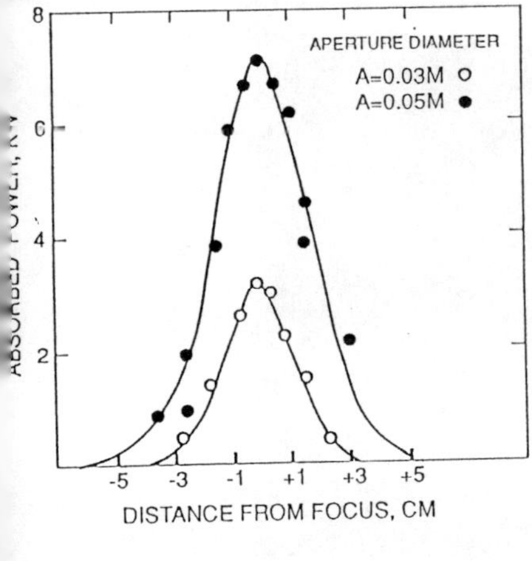

FIGURE 6: TEMPERATURE OF ZIRCONIA DISC EXPOSED TO CONCENTRATED RADIATION ALONG THE OPTICAL AXIS

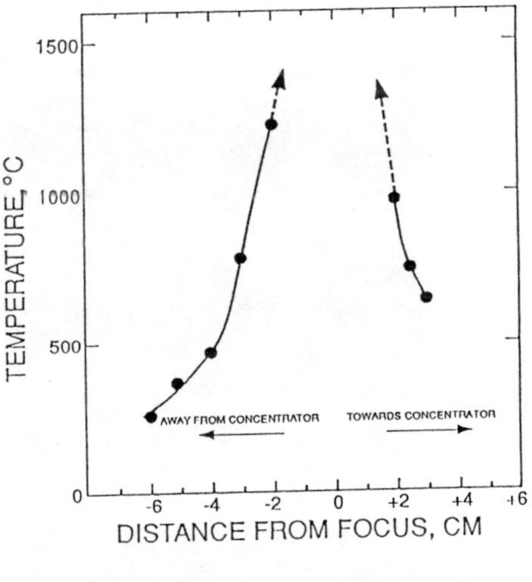

FIGURE 7: REACTOR CROSS-SECTION

1. ø200mm fused silica belljar; 2. zirconia plate;
3. cooled brass plate; 4. zirconia felt insulation;
5. porous zirconia crucible; 6. zirconia receiver cavity;
7. steam inlet; 8. H_2 enriched steam; 9. H_2 depleted steam

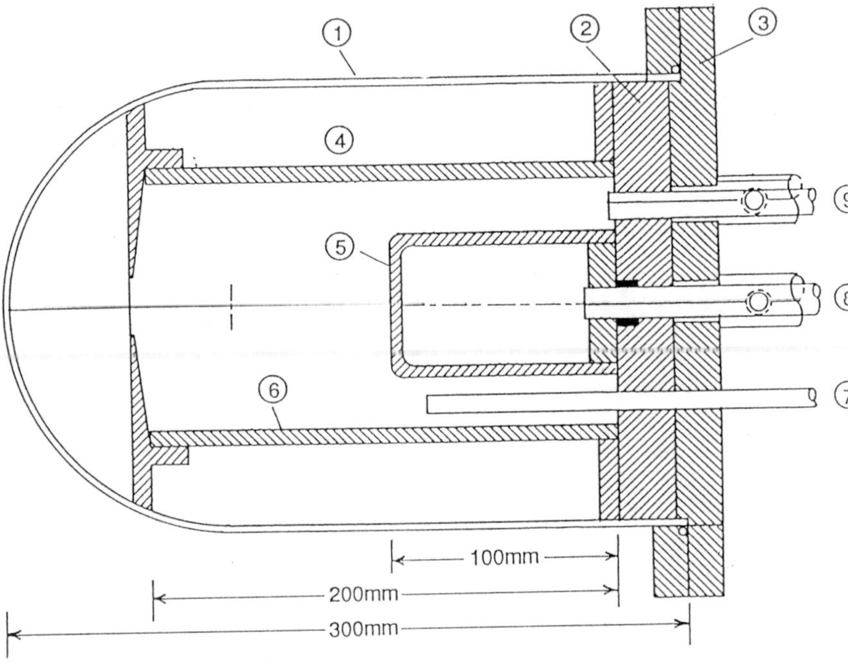

FIGURE 8: REACTOR PARTS

FIGURE 9: SCHEMATIC REPRESENTATION OF ULTRAFINE CERAMIC MEMBRANE COATING ON A COARSE SUPPORT

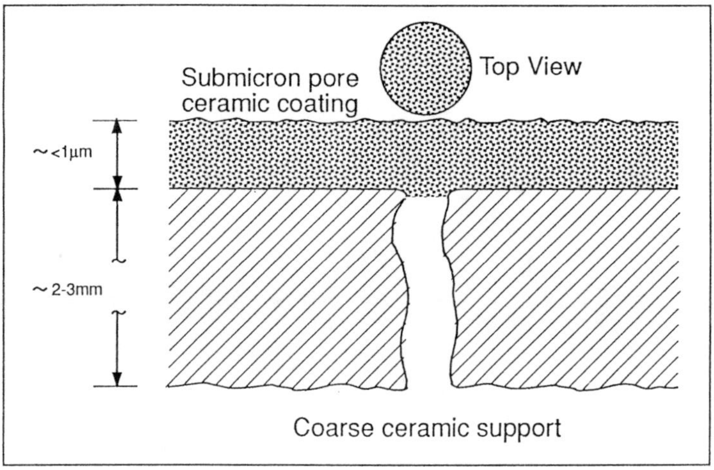

FIGURE 10: S.E.M. MICROGRAPH OF A ZIRCONIA COATING DEPOSITED ON A SILICA PLATE BY SOL-GEL PROCESS AFTER TREATMENT AT 700°C.

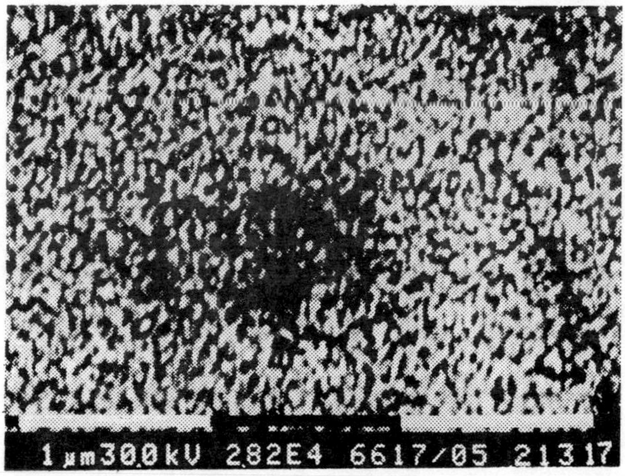

CHAPTER THREE

Hydrogen Fueled Internal Combustion Engines

DIRECT HYDROGEN INJECTION AND SPARK IGNITION SYSTEM FOR INTERNAL COMBUSTION ENGINES

Roy E. McAlister
President, American Hydrogen Association
219 South Siesta, H101
Tempe, Arizona 85281

ABSTRACT

Hydrogen burns at a higher rate than hydrocarbon fuels. Backfiring often occurs in engines that introduce hydrogen through the air intake system of combustion engines. Engines using hydrogen as a homogeneously mixed fuel are derated compared to operation on gasoline. Direct combustion-chamber injection of hydrogen and spark ignition provides a method to develop higher specific power ratings and higher thermal efficiencies than with homogeneous-charge approaches. A system for retrofitting existing internal-combustion engines is disclosed. Devices that take the place of conventional spark plugs or diesel fuel injectors are used to provide hydrogen fuel injection and positive ignition. In all modes of operation very low emissions of oxides of nitrogen occur. Operation is convenient and backfiring is prevented.

INTRODUCTION

Transportation engines produce about 80% of the atmospheric pollution in cities throughout the world. Particularly harmful emissions of carbon monoxide, oxides of nitrogen, carbon dioxide, unburned hydrocarbons, and carbonaceous particles are sourced by motor vehicles burning fossil fuels. Carbon dioxide gas concentrations in the atmosphere have accumulated to higher levels than at any time during the past 160 thousand years as the world's motor-vehicle population has risen to more than 400 million vehicles. Depletion of stratospheric ozone, acid rain, development of atmospheric warming by the "greenhouse effect" and reduced living standards have resulted. Practical solutions to these problems are offered by adopting renewable hydrogen to replace diminishing supplies of fossil fuels.

Hydrogen production from biomass wastes and solar technologies are indicated as more desirable alternatives to continued alteration of the environment by combustion of fossil fuels. Making a smooth and rapid transition to renewable hydrogen is the mission behind the technology disclosed in this paper.

Three serious difficulties with hydrogen fueled vehicles that use internal combustion engines are reported by previous investigations. [1, 2] The problems are backfiring, reduced power, and reduced range for equivalent fuel-storage volume.

Hydrogen fueled vehicles have the reputation of backfiring and knocking due to pre-ignition. However, attempts to replace diesel fuel with hydrogen have met with the difficulty of ignition failure because the compression-ignition pressure is much higher than for diesel fuels.

Homogeneous charge engines adapted to use hydrogen fuel have produced about 15 to 30 percent less power than when using gasoline. Gasoline vapors are denser than air and provide more energy release per unit volume than hydrogen. Hydrogen is about 15 times less dense than air. Hydrogen takes much more room to flow through the intake system than gasoline vapors. The result is that engines that mix hydrogen with air produce less power because there is less air and less fuel delivery to the combustion chambers.

The volume offered by a 20 gallon gasoline tank stores only about 25 percent as much energy as cryogenic liquid hydrogen. Hydride storage is more compact but the weight penalty for storing hydrogen in metallic lattices is very high.

These problems must be overcome if hydrogen is to be welcomed in a free market.

Direct Hydrogen Injection with Spark Ignition

These problems can be overcome, eliminated, or reduced by a new technology for combining direct fuel injection with spark or plasma ignition. Called the Precision Spark Injection (PSI) Technology, this approach will allow the world's existing fleet of some 400 million motor vehicles to be equipped with devices that overcome the problems of backfiring and reduced power.

Backfiring is eliminated by not having fuel present in the intake system. Only air is allowed to travel through the intake system. In the absence of fuel in the intake system previous problems such as sparks, flame leaks, and hot spots fail to cause backfiring. This improves vehicle safety. Eliminating the inventory of fuel in the carburetor and intake system greatly reduces the ability of a fire to be sustained under the hood of an automobile.

Loss of power is overcome by using the intake capacity to deliver as much air as possible to the combustion chamber. More air enters the combustion chamber than before because the air is not restricted to control the power level of the engine. With more air present more fuel can be added to produce more power than before.

PSI Technology helps overcome the range problem by increasing the thermal efficiency and thus the range of the vehicle. PSI-hydrogen Technology goes far towards solving the problem of storage by combinations of the following:

1. Increasing the combustion efficiency of the engine.

2. Increasing the part-load efficiency of the vehicle.

3. Increasing the mechanical efficiency of converted engines.

Conversion can be accomplished in the field without tear-down. [3]
Conversion of existing low-compression engines with spark ignition is accomplished by replacement of spark plugs with devices that provide direct hydrogen injection and spark ignition. Conversion of existing compression-ignition engines is accomplished by replacement of diesel fuel injectors with devices that provide precision hydrogen injection and spark ignition.

Fuel metering and ignition are electronically controlled to produce the best use of hydrogen for the power requirements of the moment. Installation of electronic controls and replacements for the spark plugs is accomplished in about the same time as an ordinary tune-up. Redesign of the engine is not necessary and required modifications are accomplished by devices that are convenient replacements for standard components.

Operation of Engines Converted to PSI-Hydrogen Operation

Figure 1 shows components of the PSI System for replacing the spark plugs of ordinary automobile engines with Precision Spark Injection devices. Intake air is allowed to enter the combustion chamber without throttling in engines converted to PSI operation. Increasing the breathing capacity of the engine results and more air is present to support combustion at all phases of operation. This greatly reduces "back-work" due to the vacuum conditions in the combustion chamber as the piston is moved against the pressure difference between crank-case pressure and the intake manifold vacuum produced by air throttling.

After substantial compression, hydrogen is delivered through the PSI plug directly into the combustion chamber and is ignited as it enters. Hydrogen burns in the compressed air with an extremely rapid flame speed. This allows late injection and late ignition near top dead center or after top dead center conditions depending upon the instantaneous piston speed.

Late injection reduces back-work due to compression of hydrogen in the combustion chamber during compression. Late ignition reduces back-work due to pressure rise due to combustion before top dead center. Power control is accomplished by metering the precise amount of fuel needed at the instant that the piston nears top dead center conditions. The amount of fuel required is less than in ordinary homogeneous charge engines because excess air insulates the combustion chamber surfaces. During deceleration there is no fuel consumption until the engine speed reaches idle or a desired hold-point.

PSI-hydrogen combustion within excess air results in greatly reduced heat losses due to radiation and conduction. Specialized hydrogen injection patterns are provided for best air utilization in each combustion chamber type. PSI plugs are designated like spark plugs for each engine retrofit. PSI-hydrogen conversion provides the following characteristics:

1. More air in the combustion chamber than with homogeneous charge operation enables more fuel to be combusted and increased power development.

2. Faster combustion rates provided by hydrogen within excess air enables injection and ignition after substantial completion of the air compression cycle. This greatly reduces back-work and eliminates detonation, back firing, and engine wear.

3. Initial combustion of hydrogen as it enters the combustion chamber is in fuel rich conditions. Combustion proceeds to excess air to provide reduced formation of nitrogen oxides per horsepower hour of operation.

4. Combustion of hydrogen within excess, unthrottled air reduces loss of cylinder lubrication because there is no intake vacuum and reduced heat losses to the combustion chamber.

5. Cold start-up of PSI-Hydrogen engines provides the same clean-burning, low-emission characteristics. During deceleration there is no fuel consumption. During other transient operations the combustion chamber receives precise amounts of hydrogen at the most optimum injection and ignition timing.

Emissions From Engines Converted to PSI-hydrogen Operation

As noted in the comparisons of Table 1, using directly-injected, stratified-charge hydrogen in place of homogenous-charge gasoline eliminates air pollution from reactive hydrocarbons and carbon monoxide in addition to eliminating sulfur compounds. Oxides of nitrogen meet the lowest projected requirements including the ultra low emissions vehicle (ULEV) contemplations of California. Directly injected hydrogen provides the long-range solution to the problem of exhaust emissions. Problems due to variations in fuel/air mixtures are eliminated.

Hydrogen burns more rapidly than hydrocarbon fuels because it is smaller and enters combustion reactions at a higher velocity than heavier molecules. This characteristic makes it possible to use mixtures of hydrogen with other renewable hydrocarbon fuels to reduce emissions of unburned hydrocarbons, oxides of nitrogen, and carbon compounds. Table 2 shows the result of mixing hydrogen with renewable methane and injecting the mixture through the PSI System. This enables existing methane production and collection systems to supply hydrogen boosted fuels for use with PSI systems to provide greatly improved thermal efficiency and reduced emissions. Hydrogen can be produced from sewage, garbage, farm wastes, and numerous other renewable solar energy conversion systems. This provides incentives for recovery of hydrogen from such wastes as city garbage and sewage rather than to allow normal decay to greenhouse gases. Agricultural wastes can produce an additional cash crop of hydrogen and while returning essential minerals to the soil.

Eliminating Combustion Engine Contributions to the Greenhouse Effect

Combustion of hydrocarbons and petrochemicals in general produce more carbon dioxide than the original amount of hydrocarbon.

$$H_xC_y + O_2 \text{ -------> } 0.5x\, H_2O + yCO_2 \qquad \text{(Equation 1)}$$

Combustion engines produce about 3 times as much carbon dioxide as the weight of propane, gasoline or diesel fuel consumed.

$$C_8H_{18} + 0_2 \text{ -------> } 9\, H_2O + 8CO_2 \qquad \text{(Equation 2)}$$

114 tons of octane gasoline -----> 352 tons CO_2

By comparison providing the same amount of energy in the form of hydrogen would produce no carbon monoxide, no hydrocarbons, no carbon dioxide, no sulfur dioxide, and considerable reductions in oxides of nitrogen. Except for a minuscule amount of oxides of nitrogen only water is produced by combustion of hydrogen in an engine.

$$H_2 + 0.5\, 0_2 \text{ -------> } H_2O \qquad \text{(Equation 3)}$$

38 tons of hydrogen ---> 342 tons of water

Smog in crowded cities has caused mandated reductions of automotive exhaust emissions of carbon monoxide, hydrocarbons, particulates, and oxides of nitrogen. Environmental damage has brought limits on electric power plant emissions of particulates, sulfur compounds, and oxides of nitrogen. Carbon dioxide has been ignored until recently. The greenhouse effect has been predicted for several decades, however only in the past decade have world leaders felt sufficiently threatened to consider dealing with pollution of the atmosphere by carbon dioxide and other greenhouse gases.

All weather on earth is driven by energy from the sun. Adding greenhouse gases to the atmosphere provides greater capacity for blocking long wave length radiation and collecting energy in the atmosphere. Solar energy in the atmosphere evaporates water from oceans, lakes, and rivers and moves it great distances to be deposited as rain and snow. Solar energy drives the winds. Seasonal variations in temperature and weather are directly correlated to the amount of solar energy that is captured by the atmosphere, continents, and oceans.

In a way the atmosphere acts as a giant heat engine. As with any heat engine, adding energy to the atmosphere allows it to do more work. More air is moved, more water is transported, and more damage is incurred during the extremes of air and water movement. We call the events storms and floods. Some special winds are tornadoes, hurricanes, and typhoons. The earliest symptom of increased energy in the atmosphere is more weather extremes including lower temperatures as well as higher temperatures.

Consider the following evidence of the ability of the atmosphere to collect more energy as a result of increased concentrations of greenhouse gases.

1. This last decade has seen the six warmest years of the last century. 1988, 1987, 1983, 1981, 1980, and 1986 were the warmest years since 1900. During this 90 year period the global temperature has risen about 0.5°C. [4, 16]

2. The first five months of 1988 were the warmest on record.[2]

3. March of 1990 set a new high-temperature record. [5]

4. North America has had over 600 destructive tornados and extremely damaging floods during the last year. During the last year records were set for highest wind velocity in a tornado. (March 13, 1990, at Heston, Kansas.) [6]

5. This year was a terrible year for destructive winds in Bangladesh and China. A devastating typhoon and flooding was followed by tornadoes. Hundreds of thousands of persons and animals were killed and countless more were left homeless as a result of this hostile weather.

In 1986 the United States consumed 74 quads of energy or 74 x 10^{15} BTU or 80 x 10^{18} J. This is equivalent to every person burning some 2,600 gallons of gasoline. Over 90 percent is sourced by oil, natural gas and coal. The largest source of photosynthetic smog pollution in cities is the transportation system. We can use PSI-hydrogen as a replacement for petrol in present vehicles to halt the needless production of smog and greenhouse gases.

The single largest source of greenhouse gas production is power plants to produce electricity. Internal combustion engines can drive alternators and assume much of the electric grid load with virtual elimination of greenhouse emissions. We can use PSI-hydrogen in decentralized stationary engine-generator sets to produce electricity and process heat. This will halt needless production of greenhouse gases from power plants. Figure 2 shows the scheme of using internal combustion engines to produce electricity and process heat.

Using PSI-hydrogen in on-site engine generators facilitates much greater utilization of energy than central power plants. On-site engines and heat exchangers can cascade heat not converted to electricity into water and space heating applications. This offers the practical advantage of doubling energy utilization compared to central power plant practices. On-site engine generators using PSI-hydrogen meet the most stringent exhaust emission standards at a fraction of the cost of modern central power plants.

Conclusions:

Hydrogen substitution for gasoline and diesel fuel solves the problem of atmospheric pollution. Gaining hydrogen from garbage, sewage, wind, tides, solar, and other renewable resources solves the problem of providing enough energy for the world's population. Transition from fossil fuels to renewable hydrogen by use of mixtures of hydrogen and methane offers significant reductions in exhaust emissions. It is in everyone's self interest to begin the transition immediately.

REFERENCES

1. S. Furuhama, "Hydrogen Engine Systems For Land Vehicles," International Journal Hydrogen Energy, Vol. 14, No. 12, pp. 907-913 (1989).

2. M.A. DeLuchi, "Hydrogen Vehicles: An Evaluation of Fuel Storage, Performance, Safety, Environmental Impacts, and Costs," International Journal Hydrogen Energy, Vol. 14, No. 2, pp. 81-130 (1989).

3. Roy E. McAlister, "Method and Apparatus for Fuel Injection-Spark Ignition System for an Internal Combustion Engine," U.S. Patent 4,066,046, (1978).

4. ARIZONA REPUBLIC, October 8, 1989.

5. NEWSWEEK, July 11, 1988.

6. DISCOVER, January 1986.

7. NEWSWEEK, July 11, 1988.

8. ARIZONA REPUBLIC, May 29, 1990.

9. KANSAS CITY STAR, March 16, 1990.

10. G. M. Woodwell, "The Energy Cycle of the Biosphere," Scientific American, Vol. 223, No. 3 (Sept. 1970).

11. C.A. Rohrman and J. Greenborg "Large Scale Hydrogen Production Utilizing Carbon In Renewable Resources," International Journal Hydrogen Energy, Vol. 2, pp. 31-40 (1977).

12. S. D. Huang, C.K. Secor, R. Ascione, and R.M. Zweig, "Hydrogen Production By Non-photosynthetic Bacteria," International Journal Hydrogen Energy, Vol. 10, No. 4 pp. 227-231 (1985)

13. S. Roychowdhury and D. Cox, M. Levandowsky, "Production of Hydrogen by Microbial Fermentation," International Journal Hydrogen Energy, Vol. 13, No. 7, pp. 407-410 (1988).

14. M. Lavitt, "Production and Excretion of Hydrogen Gas in Man," New England Journal of Medicine 281, 122 (1969).

15. Pamela Weintraub, "A New Kind of Sea Power," Discover (April 1981).

16. Glendon M. Benson, Ronald J. Vincent, and William D. Rifkin, " An Advanced 15 KW Solar Powered Free-Piston Stirling Engine," 15th Intersociety Energy Conversion Engineering Conference, paper number 809-414, published by American Institute of Aeronautics and Astronautics, New York, New York, (August 1980).

17. Richard A. Houghton and George M. Woodwell, "Global Climatic Change," Scientific American, Vol. 260, No. 4, (April 1989).

18. Harry Braun, "The Phoenix Project," Research Analysts P.O. Box 62892, Phoenix, Arizona 85082, ISBN 0-924600-00-4, (1990).

19. Raymond M. Coveney, Jr., Edwin D. Gobel, Edward J. Seller, Gisela A.M. Dreschhoff, and Ernest E. Angino, "Serpentinization and the Origin of Hydrogen in Kansas," The American Association of Petroleum Geologists Bulletin V. 71, No. 1, p. 39-48, (January 1987).

20. Cummins Power Generation, Inc., Box 3005, Columbus Indiana 47202 "Dish-Stirling System" Brochure (May, 1990).

TABLE 1: PSI HYDROGEN EMISSIONS COMPARED TO CALIFORNIA STANDARDS

STANDARDS	EMISSIONS, GRAMS PER MILE		
	RHC[1]	CO	NO_x
PSI HYDROGEN OPERATION	0.0	0.0	0.18
CALIFORNIA[2]	0.41	9.0	0.4
CALIFORNIA TLEV[3]	0.125	3.4	0.4
CALIFORNIA LEV[4]	0.075	3.4	0.2
CALIFORNIA ULEV[5]	0.040	1.7	0.2

1. RHC = Reactive Hydrocarbons
2. Present Anti-smog Standards of California
3. Transitional Low Emission Vehicle (California Air Resources Bd.)
4. Low Emission Vehicle (California Air Resources Board)
5. Ultra Low Emission Vehicle (California Air Resources Board)

TABLE 2: PSI HYDROGEN BOOSTED METHANE COMBUSTION

TEST OR STANDARD	EMISSIONS, GRAMS PER MILE		
	RHC[1]	CO	NO_x
5% HYDROGEN, 95% METHANE	0.06	1.6	0.38
50% HYDROGEN, 50% METHANE	0.03	0.4	0.21
PSI HYDROGEN CAR (100%)[2]	0.0	0.0	0.18
CALIFORNIA TLEV[3]	0.125	3.4	0.4
CALIFORNIA LEV[4]	0.075	3.4	0.2
CALIFORNIA ULEV[5]	0.040	1.7	0.2

1. RHC = Reactive Hydrocarbons
2. Essentially No SO_2 Emissions
3. Transitional Low Emission Vehicle (California Air Resources Bd.)
4. Low Emission Vehicle (California Air Resources Board)
5. Ultra Low Emission Vehicle (California Air Resources Board)

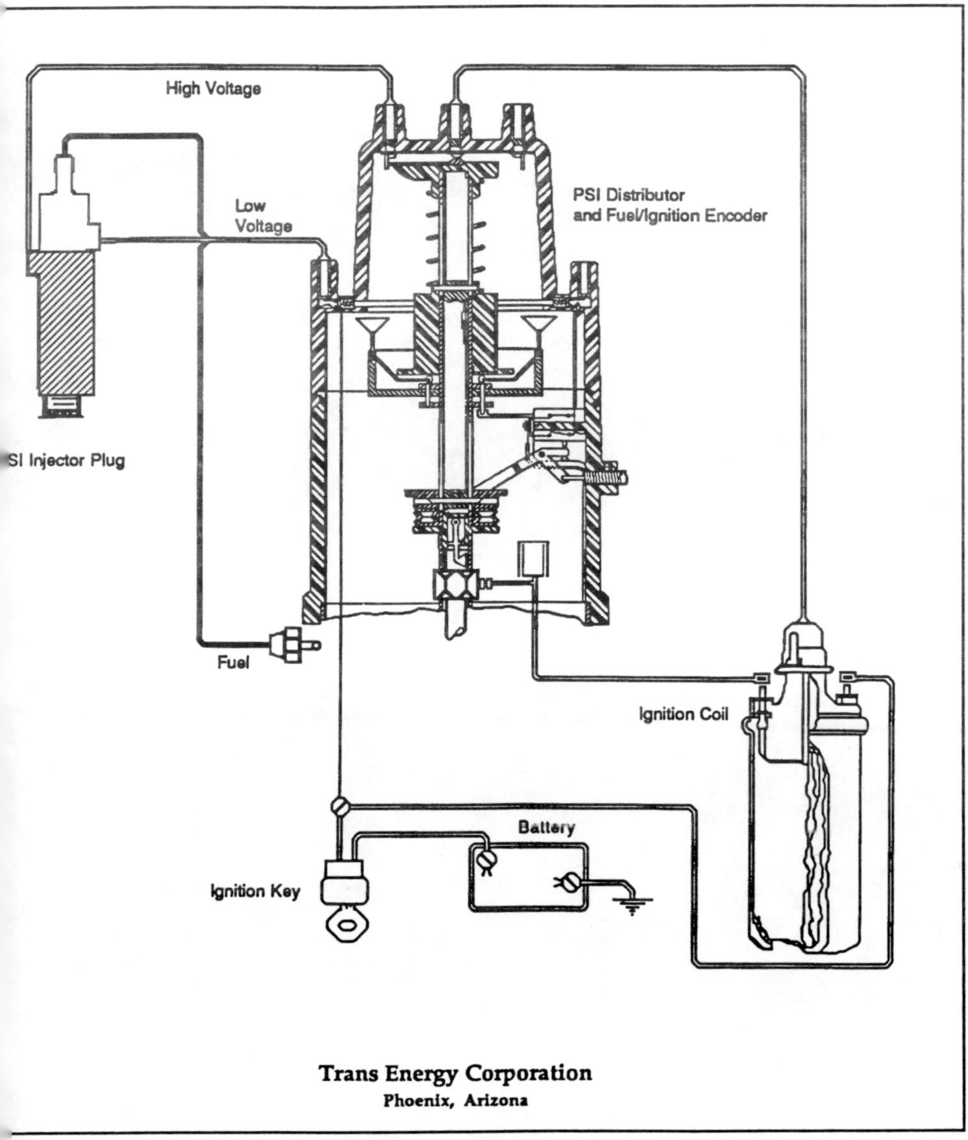

Figure 1. - Precision Spark Injection System Schematic

HYDROGEN DIESEL AND S.I. ENGINE PERFORMANCE - TARGETS FOR OTHER ENGINES TO MATCH

Drs. S.M.Lambe# and H.C.Watson+
#Research Engineer, +Reader in Transport Energy
Mechanical and Manufacturing Engineering Department
University of Melbourne, Parkville
Vic. 3052, AUSTRALIA

Abstract

This paper collates results of our hydrogen engine research over the last two decades. It is shown there is close agreement between a driving cycle performance predicted from engine test data on an s.i. hydrogen engine and measured performance of the engine installed in a car. This vehicle had a 63% improvement in energy economy over its gasoline-fuelled counterpart when tested to the U.S. city test cycle. Emissions reductions approached 100% with the exception of NOx.

Applying the same model we forecast the performance of converted diesel engines, in road driving conditions from tests on a diesel fuel ignited hydrogen engine. Diesel fuel replacement is shown to be around 85% and there is an improvement in energy efficiency. Greenhouse gas emissions are reduced by about 80%. Reductions in NOx and other emissions are also substantial.

1. INTRODUCTION

The development of the demonstration vehicle fuelled with hydrogen is the culmination of many years of research using hydrogen as a fuel. This work, begun by Watson [1], initially was directed towards fundamental understanding of engine combustion, and hydrogen was chosen for this work because of its known reaction kinetics. The work was extended in a research, single cylinder engine to study and overcome the 'backfiring' problem and these successful results were subsequently adapted to a production 4 cylinder engine [2]. Prior to transferring this technology to a multi-cylinder engine, simulation of efficiency and range capabilities of vehicles with such an engine was carried out. It is shown that there is close agreement between a driving cycle performance predicted from engine tests on an s.i. hydrogen engine and measured performance of the engine installed in a car.

The same steps are followed in the development of a hydrogen diesel engine. The steps included the conversion of a single cylinder, open chamber diesel engine to operate on hydrogen with 'pilot' diesel serving as a source of ignition [3], and then the estimation of diesel engine performance in a truck.

2. COMPARISON OF PROPERTIES OF HYDROGEN, ISO-OCTANE, AND DIESEL

From Table I it can be seen that hydrogen is quite different to conventional fossil-based fuels; namely iso-octane and diesel. The properties of prime interest are the flammability limits, minimum ignition energy, and the self-ignition temperature of a hydrogen-air mixture.

The flammability limits indicate that the engine can be operated extremely lean, allowing unthrottled operation, with a consequential improvement in efficiency due to the reduction in pumping losses and increase in mixture specific heats ratio.

The low minimum ignition energy value is the basis of the main problem when operating an engine on hydrogen. The problem manifests itself in the form of induction ignition (backfiring through the inlet tract) due to any hot regions within the combustion chamber, e.g. residual exhaust gas, spark plugs, exhaust valve surfaces, or carbon (pyrolysed oil) having sufficient temperature and energy to ignite the incoming charge before the inlet valve closes.

The high self-ignition temperature of hydrogen makes it difficult to auto-ignite in diesel engines, even at very high compression ratios. Hence an auxiliary source of ignition is required. This may be in the form of diesel fuel, injected in the normal way at a point in the cycle near top dead centre. Such an engine is commonly referred to as a dual-fuel engine.

3. SINGLE CYLINDER HYDROGEN S.I. ENGINE TESTS

The initial spark-ignition tests were performed in an ASTM-CFR engine, with hydrogen aspirated into the inlet air. Due to the flammability limit and the ignition energy of hydrogen-air mixture, no throttling was required, load being varied by hydrogen flow only. The backfire problem arose for all load conditions other than low load. A delayed port admission system was developed for hydrogen supply which overcame the backfire problem, at all speeds and loads. The engine performance in terms of efficiency and emissions has been reported in detail elsewhere [4].

4. HYDROGEN VEHICLE SIMULATION

Using the results of single cylinder tests, transfer functions were developed to predict the performance of a multi-cylinder engine on hydrogen. By assuming that engines and vehicles are scalable, it was possible to study the interactions between range and fuel storage medium: metal hydrides or cryogenic liquids.

The Vehicle Simulation Model

Fig. 1 shows the flow chart of the vehicle simulation model. Details of the mathematics may be found in reference [5]. When instantaneous speed and torque are established, fuel consumption and emissions are obtained by interpolating within appropriate numerical arrays of constant fuel flow or emission rate. Summation of the values at successive time increments leads to the total fuel or emission in the given time period. These, when divided by the distance travelled, yield values of fuel consumption in L/100km and emission rate in g/km.

The simulation reported here considers the hydrogen vehicle used in urban ares, where gaseous or liquid fuelling is likely to be most practicable with limited availability of refuelling stations. As specified in Table II, four occupants plus baggage are considered as the maximum design load, and hence vehicle ranges developed with this load will be conservative compared with a more usual occupancy of 1.3.

The vehicle concept is that of a basic vehicle structure, to which strengthening is added to support increased fuel or engine mass as the prescribed range of the vehicle or fuel type changes. An increase in any tank or transmission or engine mass has an associated 10% addition of its mass to the vehicle mass. Conservative values are selected for engine and transmission masses as functions of total engine power.

The maximum speed will not exceed 155 km/h, because of the gearing limitation, and may fall for installed powers less than about 50 kW.

To compare the energy consumption with different fuels in the simulated vehicle it is necessary to the specify driving condition. This may be a repetitive velocity-time sequence (or driving cycle) or a driving sequence which has been recorded in actual driving conditions. Here, two well recognised test cycles have been employed: the US-FTP urban dynamometer driving schedule which is also embodied in Australian Design Rule ADR27A, and the Highway driving cycle. In addition, the Australian urban cycle measured in central Melbourne traffic under congested commuter traffic conditions has been used.

Watson [6] has discussed the limitations of the model which uses steady state fuel consumption and emissions to calculate values for transient driving conditions. Analysis has indicated that most serious errors are likely to occur when simulating an engine carburetted with liquid fuel, where the transient interaction between accelerator pump fuel supply and manifold wall-wetting are likely to lead to serious departures in supplied air-fuel ratios from steady state values. With gaseous fuelling this should not be a problem.

<u>Energy Consumption</u>

The energy consumption results, projected for the vehicle detailed by the data in Table II, are presented graphically in Fig. 2. The ratio of hydrogen to gasoline energy consumed is shown over the vehicle range of 100 through to 500 km on one tank full of fuel. It can be seen that for a range of 200 km or less, hydride or liquid storage requires less than 75% of the energy needed with gasoline. For liquid fuelling or MgNi hydride only about two-thirds the energy is required.

As the range requirement increases, the energy storage requirement to overcome the aerodynamic drag for the higher speed Highway driving, results in less improvement in energy consumption than either of the urban driving conditions. The greater energy per km required in the low speed stop-go Melbourne driving, results in about 70% of the gasoline energy requirement.

In summary, for a short range urban vehicle that can be refuelled about twice per week, one third less energy will be required with hydrogen.

5. S.I. HYDROGEN CAR RESULTS

Encouraging results of our simulation work gave us the motivation to extend this work to a multi-cyinder engine. A Ford Pinto 4 cylinder 2 litre OHC with cast iron cylinder head, an engine representative of the then current automotive practice, was chosen. The performance results of this engine are shown in Fig. 3 as an engine map of thermal efficiency, when using hydrogen as fuel, also seen are the maximum torque curves for both hydrogen and gasoline and lines of constant ratio of thermal efficiency of hydrogen to gasoline. It can be seen that the peak torque curve for hydrogen fuelling is between 23 to 35% less than that for gasoline. This due to less energy being available in the fuel (see Table I), lower air volumetric efficiency when operating on hydrogen (see Table I), and the altered cam timing. Of interest is the high thermal efficiency, even at low loads, due to unthrottled air operation; and when compared with the efficiency, when operating on gasoline, there is much improved energy utilization. Below 20 to 25% of full load there is an over 70% increase in energy efficiency, with a low speed, low load maximum of over twice that for gasoline. Even at about 70% of full load there is still a 30% increase in energy efficiency. This indicates quite substantial energy savings, especially in the low speed, low load area, applicable to urban driving, and if gasoline-type maximum torque is required a larger engine can be used with consequent reductions in these energy savings.

The engine was then fitted to a 1982 Ford Cortina four cylinder, four speed manual, station wagon and emission and fuel consumption tested. Preceding this the vehicle was emission and fuel consumption tested on gasoline to be used as a comparison.

Table III records the results of tests according to the CVS-C US FTP, city cycle. This table shows that a 63% increase in energy economy has been achieved, slightly better than the modelling forecast of 56%. As might be expected NOx emissions are the only significant exhaust emission, occurring almost entirely during acceleration modes.

The vehicle is registered for road use although most of the vehicle testing has been performed in our vehicle energy and emissions testing laboratory. On the road the vehicle energy and emissions testing laboratory. On the road the vehicle has no difficulty with maintaining station in urban traffic, although the reduced power output and increased vehicle mass has resulted in a small increase in acceleration times. The vehicle requires care in clutch engagement while accelerating away from standstill as accelerator pedal controls gas flow only, the normal torque rise at part throttle is substantially reduced. Further it has very little engine braking. Both effects are due to the unthrottled operation. In general the vehicle is very easy to drive, as there is no delay or 'surge' to rapid 'throttle' opening since the gas is delivered to the inlet port.

6. HYDROGEN DIESEL SINGLE CYLINDER TESTS

A Petter's PH1W direct injection diesel engine, with specification described in Table IV, was converted to operate on the dual-fuel principle. Hydrogen was supplied in parallel with the intake air into the engine where the mixture was formed. The use of delayed port admission of hydrogen avoided the occurrence of backfiring. Minor changes to combustion chamber shape have been employed to reduce heat transfer losses. The hydrogen-air mixture was ignited by a 'pilot' charge of diesel fuel. Under high torque conditions with the compression ratio unchanged from the diesel engine's value of 16.5:1 knocking occurred. 'Solid' water injection employed to curb knock under these full load conditions [3].

With this engine a wide range of operating variables detailed in Table V were investigated. The engine performance results have been reported in detail elsewhere [3]. The mean effective pressure (mep) results at optimum injection timing (MIBT) and at full load are shown in Fig. 4. When comparing full load power output on dual-fuel (hydrogen and diesel) with that of diesel-only, it is noted that around 30% of the total air is displaced (loss in volumetric efficiency) with hydrogen-air equivalence ratios ranging between 90% and 110%.

Fig. 5 shows a 3-D plot of indicated thermal efficiencies with respect to speed and load. This plot gives an overview of the engine performance on dual-fuel in comparison with diesel-only, on the basis of thermal efficiency, over the entire load-speed matrix. If the solid surface profile (dual-fuel thermal efficiencies), is examined it can be seen that there is a region spanning from high load and high speed down to low load and low speed when the engine performs better on hydrogen and diesel as compared with diesel-only. This happens to be the normal engine operating range in vehicle installations.

7. HYDROGEN DIESEL MULTI-CYLINDER ENGINE SIMULATION

These findings encouraged us to develop transfer functions for the single cylinder data to forecast the performance of the multi-cylinder engine on dual-fuel (hydrogen and diesel). A normally aspirated Caterpillar 3208, V-8, direct injection diesel engine which is more representative of current automotive practice was selected. The specification of this engine is summarised in Table VI.

The transfer function assumes that the engine can be scaled, with power and emission rates increasing directly with cylinder volume (and engine mass) and thermal efficiencies constant irrespective of size. The resulting isopleths of fuel rates on torque-speed axes for dual-fuel operation are described in Fig. 6 for 'pilot' diesel and for hydrogen fuel, respectively. Also shown are fuel rates' isopleths over the torque and speed grid with diesel-only (baseline).

8. HYDROGEN DIESEL TRUCK PERFORMANCE FORECASTS

The same vehicle simulation model, as described earlier, was employed to forecast the on-road performance of a medium-sized diesel truck running on hydrogen and diesel (dual-fuel engine). The energy consumed and emissions rate with dual-fuel, in the simulated vehicle, are compared with diesel-only for the US-FTP city and highway cycles in Table VII. In the simulation, two types of hydrogen fuel storage systems, namely cryogenic liquid and metal hydride storage, have been considered. The test inertia shown in the table is the mass of the vehicle required to obtain the specified range of 300 km on one tank full of fuel.

The results indicate that for a cryogenic liquid tank there is a 2.7% and 2.1% energy saving running a diesel truck on hydrogen and diesel as compared with diesel-only for the urban drive cycle and the highway drive cycle, respectively. However, with hydride storage marginal energy saving (0.5%) is obtained running on dual-fuel as compared with diesel only for the urban cycle. The increased total mass obtained with hydride storage results in an energy penalty of about 3% for the highway cycle. The emission results forecast about 85% reduction in the greenhouse gas (CO_2) emission for a hydrogen-fuelled diesel truck. The NOx emissions are predicted to reduce by more than 60% for a dual-fuel truck as compared with a diesel-only.

9. CONCLUSIONS

A hydrogen car has been developed free from abnormal engine combustion and able to operate on low pressure gas supply from either hydride or liquid hydrogen storage. It is shown there is close agreement between its driving cycle performance predicted from engine tests results and measured performance of the engine installed in a car. This vehicle had a 63% improvement in energy economy over the U.S. city test cycle. Emissions reductions approached 100% with the exception of NOx.

Predictions have been made for energy consumption and emissions of a hydrogen-fuelled diesel truck. Small energy gains are obtained under urban and highway driving conditions. Greenhouse gas emissions are reduced by about 80%. There is more than 60% reduction in NOx emissions. Smoke and particulate emission reductions are also anticipated, from the single cylinder test data but have not been forecast since the latter particularly are engine design specific rather than generic. Further the pilot diesel fuel ignited vehicle, with metal hydride fuel storage, has the advantages of low heat rejection and low emissions. Therefore, it is attractive for use in densely populated urban areas and also in mines, warehouses, and other confined environments. In the short term it is these environments which should be the first to accept these virtues of hydrogen vehicles.

The energy efficiency and low pollution levels are targets that other alternative fuels will have difficulty in matching.

REFERENCES

1. WATSON, H.C., The compression ignition of fuel-oxidant mixtures, University of London, Ph.D. Thesis, 1969.
2. WATSON, H.C., MILKINS, E.E., MARTIN, W.R.B., AND EDSELL, J.D., An Australian hydrogen car, Proc. of the World Hydrogen Conference #5, Toronto, Canada, pp. 1549-1562, 1984.
3. LAMBE, S.M., WATSON, H.C., AND MILKINS, E.E., Hydrogen engine with pilot diesel fuel ignition, Proc. of the World Hydrogen Energy Conference #8, USA, 1990.
4. WATSON, H.C. AND MILKINS, E.E., Some problems and benefits from the hydrogen fueled sprak ignition engine, SAE 789212, 1978.
5. WATSON, H.C., GHOJEL, J.I., KHATIB, E.T., AND CLOSE, H.C., Heavy vehicle performance and fuel consumption - 'TABESAM' SAE-A's validated computer model, SAE-Australasia, Special Publication, March 1989.
6. WATSON, H.C., The prediction of emissions and fuel consumption in real driving conditions, SAE-A, I.Mech.E., and I.E. Aust., Symp. on Computers in Automotive Design and Analysis, Paper 78025, 1978.

TABLES

TABLE I: COMBUSTION PROPERTIES OF HYDROGEN, ISO-OCTANE AND DIESEL

Property	Hydrogen	Iso-octane	Diesel
Flammability Limits (% vol.)	4.0–75.0	1.4–6.0	0.6–6.5
A/F Stoichiometric (by mass)	34.5	15.0	14.4
Minimum Ignition Temperature (K)	858.0	~723.0	~525.0
Minimum Ignition Energy in air (mJ)	0.02	~1.35	~0.25
Heat of Combustion (lower, MJ/kg)	119.93	44.5	42.3
Burning Velocity in air (m/s)	2.65–3.25	0.37–0.43	~0.34

TABLE II: DESIGN CRITERIA SELECTED FOR THE HYDROGEN CAR STUDY

Vehicle concept	4 occupants plus baggage
Mass of pay load	340 kg
Mass of body shell and trim, etc. (less engine transmission support structure)	500 kg
Fractional mass of support structure (of engine plus transmission plus fuel tank mass)	0.10
Engine power/mass	2 kg/kW
Transmission power/mass	1 kg/kW
Vehicle power/mass	40 kW/tonne
Transmission	4 speed manual
Overall gear ratios	20.05, 9.18, 6.86, 4.01
Frontal area x Cd	0.78 m2

TABLE III: HYDROGEN CAR TEST RESULTS FOR THE US-FTP CITY CYCLE

Fuel Type	Test Inertia (kg)	Fuel Cons.	Energy Economy (km/MJ)	Emissions HC (g/km)	Emissions CO (g/km)	Emissions NOx (g/km)
Gasoline	1340	1.35 L	0.25	1.41	21.9	1.42
Hydrogen	1410	2.89 m3 (@ STP)	0.42	0.01	0.55	0.96
% Change			63	-99	-97	-32
ADR27A Requirement				2.1	24.2	1.9

TABLE IV: SINGLE CYLINDER SPECIFICATION

Engine	PETTER PH1W
Combustion Chamber	Bowl-in-piston, Low swirl
Injector	3 hole type
Injector System	Jerk pump Bryce (variable timing)
Bore	87.3 mm
Stroke	110 mm
Swept Volume	0.658 L
Compression Ratio	16.5:1
Engine Rating	6 kW at 2000 r/min

TABLE VI: MULTI-CYLINDER ENGINE SPECIFICATION

Engine	Caterpillar 3208
Combustion Chamber	Offset, bowl-in-piston
Injector	Jerk pump (var. tim.)
Bore	114.3 mm
Stroke	127 mm
Swept Volume	10.425 L
Compression Ratio	16.5:1
Engine Rating	145 kW at 2800 r/min

TABLE V: SUMMARY OF OPERATING CONDITIONS DURING THE DUAL-FUEL EXPERIMENTS

Speed (r/min)	'Pilot' Inj. Timing	'Pilot' Consumptn. (g/s)	H2% min.* (Idle)	H2% max.* (Full Load)
1000	12 deg btdc	0.0437	64%	86%
1250	11 deg btdc	0.0417	65%	89%
1500	7 deg btdc	0.0433	63%	90%
1700	5 deg btdc	0.0527	71%	89%

* H2% min. and H2% max. are the proportions of energy supplied as hydrogen

TABLE VII: HYDROGEN DIESEL TRUCK PERFORMANCE FORECASTS

Driving Condition	Fuel Type	Fuel Storage	Test Inertia for 300 km (tonnes)	Energy Consumption (MJ/km)	Emissions	
					CO2 (g/km)	NOx (g/km)
Urban	Diesel-only	Liquid	22.5	12.97	0.97	9.06
	Hydrogen & Diesel	Liquid	22.5	12.61 (-2.7)*	0.19 (-80.4)	3.04 (-66.4)
	Hydrogen & Diesel	Hydride	23.5	12.91 (-0.5)	0.19 (-80.4)	3.98 (-56.1)
Highway	Diesel-only	Liquid	22.5	12.52	0.94	8.07
	Hydrogen & Diesel	Liquid	22.5	12.25 (-2.1)	0.12 (-87.2)	3.10 (-61.6)
	Hydrogen & Diesel	Hydride	23.6	12.92 (+3.1)	0.13 (-86.2)	3.24 (-59.9)

* Numbers in parentheses indicate % change with respect to diesel-only operation

(3) Jet speed, with which hydrogen gas is injecte high pressure air is considerably lower than the spray tip speed because density of hydrogen is ve small penetration of jet is expected. In additio self-ignition temperature is about 580°C, which than that of diesel fuel (350 °C), and radiat hydrogen flame is also very small. As a result combustion process for fuel injection engine; i. e injection, mixture formation with air, igniti propagation and flame transfer from the ignited next one, is apt to be slow. Therefore, rapid should be achieved with the aid of the gas f combustion chamber.

(4) Since compression ignition of hydroge practically impossible, hot surface or spark must for its ignition. Although both ignition m feasible, spark ignition is more desirable in durability, electric energy consumption and pre preignition at start if leaked hydrogen from valve is accumulated.

LH_2-TANK, AND LH_2-PUMP FOR HIGH PRESSURE INJECTOR

One of the most important technologies fo fueled vehicle engine is on-board storage of th fuel.

The next three are candidate methods.

(i) Metal hydride (MH)

(ii) High pressure bottle (HP)

(iii) Liquid hydrogen (LH_2)

1. Comparison of the three storage methods

First, weight of each storage method is compar the same heat content of 30 liter-gasoline as Table 2. From the comparison, it is obvious heaviest among the three. And it is also notew direct fuel injection of MH stored hydrogen cylinder is difficult because hydrogen discharge insufficient for injection. Therefore, with t most effective method to enhance engine power output inject hydrogen into a cylinder while valves are In this case, amount of heat content for the mixture eased by 20 % compared with the gasoline and by 41 % with the externally formed hydrogen mixture, as n Figure 1. Two types of hydrogen injection engines en tested at Musashi Institute of Technology. The ype, injection during the first half of compression and spark ignition, is devoted as "low pressure on". Engines of this type have following eristics.

kfire never occurs because intake valve is closed injection.

injection pressure such as 1 MPa is enough.

put power increases to become 20 % more than that of e engine without preignition by means of cold n injection such as 0 $\sim$ -50 °C.

ever, when the atmospheric temperature hydrogen is d, the output power is reduced to the amount about e as gasoline's because of preignition.

second type, denoted as "high pressure injection", is on near TDC and ignition by hot surface or spark. One ge of this type is that it is free of back fire, tion and knocking. Therefore, its compression ratio raised to the range of 12 : 1 to 15 : 1 , which onds to the maximum thermal efficiency, and it can be to a large size engine.

ver, to realize this type of engine, following s should be taken.

gh pressure such as 8 MPa or more is required to hydrogen into the high pressure combustion chamber.

gh pressure hydrogen has to be prepared on board, for by means of liquid hydrogen (LH_2) pump.

drogen gas leaks very easily through a nallow gap an injection valve and its seat. Thus, extremely machining and minimal amount of lubricating oil necessary.

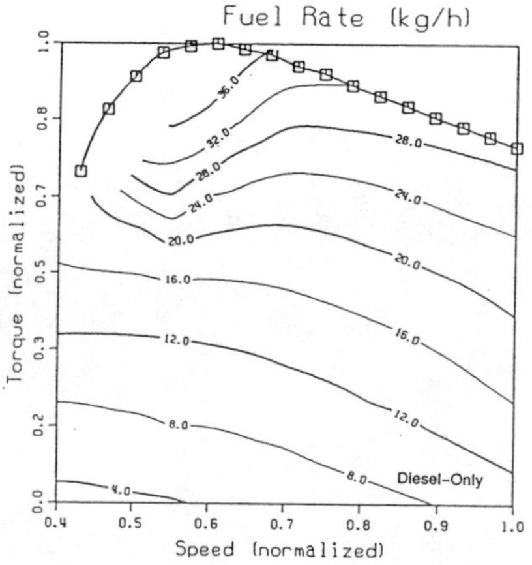

(a)

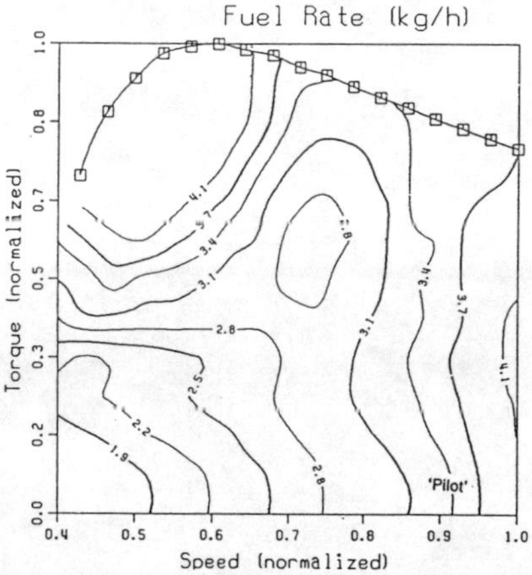

(b)

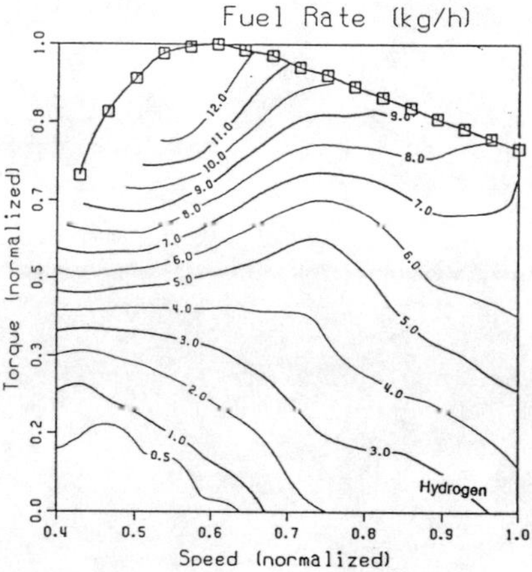

(c)

Fig. 6 Fuel rate isopleths for the V–8 diesel engine, a) measured for diesel only. Corresponding predictions for dual–fuel operation, b) 'pilot' diesel flow rate and c) hydrogen flow rate.

Hydrogen Engine Technology
" R & D at Musashi I. T. "

Shoichi Furuhama
Musashi Institute of Technology
Tokyo

INTRODUCTION

1. Hydrogen as an Alternative Fuel

It is widely believed that amount of energy consumption around the whole world will continue to increase. At the same time, it is widely acknowledged that dependence on fossil fuels in the future should be significantly reduced. The followings are two primary reasons to be substituted the new energy for fossil fuel.

(1) Fossil fuels, especially petroleum, on the earth are foreseen to be exhausted within the next century at the latest. Even before the exhaustion, the limited resource will be the major cause of regional conflicts such as the Iraqi Invasion.

(2) Since carbon is one of the principal elements of fossil fuels and alcohol fuels, their combustion products do not only include such pollutants as HC, CO, SOx, smoke and odor, but also include large amount of CO_2 (Table 1), which is recently regarded as the major cause to deteriorate the earth environment.

Hydrogen is the most promising alternative as the solution of these critical problems. Hydrogen can be produced by various types of natural energy including solar energy without using any fossil fuels. Therefore supply of Hydrogen can be limitless once its production method is established, although hydraulic power and wind power are the only practical method utilizing natural energy at the moment. Furthermore, hydrogen fueled systems discharge neither CO_2 nor pollutants listed above.

An automobile is one of modern technologies that consume a lot of petroleum (gasoline). Thus, an alternative fuel automobile will be also necessary in the future. Hydrogen seems most suitable for an automobile among various alternative fuels. Although a solar car has been successfully demonstrated, its output power is too small for practical use. A battery car has been also considered as promising, but its overweight restricts its application.

2. Problems to be solved

In order to have hydrogen practically used for automobiles as a substitute for gasoline, the following problems should be solved.

(1) Storage and carriage of hydrogen are crucial items for a vehicle. Weight ratio hydrogen and gasoline for the same calorific value is only 0.37, but hydrogen volume in normal state is 3000 times as large as gasoline in liquid state.

(2) An unit volume of stoichiometric H_2-air mixture has an only 85 % calorific value of the gasoline-air mixture. It means that the hydrogen premixture spark ignition engine has a smaller maximum output by 15% than the gasoline engine. In addition, power output of the external mixture type hydrogen engine is expected to be even lower because of backfire or preignition.

(3) Although hydrogen can be ignited by spark with an extremely low energy and in a wide inflammable range, its self-ignition temperature is so high that it is difficult to apply the compression ignition.

Presented in this paper are discussions on countermeasures against the above mentioned problems and prediction as to the best hydrogen engine for land vehicle with respect to performance and economy, which based on the experience of R & D at Musashi Institute of Technology for 20 years.

HYDROGEN INJECTION INTO THE CYLINDER FOR INTERNAL MIXTURE FORMATION

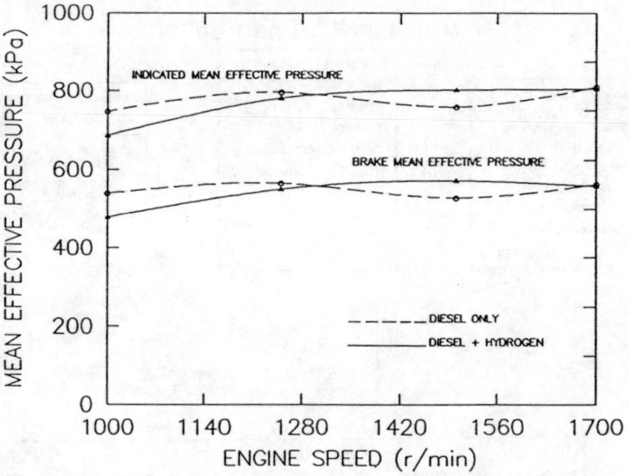

The mean effective pressures for the dual-fuel engine shown in comparison with that for diesel-only over the speed range

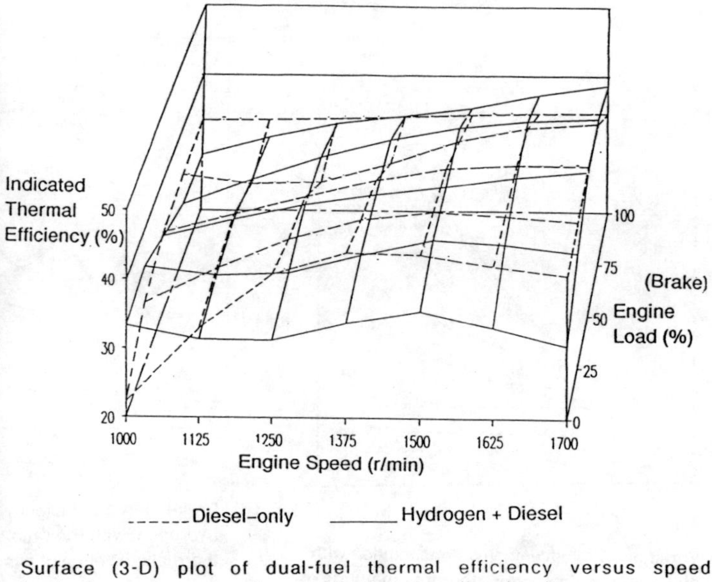

Surface (3-D) plot of dual-fuel thermal efficiency versus speed and load shown in comparison with that for diesel-only.

FIGURES

MODEL INPUTS AND OUTPUTS

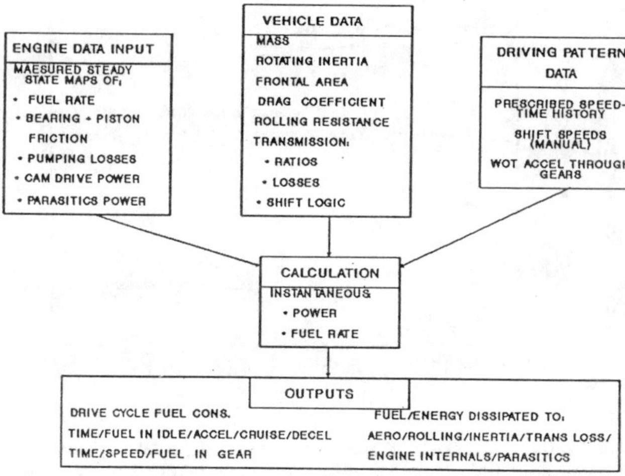

Fig. 1 Summary of vehicle model and its inputs and outputs.

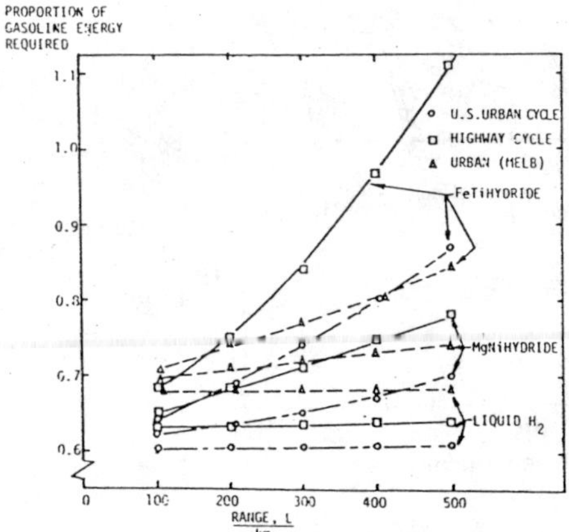

Fig. 2 Variation of relative energy consumption with range and driving pattern for three methods of hydrogen storage on board the vehicle

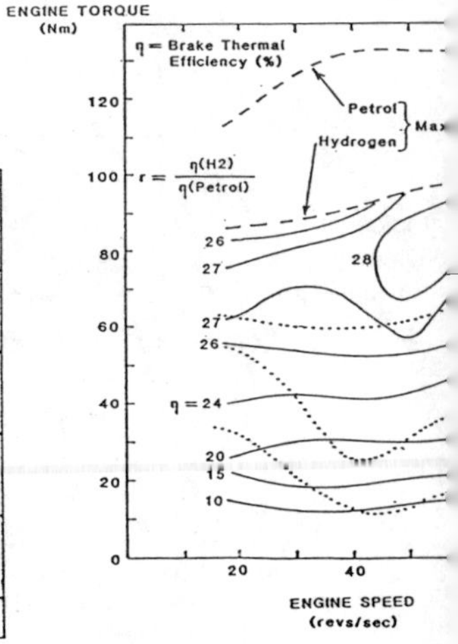

Fig. 3 Brake thermal efficiency i hydrogen, with the ratio o thermal efficiency of hydr gasoline

technology level, MH storage can not be applied to usual automobiles.

Although HP has almost the same weight as MH, higher power output can be expected because low pressure injection of HP stored hydrogen is possible. Therefore, the second storage method is suitable for limited application such as vehicles insensitive to their weight or those for short range use.

Weight for LH_2 storage shows only 1/ 10 that of the above two storage methods. In addition, LH_2 stored hydrogen can be compressed to sufficiently high pressure for fuel injection such as 8 $\sim$ 10 MPa by means of small LH_2 pump. However, high cost of LH_2 and cryogenic tank technology are disadvantages of this storage method. From viewpoints of efficiency and safety, blow-off hydrogen from the LH_2-tank must be appropriately taken care of.

2. LH_2 system for hydrogen injection engine

From the discussion above, only LH_2 storage is applicable to the transportation vehicle engine system. For practical use of LH_2-system, it is important not only to resolve its problems, but also to make efficient use of its merits.

(1) In the case of the low pressure injection engine, since injection pressure about 1 MPa is enough for the low pressure injection engine. Such injection pressure can be prepared by either of the next two methods.

a. Pressure in the LH_2 tank is raised by evaporated gas hydrogen itself. Needless to say, structure of the tank has to be designed according to this pressure.

b. While tank pressure is kept at low pressure such as 0.05 MPa, LH_2 pressure is pumped up as liquid state to the injection pressure (1 MPa). Figure 2 shows a typical injection system, where a LH_2 reciprocating pump of 15 mm bore and 15 mm stroke is driven by a DC motor. When this pump is used at 800 rpm, sufficient fuel for a 100 PS (73 kW) engine can be supplied. Delivered LH_2 immediately becomes cold gas hydrogen in the delivery pipe of exterior part of the tank, then it is heated to room temperature in the small tank, from which the hydrogen is injected into the engine cylinder through an injector. Delivery rate of LH_2 is controlled by means of electronic motor-speed control unit operated together with pressure sensor in the small tank.

(2) In the case of the high pressure injection engine, only the second fuel injection method mentioned above (b) can be applied. In addition, the special injection device is required to realize a short injection period. Figure 3 shows a new high pressure H_2 injector. High pressure diesel oil is intermittently delivered from a diesel injection pump to (1) and presses on the plunger (4) by oil pressure in the space (2). As a result, the injection valve (6), which is connected to the plunger, is opened, and high pressure H_2 is injected. At the same time, the high pressure oil is relieved through the relief valve (3). Thus, when the delivery of the oil from the diesel injection pump to (1) is ceased, pressure at (2) is lowered and the injection valve (6) is closed by the pressure of H_2 in (5). Because of this mechanism, a coil spring becomes unnecessary.

A pair of non lubricant piston and cylinder are required for LH_2 pump. Successful operation without abnormal wear was achieved polyimide material and hardened metal surfaces. Since clearance of each rubbing surfaces must be kept very small such as $1.5 \sim 3\ \mu m$ in the LH_2, a piston made of invar interior material and polyimide outer collar was developed and used against the stainless steel hardened cylinder, as shown in Figure 4. And it is a cylinder of this pump, instead of a piston, that reciprocates for preventing deformation of the long reciprocating rod.

(3) When a part of hydrogen is premixed in the intake manifold, the engine combustion can be improved. It shortens the ignition delay of injection fuel, and combustion pressure vibration can be reduces, as shown in Figure 5.

Since blow-off hydrogen from LH_2-tank can be used as a part of the premixed fuel, no blow-off loss during the engine operation can be expected. However attentions must be paid to the fact that increase of the premixed hydrogen beyond a certain limit causes the preignition.

(4) When high pressure hydrogen from the LH_2 pump is supplied to a reciprocating gas expander, power output can be obtained from its crankshaft. The performance of an expander as shown in Figure 6 was tested. According to the test results, it was first confirmed that sufficient driving power for LH_2-pump was obtained from the atmospheric heat through the expander. It was also found that necessary amount of high pressure (8 MPa) hydrogen to drive LH_2 pump is 1/8 to 1/6 of total engine fuel, hence all of the expanded hydrogen can be used as premixture.

IGNITION METHODS FOR HYDROGEN JET

1. Spark ignition for low pressure injection

For low pressure injection type, where hydrogen is injected during the first half of compression stroke, spark ignition is suitable. This is a pre-mixture type engine and large power output becomes possible, if pre-ignition can be well controlled because amount of induced air is increased. However, in practice, pre-ignition occurs at richer mixture ratio than λ (excess air ratio) = 1.2, at which condition only about the same out put as gasoline engine or smaller NH_1 can be obtained. If temperature of injection hydrogen is low such as 0 ~ -50 °C, the pre-ignition can be prevented and the maximum power can be increased to NH_2, as shown in Figure 7.

2. Hot surface ignition for high pressure injection

According to the experimental results of the hot surface ignition method, although this method is found to be very useful in terms of ignition stability if the surface temperature is higher than about 900 °C , the following problems are to be resolved.

(i) Life time is not long enough for practical use.

(ii) Electricity for heating is not negligible, especially a large capacity battery must be prepared.

(iii) At engine starting, pre-ignition or backfire tends to occur because of hydrogen leakage through the injection valve.

3. Spark ignition for high pressure injection

Recently, spark ignition was applied to the high pressure injection engine as a countermeasure to the above problems. It was found out that the spark ignition was more suitable to the hydrogen injection jet than to the liquid fuel spray, because ignition by electric spark has advantages such as very small ignition energy and very wide range of ignitable mixing ratio.

Visualization of its ignition is shown in Figure 8. The tip of the hydrogen jet was most easily ignited as (a). As the ignition timing was delayed, the ignition became more unstable as (b), and the bottom part of the jet could not be ignited as (c). It was also found out from this visualization that smaller distance between the injection hole and the spark gap enabled the jet tip to reach the gap earlier, made the ignition delay shorter, and realized more gradual pressure rise. This result indicates that in the case of applying the spark ignition to the hydrogen injection type engine, precise positioning and carefully adjusted timing will be necessary.

MUSASHI-8, THE LATEST LH_2-CAR Of MUSASHI INSTITUTE OF TECHNOLOGY

Musashi-8 is the 8th hydrogen fueled car designed and developed at Musashi Institute of Technology. Figure 9 shows its photo. The main purpose of Musashi-8 was to demonstrate the most recent technology achievement as a hydrogen fueled passenger car at the 8th WHEC held in Hawaii in July 1990. The engine and the fuel supply system of Musashi-8 were designed based on the most recent research and development results discussed above. And fabrication of the entire car was completed under support of Nissan Motor Co.. The main parts of it are as following:

1. Combustion chamber

The base engine was a 4 cylinder, 4 stroke DI-diesel engine manufactures by NISSAN DIESEL. Its bore, stroke, compression ratio and stroke volume were 96 mm, 102 mm, 18.5 : 1 and 3 liter, respectively. The modified engine for hydrogen fuel use was equipped with piston cavity, a hydrogen injection valve and a spark plug, as shown in Figure 10.

During the hydrogen engine modification, compression ratio (ε) was reduced to 13.5 : 1, and injection nozzle, which had 8 holes, were installed. It was because the previous study indicated that these modifications improved thermal efficiency and reduced NO_x emission.

Hydrogen injection pressure was increased from the previous value of 8 MPa to 10 MPa in order to improve mixture formation of hydrogen and air. A distributor-less spark ignition system, in which each spark plug has its own ignition coil, was selected, which enabled ignition timing to be arbitrarily chosen. The distance between the spark gap and one of the injection nozzle hole was decreased to 3 mm by using an custom made plug with an extended electrode. In addition, a crank angle sensor, which was driven by a crank-shaft and a pulley, was installed, and the sensed crank angle was used as computer input.

Cavity, machined at piston crown, formed a combustion chamber, and 8 hydrogen jets were injected into the cavity. Since it had been found out that the cavity shape strongly affected engine performance, the optimum shape was chosen. It was 36 mm in opening diameter of the cavity instead of the original 55 mm. However, it seemed that further improvement would be necessary.

Temperature for ignition was not affected by compression ratio, thus the ε value should be determined to maximize the engine performance and to decrease the maximum pressure. Lower maximum pressure improved fuel jet penetration and mixture formation. For the Musashi-8 engine, the ε value 13.5 : 1 was chosen as the experimental results indicated.

2. The effect of injection timing and ignition timing

Spark timing should be selected to be optimum under all operating conditions. In other words, there are two variables to be controlled. Results in Figure 11 show effects of different ignition timing under the operating conditions where engine speed, BMEP and injection beginning θ_i were kept at constant values of 1000 rpm, 0.45 MPa and 6° CA BTDC, respectively. The upper figure is for spark timing θ_s = 6° CA BTDC, while the lower figure is for θ_s = 3° CA BTDC. The figure in right-hand side illustrate the measured pressure inside the cylinder for successive cycles, and they indicated cycle-to-cycle variation. When θ_s was equal to θ_i , the pressure profile was smooth as a result of stable ignition. On the other hand, in the lower figure, where θ_s was later than θ_i by 3 degree, pressure rise in the initial combustion storage was very steep and pressure vibration was also observed. In addition, misfire often occurred in this case.

In the course of Musashi-8 development, it was revealed that θ_s should be electronically controlled to be equal to θ_i within 0.05 ms in order to realize comfortable driving on the road. This was one of the most difficult problems to overcome before Musashi-8 was able to be successfully demonstrated.

The LH_2 tank for Musashi-8 was made of stainless steel plate and it had a double wall vacuum space and super insulation, as shown in Figure 10. The tank, whose weight and content volume were 60 kg and 100 liters, respectively, was loaded inside the rear trunk.

The LH_2 pump was a reciprocating type, and its bore and stroke were both 15 mm. The pump was driven by a 600 W DC motor.

3. Test drive

The body of Musashi-8 was a sport car type, as shown in Figure 9. Musashi-8 achieved successful running on the test course of NISSAN MOTORS in Oppama, Yokosuka in July 5, 1990. At this test, all parts of it operated smoothly under the sufficient electronic control, and the maximum speed 130 km/h was attained. The test was reported in many TV programs, newspapers and journals in Japan and abroad. Then from July 22 to 26, Musashi-8 was exhibited at WHEC-8 held in Hawaii. Its demonstration drive was also conducted at University of Hawaii.

SAFETY OF HYDROGEN

From the author's 20 year experience of hydrogen research, it may be concluded that hydrogen is in general as dangerous as gasoline or propane. In fact, among a number of current and previous research activities on hydrogen engines and hydrogen fueled vehicles, there have been no big accident reports. However, it should be noted that each fuel has different characteristics with respect to its safety.

From the following viewpoints, hydrogen is safer than other fuels.

(i) Self ignition temperature of hydrogen is high, therefore hydrogen can be seldom ignited without a high temperature igniter.

(ii) Hydrogen has very small density and high diffusively, therefore even if leakage of hydrogen should occur, the leaked hydrogen immediately diffuses up and can be rarely ignited in the open air. For example, when the UCLA LH_2 vehicle was turned over and its damaged LH_2 tank blew out the stored hydrogen, no fire was caused. It was also reported by American National Bureau of Standard in 1970 that concentration of hydrogen and that of methane were compared, inside an automobile room in the case, where both gases leaked from the tank inside a rear trunk, and that their behaviors to reach their ignition limits were almost the same.

(iii) Hydrogen flame is colorless and does not emit radiative heat, whereas flames of oil and coal emit a strong radiative heat with red color. Therefore, heat does not transfer downward or sideboard from the flame, which narrows the dangerous area. While it is well known that 36 people died of the famous Hindenburg fire at Lake Hurst NJ. in 1937, the few remember that 65 people survived at the accident. Those 65 people were either below or on the sides of the flame.

(iv) Hydrogen gas can be converted to H_2O with the oxidizing catalyst, which works at room temperature. This is especially interesting to the safety device for blow-off hydrogen gas from the LH_2 tank in parking garage.

On the other hand, hydrogen is more dangerous in the following aspects.

(i) Hydrogen can be ignited with very small energy if there is a high temperature spots as an electric spark.

(ii) Hydrogen can be ignited even at very lean mixing ratio.

(iii) Hydrogen can leak out through a very narrow gap.

CONCLUSIVE REMARKS

In this paper, it is presented that research and development results on the hydrogen vehicles, which have been carried out at Musashi I.T. for 20 years.

The latest LH_2-car Musashi-8 has a LH_2-tank, high pressure 100 bar, LH_2-pump for high pressure H_2-gas injection into the combustion chamber, and spark ignition with the optimum timing against the injection timing. The author believes that this system is more practical and superior than the other types which apply the pre-mixing and storage of MH or HP. Because it achieves the high output power without any abnormal combustion, and also a light weight of storage tank.

However, listed below are items of R & D required in future.

(1) To make more easy use of cryogenic, low viscous LH_2.

(2) To improve LH_2-tank insulation.

(3) To increase LH_2-pump efficiency and reliability.

(4) To optimize combustion chamber and gas flow pattern in it.

LIQUID HYDROGEN CAR WITH HIGH PRESSURE HYDROGEN INJECTION AND SPARK IGNITION ENGINE

Katsuyoshi Koyanagi, Masaru Hiruma and Shoichi Furuhama

Musashi Institute of Technology
Department of Mechanical Engineering
Engine Research Laboratory

1-28-1 Tamazutsumi Setagaya-ku,
158, Tokyo,
JAPAN

ABSTRACT

Hydrogen automobile research group at Musashi Institute of Technology, Engine Research Laboratory, has been carried out research and development on the hydrogen engine. Recently, the 8th test car, Musashi-8, was accomplished. This hydrogen car has a newly developed engine. Hydrogen was intermittently injected into the cylinder near TDC by the injector and ignited by the spark plug.

The combustion form of this engine was different from that of ordinary gasoline engines and diesel engines. It was reason that, in the combustion chamber, there is exactly the polarization of injected hydrogen, the only one ignition point and the ignition period. It was pointed out that injection timing, ignition timing and their relative timing had significant influences on ignition and combustion behaviors. In this study, measured combustion pressure diagram and experiment of the modified combustion chamber for visualization indicated that rigorous control of both injection ignition timing was necessary for stablo combustion and high thermal efficiency.

INTRODUCTION

A hydrogen injection and hot surface ignition engine, which was also developed previously at Musashi I. T.[1], showed such difficulties as short life of the hot surface, relatively large electric power consumption, and occasional misfire and back fire at start-up.

In order to overcame these difficulties, spark ignition, instead of hot surface ignition, was selected for Musashi-8. According to the previous study[2], the spark timing was required the delicate and accurate timing control, absolutely and relatively with the injection timing.

LIQUID HYDROGEN FUELED CAR "MUSASHI-8"

Figure 1 shows Musashi-8 and its engine room. A liquid hydrogen tank and a liquid hydrogen pump were loaded in the rear portion of the Musashi-8. The liquid hydrogen pump, which was driven by a DC motor, compressed liquid state hydrogen to 10 MPa. Since accurate control of injection and ignition was necessary, as discussed above, constant injection pressure was desirable. In order to keep the injection pressure constant, the motor speed should be adjusted according to the hydrogen amount required by the engine. The high pressure liquid hydrogen was heated to become high pressure gas, which was directly injected into the engine cylinder near TDC by means of an oil pressure driven injection valve. A spark plug was selected for its ignition. In order to realize stable ignition during the test drivo , ignition timing, injection timing and injection pressure should be rigorously controlled. The optimum design of the combustion chamber and compression ratio were adapted to obtain the stable and higher performance.

COMBUSTION OBSERVATION BY SCHLIEREN METHOD

Constant volume combustion chamber

This experiments were conducted in a constant volume combustion chamber with schlieren method in order to optimize the spark gap position and timing which could realize stable ignition with a short ignition delay. Ignition for an actual engine took place in a much larger region than for a constant volume combustion chamber due to high pressure, high temperature and gas flow, although a constant volume combustion chamber can be used for understanding the fundamentals.

Figure 2-A shows two typical results. In the case of (a), where the tip of the hydrogen jet was ignited, ignition lag τ was short and stable combustion was obtained. On the other hand, in the case of (b), ignition lag τ was long because spark was given after the tip of the jet passed by. Very rapid flame propagation under such an ignition condition was demonstrated clearly. Ignition instability was observed under this ignition timing.

Figure 2-B shows the effect of distance X on the ignition lag τ. The lag τ can be divided into two periods; τ_o and τ_s, where τ_o is from injection to jet arrival at the spark gap, and τ_s is from arrival to ignition. It was clearly demonstrated that while τ_o was a function of X, τ_s was not affected by X. From this measurement, it was found out that the optimum distance X, under this condition, was about 5 mm. In shorter than this distance, ignition was became instability.

Bottom View Type Optical research engine

The second step, the schlieren method was applied to the optical research engine which was similar to the actual engine, and effect of the air flow and process of the flame propagation was visualized by a high speed camera. Figure 3-A shows the optical system in this experiment.

The bottom view type optical research engine was modified single cylinder 4 stroke diesel engine. Its bore, stroke and stroke volume were 91.5 mm, 95 mm and 625 cc, respectively. The high speed photograph was taken through the observation window whose visible diameter was ϕ50 mm and located at the bottom of the piston cavity. Figure 3-B shows the combustion chamber observed.

Since density of hydrogen gas was very small, the photographs, especially hydrogen jets, turned out to be difficult to interpret. Figure 4-A shows the effect of spark timing on combustion pressure by indicator diagram, and Figure 4-B shows the comparison of flame propagation by the high speed photographs. Each (A) and each (B) were corresponded. The result (A) was obtained when injection began at θ_i=16 °CA BTDC and spark began at θ_s=15 °CA BTDC, while the result (B) was obtained under the same injection timing, but when spark began later than that for (A) by 1 ° CA. The number indicated below each photographs are corresponding time the injection begins.

In base cases, although injection valve was lifted at 18 °CA BTDC, hydrogen jet came out of the nozzle port 2 degree (0.8 ms) later than the valve lift. Injection lag was clearly pointed out by these schlieren photographs.

If electric spark was given at the tip of the hydrogen jet stream, as shown in (A), stable ignition was obtained, and as a result, ignition lag became short and combustion pressure rise became gradual. In addition, it was demonstrated that the initially ignited flame was propagated to the neighboring jets only in the swirl direction.

On the other hand, in the case of (B), where ignition timing was delayed by only 1 CA(0.4 ms), spark was given after the jet tip passed by. Its pressure diagram and its photographs showed steep rise of the combustion pressure. According to the schlieren photographs, the flame in (B) spread out more abruptly than in (A) because larger amount of inflammable mixture was formed before ignition in the case of (B). It was also observed that the initial flame propagated radically to the neighboring jets in (B). This was clearly different from the propagation in (A), where smaller amount of the formed inflammable mixture before ignition resulted in gradual combustion.

This visualization experiment revealed in the case of spark ignition, the optimum ignition timing lay in the very small range and that highly accurate electronic control was necessary to obtain stable combustion with short ignition lag and to realize gradual flame propagation.

ENGINE FOR HYDROGEN CAR MUSASHI-8

Combustion chamber

The modified engine for hydrogen fuel use was 4 cylinder 4 stroke DI-diesel engine. Its bore, stroke, compression ratio and stroke volume were 96 mm, 102 mm, 13.5 and 2.953 liters, respectively. It was equipped with piston cavity, a hydrogen injectior which had 8 holes nozzle and a spark plug, as shown in Figure 5.

The effect of injection timing and ignition timing

Injection timing was determined by a centrifugal type injection advance timer attached to the injection pump for opening the injection valve.

Result in Figure 6 shows effects of different ignition timing under the operating conditions where engine speed, BMEP and θ_i were kept at constant values of 1000 rpm, 0.45 MPa and 6° CA BTDC, respectively. The upper figure is for ignition timing θ_s=6 CA BTDC, while the lower figure is for θ_s=3° CA BTDC. The figure in right-hand side illustrate the measured pressure inside the cylinder for successive cycles, and they indicate cycle-to-cycle variation. When case (1), the pressure profile was smooth, and as a result, ignition was stable. On the other hand, case (2), pressure rise in the initial combustion stage was very steep and pressure vibration was also observed. In addition, misfire often occurred in this case. As discussed in the visualization experiment, such difference in combustion phenomena was mostly because the ignition location of the hydrogen jet differed. From this Musashi-8 engine experiment, it was confirmed that rigorous control was crucial.

CONCLUSION

(1) In direct injection spark ignition engine, the combustion behavior was demonstrated by combustion chamber visualized.

(2) Optimum control for the timing of injection and ignition was realized under all operating condition.

REFERENCE

(1) T. Fukuma, T. Fujita, P. Pichainarong and S. Furuhama, Hydrogen Combustion Study in Direct Injection Hot Surface Ignition Engine, SAE paper 861579 (1986)

(2) H. Kobayashi, T. Matsushita, S. Furuhama, Studies on Ignition Methods for Hydrogen Engines with Injection Right Before TDC, JSME, 88-0355, 1988

FIGURE 1 PHOTOGRAPHS OF MUSASHI 8

A. Musashi 8

B. Engine room

FIGURE 2 CHARACTERISTIC OF SPARK IGNITION

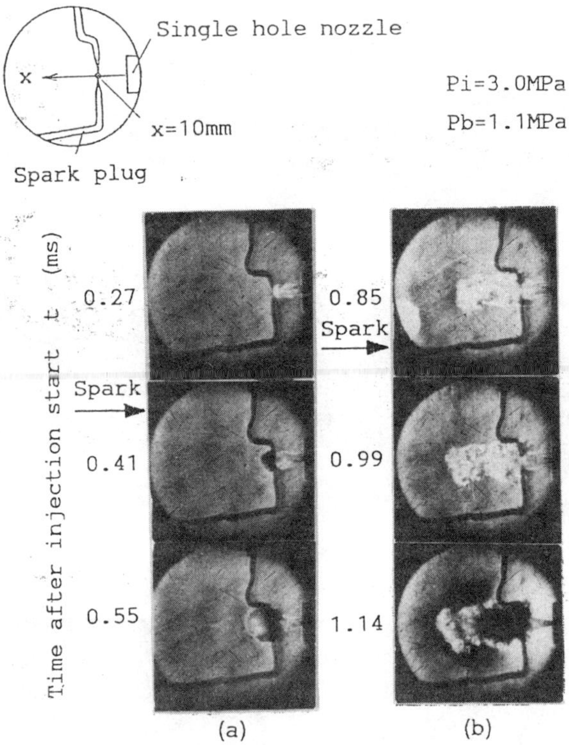

A. Effect of spark timing

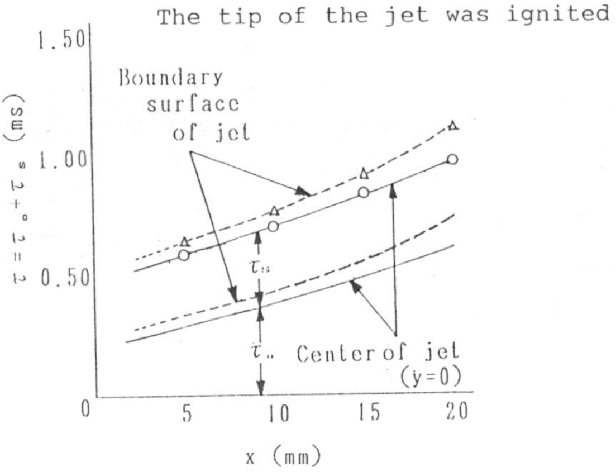

B. Effect of distance X ignition lag τ

FIGURE 3 OPTICAL SYSTEM

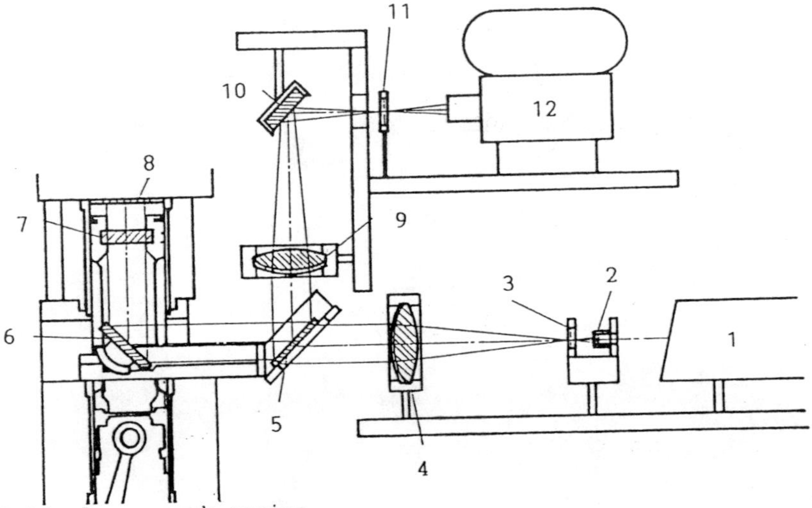

Optical research engine

- 1: Ar ion laser (4W)
- 2: Lens
- 3: Pinehole
- 4: Lens
- 5: Beamsplitter
- 6: Mirror
- 7: Quarts window
- 8: Mirror (Cylinder head, In.valve, Ex.valve)
- 9: Lens
- 10: Mirror
- 11: Knife edge
- 12: High speed camera

A. Optical system

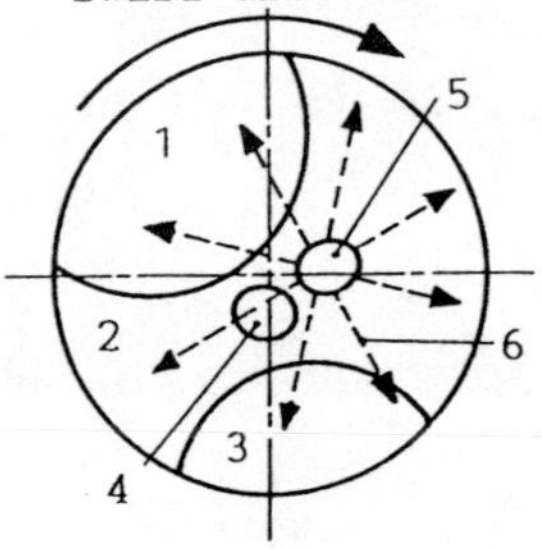

- 1: Intake valve
- 2: Cylinder head
- 3: Exhaust valve
- 4: Spark plug
- 5: Injection nozzle (8 holes)
- 6: Hydrogen jet

B. Explanation for observed combustion chamber

FIGURE 4 RESULT OF OPTICAL RESEARCH ENGINE (500rpm)

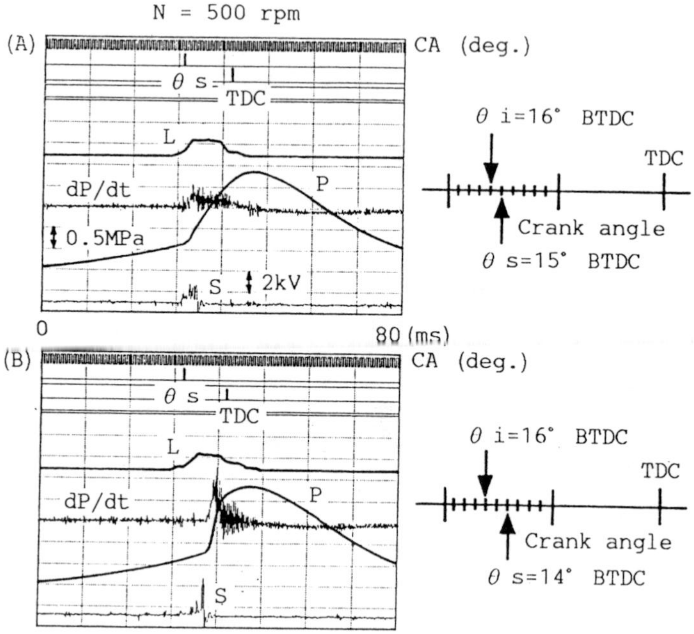

A. Effect of spark timing on combustion pressure

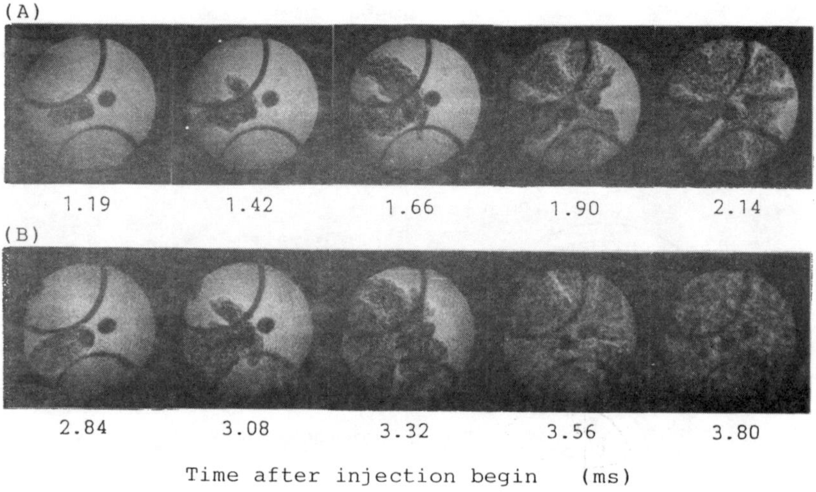

B. Comparison of flame propagation

FIGURE 5 COMBUSTION CHAMBER OF HYDROGEN ENGINE AND SPARK PLUG

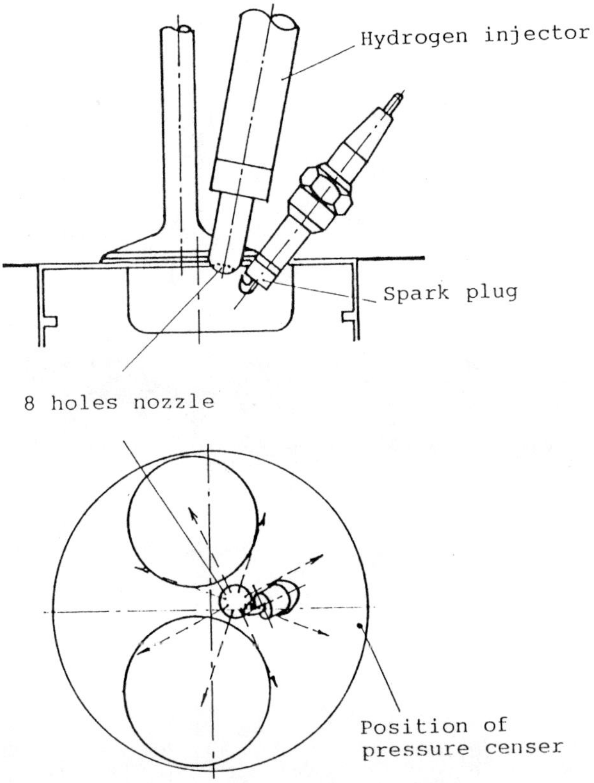

A. Combustion chamber of hydrogen engine

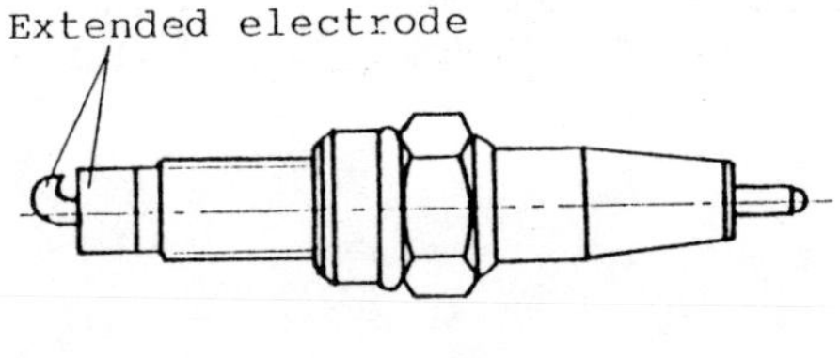

B. Spark plug of hydrogen engine

FIGURE 6 INDICATOR DIAGRAM SHOWING EFFECT OF SPARK TIMING ON THE CYCLE VARIATION (BMEP=0.45MPa at 1000rpm)

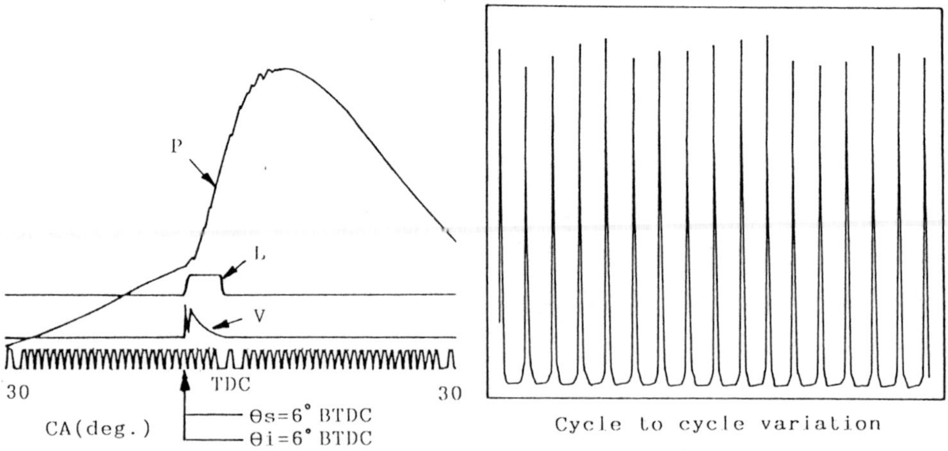

(1) Θi = 6° BTDC , Θs = 6° BTDC

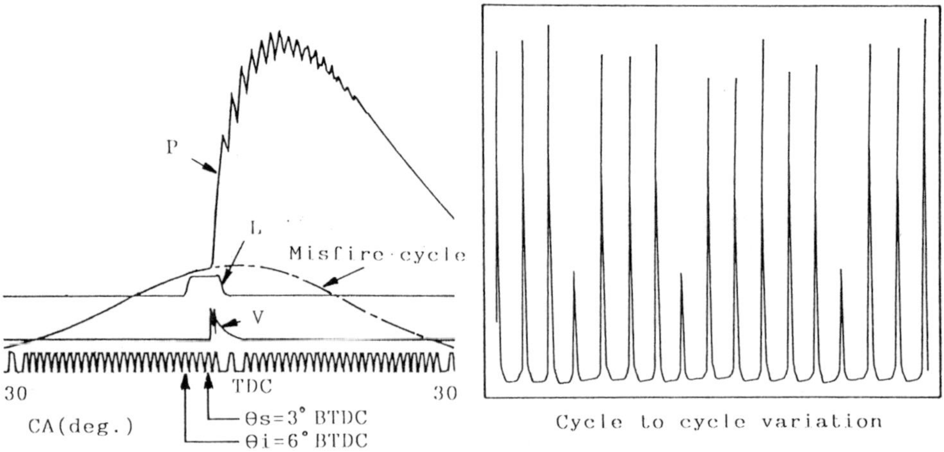

(2) Θi = 6° BTDC , Θs = 3° BTDC

CHAPTER FOUR

Hydrogen Fuel Cells

THE PROTON EXCHANGE MEMBRANE FUEL CELL
- A STRONG CANDIDATE AS A POWER SOURCE FOR ELECTRIC VEHICLES

D.H. Swan, O.A. Velev, I.J. Kakwan,
A.C. Ferreira, S. Srinivasan and A.J. Appleby

Center for Electrochemical Systems and Hydrogen Research
Texas Engineering Experiment Station
Texas A&M University System
238 Wisenbaker Engineering Research Center
College Station, TX 77843-3402
USA

ABSTRACT

Electric vehicles will play a significant role in solving energy and environmental problems. The new clean air laws in California mandate the sale of zero emission vehicles by 1998, (2% of total vehicle sales increasing to 10% by the year 2005). The only vehicle power systems that can meet the zero emission requirement are battery or fuel cell power systems with on-board hydrogen. Initial sales requirements will be met by battery powered vehicles for use in an urban setting (range of less than 150 miles/day). Long range general purpose vehicles will require fuel cells. Of all fuel cell types, the proton exchange membrane fuel cell (PEMFC) system is presently the strongest candidate. The PEMFC uses platinum as a catalyst and at present electrode loading levels will require substantially more per vehicle than is presently used in catalytic converters. This paper highlights the progress made in air breathing, low platinum loading PEMFCs at the Center for Electrochemical Systems and Hydrogen Research (CESHR). The effects of platinum loading, and operating pressure are described and an analysis of air compression energy and its net affect on fuel cell performance is presented.

INTRODUCTION

The energy crisis in 1973 stimulated the development of batteries and fuel cells as power sources for electric vehicles. The R&D expenditure since 1973 for developing advanced batteries for electric vehicles has exceed that of fuel cells by at least a factor of ten. Recent fuel cell developments indicate that the PEMFC may be a the first fuel cell to be commercialized for vehicle use. Even though the attainable power and energy densities in PEMFCs may be sufficient to serve as a sole power source for electric vehicles, a hybrid propulsion system (i.e. a PEMFC with advanced batteries) appears more attractive due to lower capital cost[1].

The General Electric Company invented PEMFCs (originally referred to as solid polymer electrolyte fuel cells). Historically, this fuel cell was the first to find an application (auxiliary power source in NASA's Gemini space vehicles, 1960s). Due to the low performance and stability of the proton conducting membrane at that time (polystyrene sulfonic acid), the PEMFC was displaced by alkaline fuel cells for the subsequent space missions (Apollo, Space Shuttle). General Electric continued to made advances in the 1970s by replacing polystyrene sulfonic acid membranes with DuPont's Nafion membranes. High levels of performance were attained[2].

The organizations which were responsible for making major strides in the 1980s, are Ballard Power Systems, Inc. (BPSI) and Ergenics Power Systems, Inc. (EPSI). BPSI demonstrated high power density electrodes and built 5 kW prototype units for military applications. Recently they have been engaged in designing and constructing an array of 5 kW modules (120 kW gross power) for a transit bus propulsion system. EPSI incorporated novel humidification and water removal systems in 200 W and 1 kW systems and utilized hydrides for supplying hydrogen to the electrochemical cell stack.

The BPSI and EPSI PEMFCs contain high platinum loadings (about 4 mg/cm^2) in each electrode (8 mg/cm^2 of membrane area). A novel method was developed at Los Alamos National Laboratory (LANL) to utilize low platinum loading electrodes in PEMFCs and attain performances close to those in PEMFCs with high platinum loading electrodes. The attainment of high energy efficiencies and high power densities in fuel cells with low platinum loading electrodes, first at LANL and then at CESHR was made possible using the following procedures: (i) impregnation of a proton conductor into the electrode structure to enhance the three-dimensional reaction zone just as when using a liquid electrolyte; (ii) hot-pressing of the proton conductor-impregnated electrodes to the proton conducting membranes at a temperature close to that of the glass transition temperature; (iii) optimize humidification of reactant gases; (iv) operation at elevated temperatures and pressures; (v) localization of platinum near the front surfaces of the electrodes by using thinner active layers and depositing a very thin layer of platinum on the front surface; and (vi) use of membrane with higher conductivities and water retention characteristics.

CESHR AIR BREATHING FUEL CELL DATA

The CESHR fuel cell experimental data and analysis is presented in the following text and Figures 1 through 7. All data was taken using single test cells with 5 cm^2 electrodes. The electrode membrane assemblies were fabricated using porous gas diffusion electrodes and proton conducting polymer membranes (Dow Chemical, 125 μm thickness). The fabrication method and a schematic of the test cell is provided in reference[3]. The fuel cell gases were humidified and temperature regulated. The performance evaluation was conducted using a computer controlled power supply (HP 6033A) and a data acquisition unit (HP 3421A) interconnected through a GPIB bus.

The fuel cell data is presented in the form of polarization and performance curves. The polarization curves show the single cell voltage and electrode current density. The performance curves present the electrode power density (voltage x current density) and the chemical to electrical efficiency based on the higher heating value of hydrogen.

Electrode Platinum Loading

In Figure 1 polarization data is present that compares high and low platinum loading electrodes on air and oxygen. Oxygen is not practical for fuel cell use on an electric vehicle (due to storage) but is given for comparison. The platinum loading presented are 10.0 and 0.45 mg/cm^2 of electrode area. This translates to 20.0 and 0.9 mg/cm^2 of membrane material and is considered as being very high and low values.

The polarization curve is typically characterized by three regions, an initial region followed by a linear region, and finally a mass transfer limited region. The initial region shows an initial steep drop in the cell potential due to slow cathode kinetics. With sufficient voltage drop (0.1 to 0.15 volts) the kinetics are high and are no longer limiting. The linear region is characterized by a linear voltage drop primarily due to ionic resistance in the electrolyte. As the current density further increases, the polarization curve enters the mass transfer limited region. Cell potential drops off rapidly primarily due to the inability of oxygen to reach reaction sites fast enough. This inability may be caused by a combination of an oxygen gradient through the electrode, nitrogen blanketing and or process water blockage.

Figure 1 shows that oxygen and higher platinum loading increases the cell voltage at a given current density. Consider the two oxygen curves, the high platinum loading curve is offset approximately

60 mV higher than the low loading curve, neither curve shows mass transfer limitations (data was not taken to the reach this region). The air curves at the beginning of the linear region have an offset of approximately 60 mV but the spread increases to > 100 mV as the potential drops due to mass transfer limitations. This may be explained by the high platinum loading having more reaction sites available, thus mass transfer limitations occur at a high current density. Severe mass transport problems with air are encountered at current densities greater than 1.0 A/cm^2. Within the linear region the difference between air and oxygen for the same platinum loading is approximately 60 mV.

Figure 2 presents the Figure 1 data in a performance curve form. The ordinate is electrode power density in W/cm^2 and is an indication of how compact the fuel cell could be constructed. A higher electrode power density means less surface electrode area will be needed to achieve a given power level. The abscissa is electrode efficiency based on the higher heating of hydrogen. Higher efficiency translates to greater fuel economy thus a smaller on-board hydrogen storage system and or a longer vehicle range. It should be noted that lower efficiency translates also to greater heat generation within the fuel cell.

The curves indicate that there is a trade off between power density and efficiency. A higher power density is achieved at a lower efficiency, a peak is reached where the power density and efficiency declines. At no time would it be advantageous to operate the fuel cell beyond the peak power density.

The figure shows that oxygen and higher platinum loading increases the fuel cell power density and efficiency. The high platinum loading air case increases the peak power density by approximately 33% (0.3 to 0.4 W/cm^2). However this is accomplished at great cost to platinum utilization, 48.8 g of platinum/kW compared to 2.9 g/kW for the low platinum loading case (factor of 17). Considering a hybrid vehicle (fuel cell battery combination) may require 25 kW of fuel cell power this translates to 1220g and 72.5 g respectively. The average conventional vehicle contains 3 g of platinum.

Using an example design power density of 0.25 W/cm^2 the higher platinum loading increases efficiency from 43.5% to 49% thus resulting in a high fuel economy. Cyclic voltametric was used to determine the elctrochemically active surface area of the 0.45 mg Pt/cm^2 electrode and it was found to be about 15% to 20% leaving significant opportunity for further increases in performance[4].

<u>Operating Pressure</u>

In Figure 3 polarization curves are presented for operating pressures of 1 and 5 atmospheres utilizing low platinum loading electrodes. Pressurization significantly increases the performance when operating on air or oxygen. Consider the oxygen curves, the voltage increase with pressure at 400 mA/cm^2 is approximately .09 volts increasing in difference to 0.13 volts at 1800 mA/cm^2. This phenomena is particularly pronounced in the air curves where the voltage spread occurs rapidly. The 5 atmosphere air case achieves 0.6 volts at 1000 ma/cm^2 while the 1 Atm curve can only produce 475 mA/cm^2, greater than a factor of 2 decline in current density. It would seem logical that to achieve high performance from a fuel cell, the gases should always be compressed. For oxygen breathing fuel cells this is not a problem as the oxygen will be generally available in a compressed state. However, for an air breathing electric vehicle fuel cell the pressurization will require energy. The following section analyzes this energy requirement and its effect on fuel cell performance.

<u>Energy of Air Compression</u>

By compressing the air going into a fuel cell stack the partial pressure of the oxygen is increased and the stack performance is improved. However the energy for compression must be supplied by the fuel cell system, and the net fuel cell performance is therefore less than the stack performance. The air compression process can be adiabatic or isothermal and part of the compression energy may be recovered by an expander such as a turbine. To simplify the following analysis, it is assumed the compression to be performed by an ideal adiabatic compressor with no energy recovery upon expansion. All values are calculated on a per cm² basis for the electrode (specific values).

The specific power needed to operate the compressor (w_C/cm² of electrode area) is proportional to the reaction rate, (the amount of air needed, $\dot{m}_{air}$, g/hr) and the pressure ratio (P_2/P_1). The stoichiometric mass fuel ratio of oxygen to hydrogen is 8 to 1. Air contains 23.18% oxygen by mass, 21.99 % by volume. Therefore the stoichiometric mass fuel ratio of air to hydrogen is 34.5 to 1. However due to nitrogen blanking of the electrode the fuel cell will probably be operated between 1.5 and 3.0 stoichiometric mixtures of air.

The specific amount of air mass needed for the electrode (g/hr/cm²) is dependent on the current density (i, amps/cm²). The mass needed can be determined by Faradays' Law - 96487 coulombs per gram equivalent of consumed hydrogen. Faradays' Law translates to 26.8 amp hrs/g of H_2, or 26.8 amp hrs/34.5 g of air assuming a 1 stoichiometric mass fuel ratio. Inverting this value results in 1.287 g of Air/hr/i per stoichiometric mass fuel ratio. The specific rate of air use as a ratio of current density is as follows:

(1) $$\frac{\dot{m}_{air}}{i} = \frac{1.287}{3600} \times \text{\# of Stoich. Mixtures} \qquad \frac{\text{g/second}}{\text{amp/cm}^2}$$

The specific energy to adiabatically compress Air is as follows:

(2) $$E_{1\to2} = C_p\, T_1 \left[\left(\frac{P_2}{P_1}\right)^{\frac{k-1}{k}} - 1 \right] \qquad \frac{J}{g}$$

Where
- $E_{1\to2}$ Energy of Adiabatic Compression
- C_p Specific Heat (Air 1.004 J/(g K))
- T_1 Inlet Air Temperature
- P_1 Inlet Air Pressure
- P_2 Fuel Cell Air Pressure
- k Specific Heat Ratio (Air 1.4)

Multiplying the specific rate of air use, $\frac{\dot{m}_{air}}{i}$ and the specific energy of compression results in specific power of compression for the electrode (w_C/i):

(3) $$\frac{w_C}{i} = \frac{1.287}{3600} \times \text{\# of Stoich. Mixtures} \times C_p\, T_1 \left[\left(\frac{P_2}{P_1}\right)^{\frac{k-1}{k}} - 1 \right] \qquad \frac{W}{\text{amp/cm}^2}$$

The electric power to operate the compressor (w_C) comes from the fuel cell. Thus the gross specific fuel cell power (w_{FC}) is the summation of the specific electric load power (w_{FCL}) and specific compressor power (w_C)

(4) $$w_{FC} = w_{FCL} + w_C = V_{cell} \times i \qquad W$$

Consider that the compressor is in series with the electrical load as in Figure 4. Both the electrical load and compressor received the same specific current (i) with the respective voltage drops for the compressor (v_C) and the electrical load (v_L) being in proportion to the specific powers w_{FCL} and w_C:

$$(5) \qquad w_{FC} = w_{FCL} + w_C = i \times (v_L + v_C) \qquad W$$

The specific work of compression is simply the product of the specific amps and the voltage drop due to the compressor:

$$(6) \qquad w_C = i \times v_C \qquad W$$

Rearranging and applying the specific energy of compression equation results in the following:

$$(7) \qquad v_C = \frac{w_C}{i} = \frac{1.287}{3600} \times \text{\# of Stoich. Mixtures} \times C_p\, T_1 \left[\left(\frac{P_2}{P_1}\right)^{\frac{k-1}{k}} - 1 \right] \qquad V$$

It is important to note that the effective compressor voltage v_C is a function of the number of stoichiometric mixtures, the initial air temperature T_1 and the pressure ratio P_2/P_1 only. Thus the effect of air compression can be simply presented on a polarization diagram as a reduction in voltage. The reduced voltage curve then may be used to determine the impact air compression has on power density and efficiency. Note that the analysis considered the compressor in series with the electrical load for ease of explanation, the effective result in equation #7 also applies to a parallel placement of the compressor.

Figure 5 presents the results of equation # 7. The effect of stoichiometric mixtures is to linearly increase the effective compressor voltage while the pressure ratio effect is a power function. Utilizing the analysis, the air curves of Figure 3 are presented in Figure 6. The 1 atmosphere case is unaffected while the 5 atmosphere curve is simply reduced by 0.125 V to represent the result of adiabatically compressing 2 stoichiometric mixtures. The figures shows that the 1 atmosphere condition actually out performs the 5 atmosphere condition up to 0.62 V/425 mA where the two curves cross.

Figure 7 presents the Figure 6 data as performance curves. The curves show that when compressor energy is not considered, pressurization significantly increases both power density and efficiency e.g., 1 atmosphere, 0.31 W/cm² and 35 % efficiency, 5 atmosphere, 0.64 W/cm² and 36 % efficiency. The fuel cell stack efficiency gain with pressure is most pronounced by considering a design power density of say 0.25 W/cm² At this condition the 1 atmosphere case has an efficiency of 43.5% while the 5 atmosphere case has an efficiency of 51.5%. This translates to an approximate fuel saving of 20%.

However when adiabatic compressor energy for 2 stoichiometric mixtures is considered, the peak power is reduced from 0.64 W/cm² and 36% efficiency to 0.49 W/cm² and 28% efficiency, a loss in power density and efficiency. The efficiency drop may be most significant as it translates to a larger on-board hydrogen storage system. For the example design power density of 0.25 W/cm² the efficiency drops to 42%, a lower efficiency than the 1 atmosphere case.

Air compression improves platinum utilization, Comparing the 1 and 5 atmosphere cases, the platinum utilization increases from 2.9 g of platinum/kW to 1.8 g/kW (factor of 1.6 reduction). Considering a hybrid vehicle (fuel cell battery combination) may require 25 kW of fuel cell power, this translates to 72.5 g and 45 g respectively. The average conventional vehicle contains 3 g of platinum and thus these values are still an order of magnitude greater. However future research will probably remove this barrier by incremental improvements in platinum utilization or its outright replacement as a catalyst.

CONCLUSIONS

Mandates for zero emission vehicle are stimulating the development of fuel cells as a power systems for electric vehicles. The PEMFC is a favored technology due to its use of a solid electrolyte, cold start capability, high power density, and efficiency characteristics. This paper presented performance data on the effects of electrode platinum loading and operating pressure.

The use of platinum as a catalyst is a major obstacle to the wide spread use of PEMFCs for transportation. Present low loading electrodes would use an order of magnitude more platinum per vehicle than is presently used in catalytic converters. Future research will probably remove this barrier by incremental improvements in platinum utilization or its outright replacement as a catalyst. Air compression increases the fuel cell power density and thus the platinum utilization is improved. However the energy for compression must be supplied by the fuel cell system, and the net fuel cell performance is therefore less than the stack performance. Through analysis it was found that the theoretical effect of air compression can be simply related to a constant reduction in stack voltage. The reduced voltage curve then may be used to determine the impact air compression has on power density and efficiency.

The trade off between power density, efficiency and platinum utilization is complex and will depend on the vehicle mission requirements. The optimum combination will also depend on the cost of hydrogen and its associated on-board storage system, and finally the availability of platinum and its cost. The future development of alloy electrocatalysts and the optimization of electrode structure will in all probability enhance the performances. From a systems point of view a variable speed *smart* air compressor may be necessary to balance the conflicting goals of power density and efficiency.

ACKNOWLEDGMENTS

Support for this work was provided through a grant No 249 of the Energy Research in Application Program of the Higher Education Coordinating Board of the State of Texas. The authors are grateful to Dow Chemicals for providing the proton exchange membrane for these studies.

REFERENCES

1 D.H. Swan, "Fuel Cell Powered Electric Vehicles", Society of Automotive Engineers, SAE Paper #891724, Appears in SAE Publication SP-793 "Recent Advances in Electric Vehicle Technology", (1989).

2 A.J. Appleby and F.R. Fowkes, *Fuel Cell Handbook, Van Nostrand Reinhold, New York, 1988.*

3 S. Srinivasan, D. Manko, H. Koch, M.A. Enayetullah, A.J. Appleby, J. Power Sources **29**, 367 (1990)

4 H. Koch, A. Nandi, N.K. Anand, O. Velev, D. H. Swan, S. Srinivasan, A.J. Appleby, 178th meeting of the Electrochemical Society, Seattle, WA October 14-19, 1990 Extended Abstracts No 115.

FIGURE 1: POLARIZATION CURVE - AIR AND OXYGEN AT HIGH AND LOW ELECTRODE PLATINUM LOADING

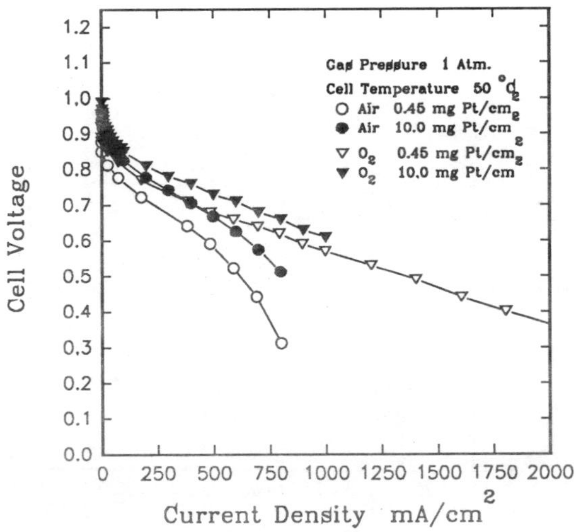

FIGURE 2: PERFORMANCE CURVE - AIR AND OXYGEN AT HIGH AND LOW ELECTRODE PLATINUM LOADING

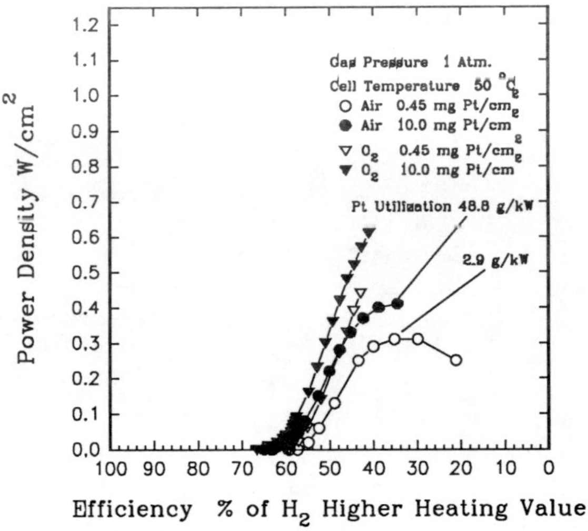

FIGURE 3: POLARIZATION CURVE - AIR AND OXYGEN AT 1 AND 5 ATMOSPHERES PRESSURE

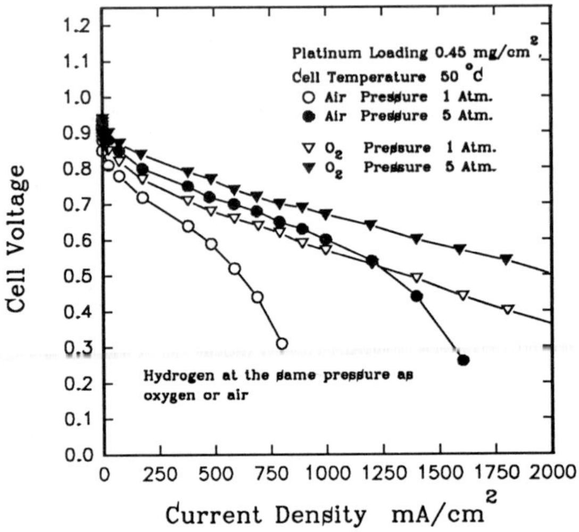

FIGURE 4: FUEL CELL SCHEMATIC

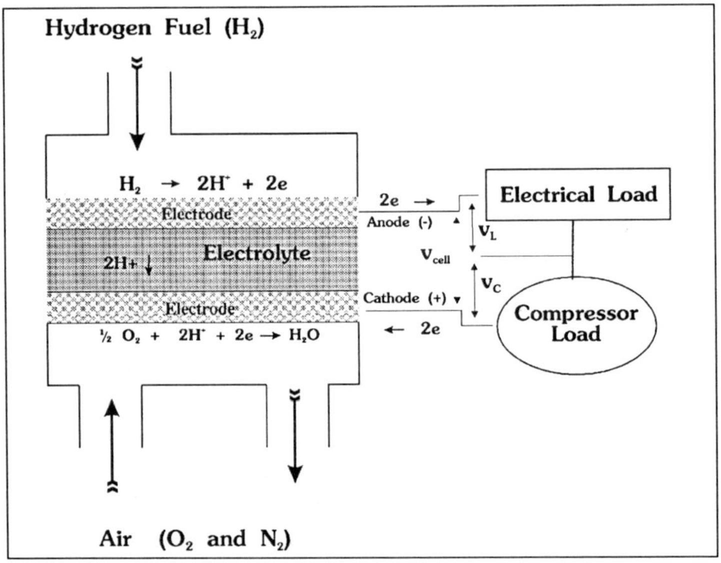

FIGURE 5: EFFECTIVE VOLTAGE LOSS DUE TO ADIABATIC AIR COMPRESSION

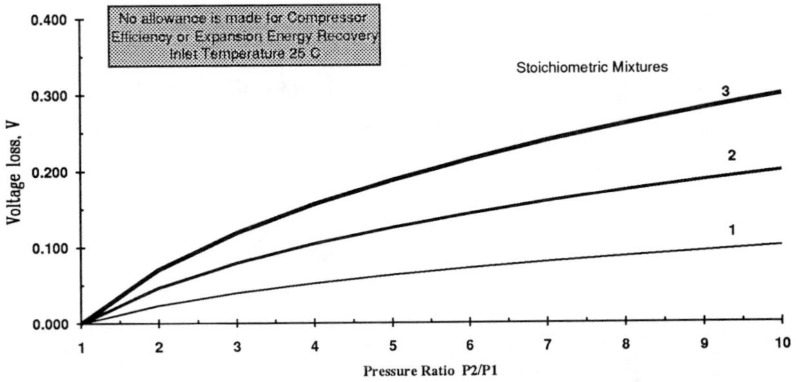

FIGURE 6: POLARIZATION CURVE - AIR AT 1 AND 5 ATMOSPHERES PRESSURE AND EFFECT OF ADIABATIC COMPRESSION

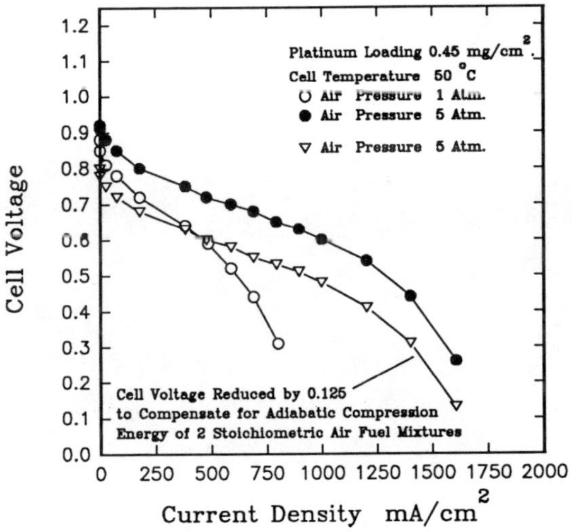

FIGURE 7: PERFORMANCE CURVE - AIR AT 1 AND 5 ATMOSPHERES PRESSURE AND EFFECT OF ADIABATIC COMPRESSION

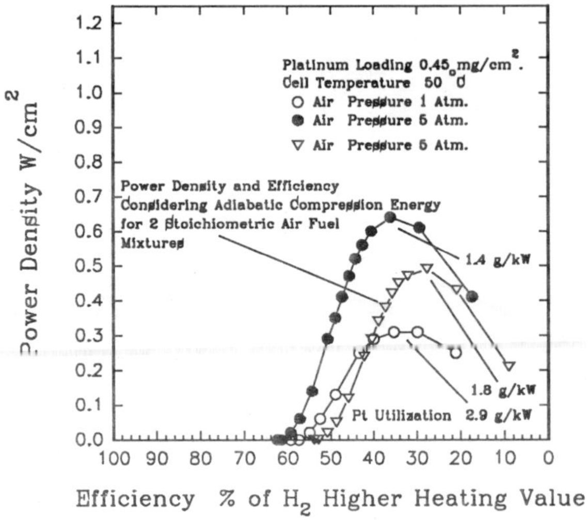

OPERATING CHARACTERISTICS OF A SOLID POLYMER FUEL CELL

J. C. Amphlett, M. Farahani, R. F. Mann
B. A. Peppley and P. R. Roberge

Royal Military College of Canada
Dept. of Chemistry and Chemical Engineering
Kingston, Ontario, Canada.
K7K 5L0

ABSTRACT

Various aspects of fuel cell efficiency are discussed including ideal efficiency, efficiency based on fuel utilization, and net efficiency after parasitic power losses. Typical operating efficiencies are calculated for a hydrogen/air power system. Variation in operating performance with various operating parameters has been studied. Pressure, temperature, mole fraction hydrogen and mole fraction oxygen were all found to effect the fuel cell performance. These effects can be explained in terms of reaction kinetics at the electrode interface. Fuel cell efficiency decreases rapidly for oxygen levels below 0.2 mole fraction. The excess oxidant flow will have an indirect effect on the average effective oxygen concentration at the cathode. Excess fuel and excess oxidant for pure hydrogen and oxygen did not have a significant effect on single cell performance. This indicates that the mechanisms of humidification and water removal within the cell are stable over the range of flows investigated.

INTRODUCTION

Solid polymer fuel cells (SPFC's), also known as polymer electrolyte membrane (PEM) fuel cells, consist of a proton-conducting membrane such as a perfluorosulphonic acid polymer, sandwiched between two platinum-impregnated porous electrodes which have been made hydrophobic by coating with an appropriate compound. The fuel, either hydrogen or a hydrogen rich gas mixture, diffuses through the anode to a catalytic site where it reacts giving up electrons to the external circuit. The hydrogen ions or protons which remain are solvated with a certain number of water molecules and diffuse through the membrane to the cathode. At this electrode oxygen molecules accept electrons returning from the external circuit and react with the hydrogen ions to form water.

Polymer electrolyte membrane fuel cells offer a number of advantages over liquid electrolyte fuel cells. These advantages include low temperature operation, no corrosive fluid spillage hazard, lower sensitivity to orientation, and fast start up. With the improvements in power density and durability which have been achieved in the past several years, SPFC's are now being considered for a wide variety of applications ranging from remote land based power stations to single passenger and mass transit vehicles[1].

FUEL CELL EFFICIENCY

The efficiency of a fuel cell, unlike an internal combustion engine, is not subject to the thermodynamic limitations associated with heat engines. The ideal efficiency of a fuel cell can be expressed as,

$$\epsilon_{id} = \frac{\Delta G}{\Delta H} \tag{1}$$

where

$$\Delta G = \Delta H - T\Delta S \tag{2}$$

For a hydrogen/oxygen fuel cell $T\Delta S$ is small relative to ΔH and the ideal efficiency is close to unity.

The electrochemical efficiency of a cell, however, is reduced by two factors; i) IR losses and ii) kinetic losses. IR losses are associated with the internal resistance of the cell which is affected by the conductivity of the membrane and the effectiveness of the electrical contact between the electrodes, catalyst and collector plate. These properties are mainly a function of the construction of the cell, however they are also affected by the humidification of the reactant gases. Kinetic losses are associated with the efficiency of the electro-catalyst, the operating temperature and pressure, and the concentration of hydrogen or oxygen in the reactant gases. Internal water management also plays an important role with respect to kinetic losses as water hold-up in the cell can potentially flood the electrode and block the electro-catalyst sites. A further factor in kinetic losses is poisoning of the catalyst by undesirable impurities in the fuel gas.

These factors lead to a unique set of operating characteristics distinctly different from the characteristics of a heat engine. Furthermore solid polymer fuel cells are unlike liquid electrolyte cells in that there is no need for ongoing management of the electrolyte concentration and composition, factors which can be quite significant to the efficiency and endurance of fuel cells such as phosphoric acid cells [2].

The effects of various operating conditions on the output voltage of a Ballard Power Systems solid polymer electrolyte fuel cell have been studied by our group. The variation in fuel cell performance with temperature, pressure, %excess fuel and oxidant, and concentration of hydrogen and oxygen in the fuel and oxidant streams, respectively, has been modelled using a parametric approach. These results are reported elsewhere [3].

In this work the results of this previous study have been used to show the net effect of each of these operating conditions on the efficiency of a fuel cell power plant.

PARASITIC POWER LOSSES

Compressor Power Requirements

Compressor power for the purposes of this study is calculated based on typical isothermal efficiencies of 50 to 70% [4]. The theoretical isothermal power requirement is calculated using the equation

$$P\ (kW) = 1.0133\times10^5\ p_1\ v_1\ \ln\left(\frac{p_2}{p_1}\right) \tag{3}$$

and the actual power required is calculated using efficiencies of 50, 60 and 70%. This simplified method of calculating compressor power requirements was chosen for the purpose of demonstrating the magnitude of the parasitic losses. Isentropic calculations would be more accurate and would be required for actual design calculations.

In practice the compression efficiency will depend very much on the application. Small compressors, such as would be used in low power remote generators, would have a lower maximum efficiency but would probably operate in a steady state mode close to this maximum. Medium power 20 - 120 kW systems, such as required for vehicle applications, would use larger more efficient compressors however the variable load demands associated with acceleration and idling conditions would require the compressor to operate in ranges which would be less efficient. High power (greater than 1 MW) systems for marine applications and stationary power plants would be able to use multiple compressors to maintain maximum efficiency at each power level.

Cooling System Power Requirement

The power requirements for cooling a fuel cell power system will again be application dependent. Small systems have a large surface area to active area ratio for air cooling and will require less additional cooling. Larger systems in enclosed spaces, such as marine applications, will require larger cooling systems and therefore incur a greater parasitic power loss; however, larger pumps are more efficient and larger systems can divide the load between several pumps to maintain optimum efficiency at varying load levels.

The effect of fuel cell operating temperature on cooling system power requirements will depend on the available ambient heat sink. For marine applications there will generally be an adequate temperature difference between the fuel cell cooling water and sea water. For land vehicle applications the temperature difference between the fuel cell and the ambient air could be a more significant factor. In this situation, higher fuel cell operating temperatures could result in less heat exchange surface being required and hence lower pumping losses.

Because of the potential variations in cooling system configurations no attempt has been made in this paper to account for the potential parasitic power losses associated with cooling requirements. Generally they are expected to be much less significant than compressor power losses.

Net Efficiency of Solid Polymer Fuel Cell Power System

Figure 1 shows typical efficiency curves for a solid polymer fuel cell stack operating on hydrogen and air. The electrochemical efficiency is a measure of the ideal efficiency as defined by equation 1. As can be seen the electrochemical efficiency is highest at low power and decreases as the current density increases.

The Ballard fuel cell can be operated for long periods of time at nearly zero excess hydrogen. However it can not be operated under "all" operating conditions with zero excess hydrogen. Ballard therefore recommends operation with some excess hydrogen for operational flexibility.[5] This excess is typically 5 to 20%. The fuel efficiency is therefore a measure of the power output achieved versus the lower heating value of the hydrogen fed to the system. For the purposes of this example an excess of 15% was used. Lower excesses should be possible as will be discussed below. A 15% excess however is considered the maximum excess of hydrogen which would be required. Under certain circumstances it could be possible to incorporate a recycle pump which would recover the excess hydrogen exhausted from the anode. This would reduce the effective excess to near zero and make the fuel efficiency near to the electrochemical efficiency. Recycling of fuel, however, could be limited by the build-up of impurities and inerts in the recycle stream.

Net efficiency, after parasitic losses, is more difficult to estimate. Parasitic losses, as described above, are very application specific. The main parasitic loss for a hydrogen-air system is the compressor power. Air pressures from just above atmospheric to over 2 atm abs. are required, depending on the application, to maintain adequate oxygen concentrations at the cathode. Compressor power will also depend on the %excess oxidant required. Excess air flow is necessary in order to maintain adequate average oxygen concentrations and also to remove product water of reaction. For the purposes of the example described by Figure 1 an air pressure of 4.4 atm abs. (50 psig) at an excess of 100% was used. Again these are considered the maximum necessary conditions and would only be required in situations requiring high power densities. Equation 3 was used to estimate compressor power for these conditions. The range of estimated net efficiency as calculated using the method described above is in good agreement with similar results published by Ballard Power Systems[6].

As can be seen the net efficiency is not shown below 20% maximum power. In this region there will be a large variation in the net efficiency depending on the compressor configuration being used.

EXPERIMENTAL APPARATUS AND PROCEDURE

The PEM fuel cell and stack studied in this paper were manufactured by Ballard Power Systems of N. Vancouver, B.C., Canada. The design is termed the Mark IV. The active cell sections and the flow distribution sections are sandwiched between endplates as in a filter press. All the cells studied by our group used Nafion™ 117 membranes. The single cell used a traditional platinum based catalyst while the 12-cell stack used an experimental CO-tolerant catalyst. The cells were found to be easy to operate and quite stable during changes in operating conditions.

The testing equipment used to study the performance of the fuel cells and stack consists of a pressure regulated gas manifold system, a temperature controlled water recirculation system, a high current capacity resistive load bank and a computer controlled alarm and data acquisition system.

The gas distribution system can be configured to provide pure or diluted hydrogen to the anode and pure or diluted oxygen to the cathode. The load bank consists of 1000 watt resistors which were selected in order to achieve very steady current density. The data acquisition system monitors output voltage and current, cell temperature, and fuel and oxidant flows in and out. The gas pressures are accurately maintained by in line regulators and the pressures are manually read from dial gauges on the gas supply lines.

The procedure used for measuring the effect of the various operating parameters consisted of establishing a steady state performance at a baseline set of conditions and then making a step deviation in the operating variable being investigated. When a new steady state was established the cell voltage at the new operating point was recorded. In this way the effect of each operating variable was measured as a deviation from baseline performance.

The baseline conditions for the study were;

Current Density :	400 ASF (420 mA/cm^2)
Pressure :	2.6 atm abs. (24 psig) fuel and 2.6 atm abs. (24 psig) oxidant
Temperature :	80 °C
Excess O_2 :	50%
Excess H_2 :	15%
Oxidant :	pure O_2 (ie. mole fraction = 1)
Fuel :	pure H_2 (ie. mole fraction = 1)

RESULTS AND DISCUSSION

Pressure

Figure 2 shows the variation in fuel cell efficiency with pressure at 400 ASF. Over the range from 2.4 to 3.4 atm abs.(20 - 35 psig) the electrochemical efficiency increased from 61.6 to 63.4% while the fuel efficiency increased from 53.0 to 54.6% at a fuel excess of 15%. One explanation for this behaviour in terms of reaction kinetics is that increasing the pressure increases the oxygen concentration which will increase the rate of reaction and hence reduces kinetic losses.

The experimental results shown in Figure 2 were correlated using a logarithmic equation of the form

$$\epsilon = \beta_0 + \beta_1 \ln(P) \quad (4)$$

Based on this equation the efficiency will increase more slowly as the pressure continues to increase. For our test program the oxygen was supplied by high pressure tanks and therefore there were no compressor losses. However, for a system where the oxidant was being supplied by a compressor, at some point the increase in parasitic power demand will equal the increase in electrochemical efficiency and further increases in pressure will no longer produce a net increase in efficiency.

Temperature

Figure 3 shows that the electrochemical efficiency at 400 ASF increased from 60.3% at 60°C to 62.2% at 80°C. Again, based on reaction kinetics, increasing the operating temperature should increase the rate of reaction and hence reduce kinetic losses. It should be noted that the fuel cell performance was stable over the entire temperature range. This means that the membrane did not dehydrate at the higher temperature or flood at the lower temperature despite a significant change in the rate of evaporation over this temperature range.

Excess Fuel Flow

As mentioned above, there generally will be an excess flow of fuel to the anode for stable cell performance. Figure 4 shows that there was no change in electrochemical efficiency over the range from 5 to 24 % excess. These results are very significant to the performance of multiple cell stacks.

The fuel supply lines to the anodes in a stack are normally connected in parallel. This means that any variation in flow resistance from cell to cell will result in a variation in %excess fuel. Very precise machining and cell construction will minimize this variation. However, there will always be some residual variation in cell geometry due to random factors such as catalyst distribution on the anode and axial sealing pressure along the stack. If cell performance was sensitive to variations in fuel supply it would be very difficult to achieve stack performance equal to that of a single cell. The results shown in Figure 4 indicate that for a stack operating on hydrogen at a 15% excess there could be a variation of ±67% (ie. from 5 to 25% excess hydrogen) in flow to any cell in the stack without a significant loss in electrochemical efficiency. Multiple cell stack performance equal to single cell performance has been reported by Ballard Power Systems[6] and has been confirmed in previous studies by our group[7].

Figure 4 also shows that the fuel efficiency decreases proportionally from 59% at 5% excess fuel to 49% at 24% excess fuel. Operating at as low an excess fuel rate as possible is obviously desirable. Recycling the exhaust hydrogen, when the build-up of inerts is tolerable, is also a possible method of reducing the loss in fuel efficiency associated with excess fuel flow.

Excess Oxidant Flow

As described in the Introduction, the water of reaction forms at the cathode. One means of removing this water is by exhausting it with an excess flow of oxidant. If the oxidant exhaust is stopped the cathode will flood and eventually the reaction will stop, shutting down the cell.

Figure 5 shows that the electrochemical efficiency did not change over the range of %excess oxidant from 50 to 135%. A 50% excess of oxygen at 400 ASF corresponds to an inlet oxygen flowrate of 2.25 cm^3/min(STP) per cm^2 of active area. These results show that at even this low a flowrate there was sufficient water removal capability to keep the cathode from flooding. As discussed above, this operating stability is very significant with respect to flow distribution in multi-cell stacks.

For systems operating on pure oxygen, such as submersibles and space craft, using as low an excess as possible and also recycling the oxidant stream is desirable in order to increase oxidant utilization and hence reduce oxygen storage requirements.

For power systems utilizing a compressor to supply air as the oxidant, the excess flow will have a direct effect on parasitic power losses as can be seen by equation (3). If water removal was the only factor at the cathode affecting the fuel cell performance the results shown in Figure 5 would suggest that a stoichiometric supply of air would be sufficient since the accompanying nitrogen would be more than adequate for water removal. However, when air is the oxidant, the %excess flow will also have an affect on the average concentration of oxygen at the cathode. The significance of this will be discussed below.

Concentration of Reactants

The variation in electrochemical efficiency with logarithmic mean oxygen concentration is shown in Figure 6. It is important to note that for all these results oxygen is supplied at 50% excess and the current density is 400 ASF. As can be seen, the cell efficiency begins to decrease rapidly below 0.2 mole fraction oxygen. The cell efficiency was correlated with $\ln(y_{O2})$ using least squares. The best fit equation was found to be

$$\epsilon = 64.45 + 4.5 \ln(y_{O_2}) , \quad r^2 = 0.97 \tag{5}$$

As can be seen in Figure 6, equation 5 provides a reasonably good fit of the data. The logarithmic mean oxygen concentration as used in equation 5 is calculated as

$$y_{O_2} = \frac{y_{in} - y_{out}}{\ln(y_{in}/y_{out})} \tag{6}$$

where y_{in} is the mole fraction of oxygen fed to the cell and y_{out} is the mole fraction of oxygen in the exhaust. The diluent in all cases was nitrogen.

The effect of oxygen concentration on cell efficiency can again be understood in terms of reaction kinetics at the cathode, since the rate of reaction is proportional to reactant concentration. The results shown in Figure 6 are particularly significant since they show that cell efficiency varies significantly near the mole fraction of oxygen in air, dropping steeply as y_{O2} decreases below 0.21 (ie. air)

As mentioned above, the %excess oxidant flow will indirectly effect the concentration of oxygen at the cathode. The mole fraction of oxygen in the exhaust is simply the moles of oxygen fed to the cell less the consumption in the cell all divided by the total moles of gas in the exhaust. This can be expressed as

$$y_{out} = \frac{y_{in}\ (\%E\ /100)}{(1-y_{in}) + (\%E\ /100)} \tag{7}$$

where %E is the percentage excess of oxygen fed to the cell. For air Equation 7 reduces to

$$y_{out} = \frac{0.209\ (\%E\ /100)}{0.791 + (\%E\ /100)} \tag{7a}$$

Substituting equation 7 into equation 6 shows that the average mole fraction of oxygen at the cathode will increase as the %excess air flow increases.

An estimate of the effect of hydrogen concentration was obtained from earlier experimental work on the 12-cell stack[7]. The electrochemical efficiency for a fuel gas mixture of 75% H_2 and 25% CO_2 was approximately 95% of the efficiency for pure H_2/O_2 at 400 ASF. This estimate is in good agreement with similar results published by Ballard Power Systems [8].

CONCLUSIONS

The variation in operating performance with a number of different operating variables has been studied. Pressure, temperature, mole fraction hydrogen and mole fraction oxygen were all found to effect the fuel cell performance.

These effects can be explained in terms of reaction kinetics at the electrode interface.

Fuel cell efficiency decreases for oxygen levels below 0.2 mole fraction. The excess oxidant flow will have an indirect effect on the average effective oxygen concentration at the cathode

Excess fuel and excess oxidant for pure hydrogen and oxygen did not have a significant effect on single cell performance. This indicates that the mechanisms of humidification and water removal within the cell are stable over a range of flows.

NOMENCLATURE

P	- power (kW)
p_1, p_2	- pressure atmospheric and fuel cell supply respectively (atm abs.)
v_1	- volumetric flowrate at inlet conditions ($m^3\ s^{-1}$)
y_{in}, y_{out}	- mole fraction of oxygen
ϵ	- efficiency
r^2	- correlation coefficient

ACKNOWLEDGEMENT

This work is supported by the Canadian Department of National Defence through funding supplied by the Chief of Research and Development.

REFERENCES

1. R.Kumar, M.Krumpelt, B.Misra, "Fuel Cells for Vehicle Propulsion Applications: A Thermodynamic Systems Analysis", Proc. Intersociety Energy Conversion Conference, 1989, pp 1601-1606.

2. Jerome M. Feret, "Westinghouse Air Cooled PAFC Technology", J. of Power Sources, **29** (1990) 59 - 69

3. J.C.Amphlett, M.Farahami, R.F.Mann, B.A.Peppley and P.R.Roberge, "The Operation of A Solid Polymer Fuel Cell: A Parametric Model", Proc. 26th Intersociety Energy Conversion Conference (In Press)

4. Max S. Peters and Klaus D. Timmerhaus, Plant Design and Economics for Chemical Engineers, 4th Ed., McGraw-Hill Inc., 1991.

5. N.E.Vanderborgh, J.R.Huff and J.Hedstrom, "Heat and Mass Transfer Effects in PEM Fuel Cells", Proc. 24th Intersociety Energy Conversion Conference, 1989, 1637-1640.

6. Ballard Power Systems, "The Ballard Fuel Cell", Information Brochure, 700.306.008.

7. Electrochemical Power Sources Group Annual Report , Chief of Research and Development, Canadian Dept. of National Defense, FE Grant RMC003.

8. D.Watkins, K.Dircks, D.Epp and J.Blair, "Canadian Solid Polymer Fuel Cell Development," 5th Annual Battery Conference, 1990.

9. S.Furuhama, "Hydrogen Engine Systems for Land Vehicles", Int. J. Hydrogen Energy, **14** (1989) 907-913

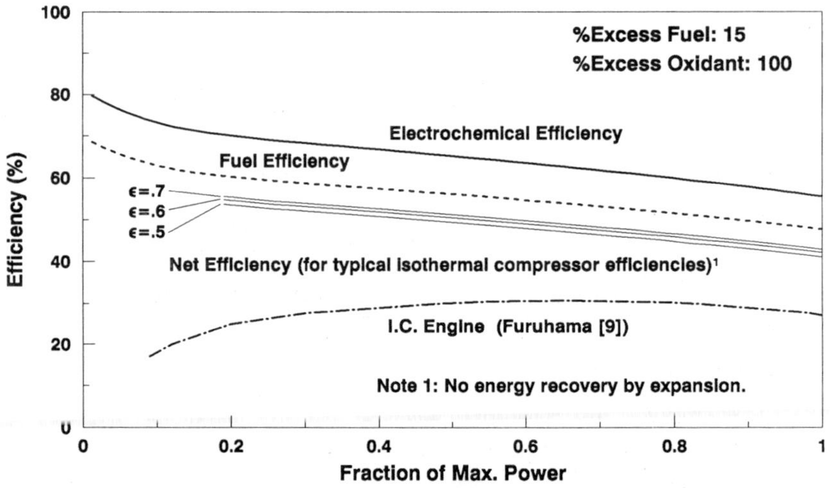

Figure 1: Typical Efficiencies of Fuel Cell Stack Operating on Pure Hydrogen and Air compared to Internal Combustion Engine

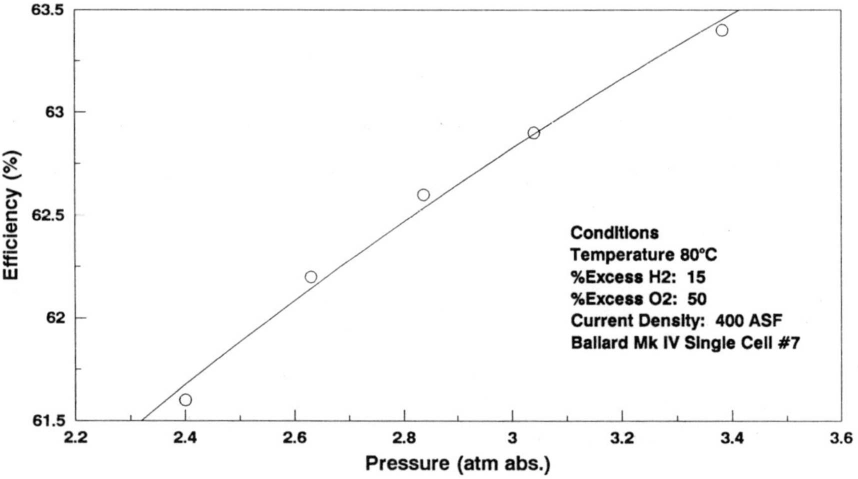

Figure 2: Variation in Electrochemical Efficiency with Pressure for Single Cell Operating on Hydrogen and Oxygen

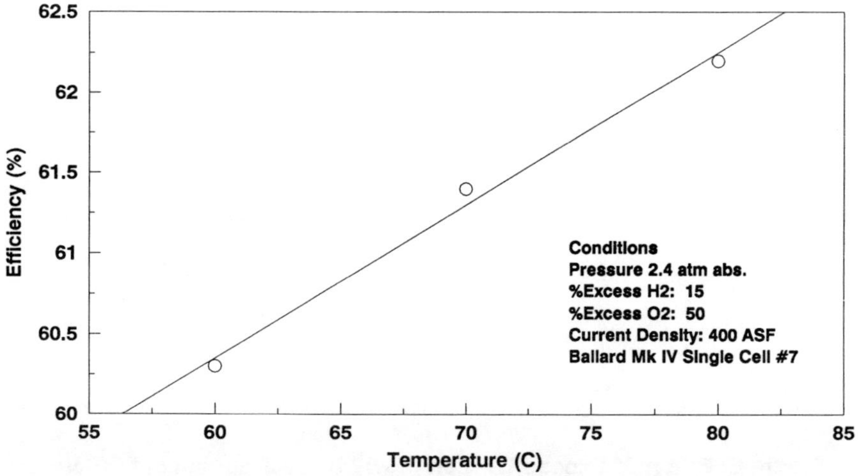

Figure 3: Variation in Electrochemical Efficiency with Temperature for Single Cell Operating on Hydrogen and Oxygen

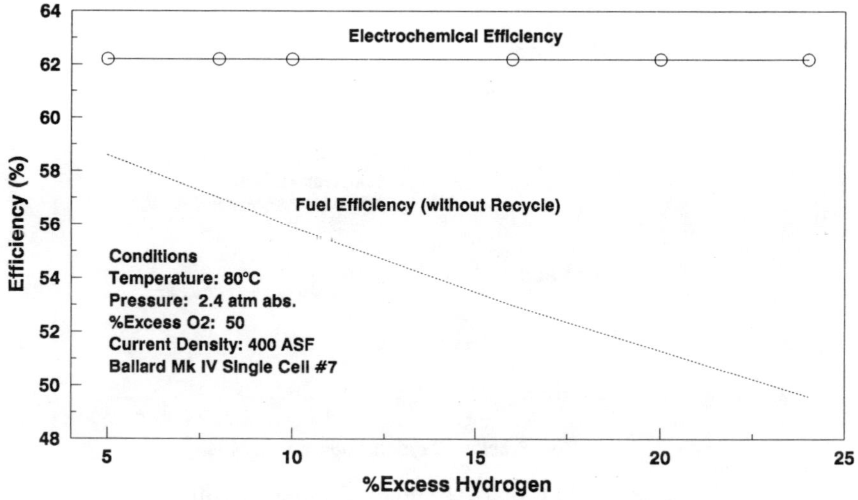

Figure 4: Variation in Efficiency with %Excess Hydrogen for a Single Cell Operating on Hydrogen and Oxygen

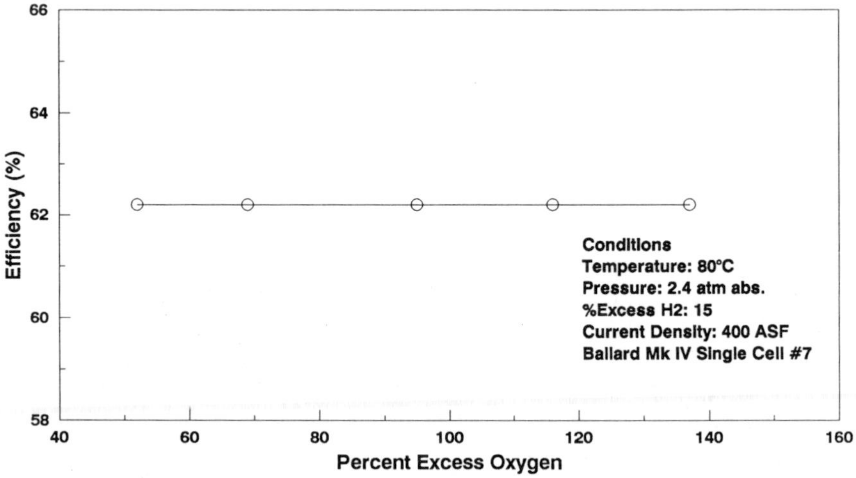

Figure 5: Effect of Excess Oxygen Flow on Electrochemical Efficiency of Single Cell Operating on Hydrogen and Oxygen

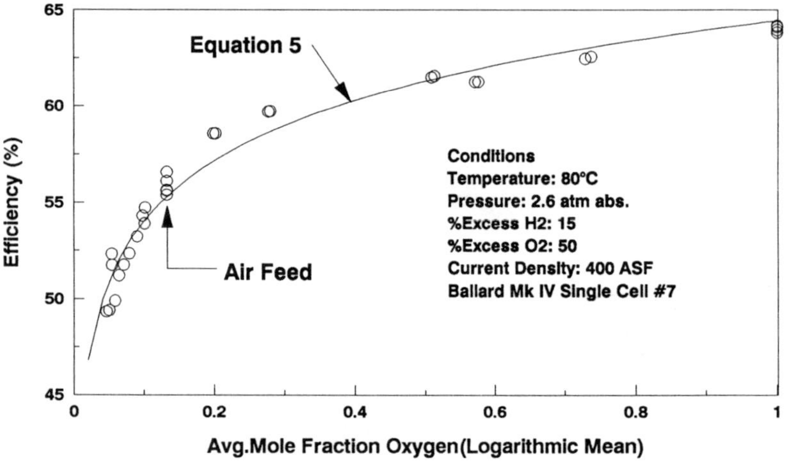

Figure 6: Variation in Electrochemical Efficiency with Log Mean Oxygen Concentration at the Cathode

MOLTEN CARBONATE FUEL CELL DEVELOPMENT AT M-C POWER

T.G. Benjamin, R.M. Laurens, and E.H. Cámara
M-C Power Corporation
8040 South Madison Street
Burr Ridge, Illinois 60521, U.S.A.

and

L.G. Marianowski
Institute of Gas Technology
3424 South State Street
Chicago, Illinois 60616, U.S.A

ABSTRACT

M-C Power's (MCP) molten carbonate fuel cell (MCFC) program is designed to culminate in delivery of market entry power plants in 1997. The path to commercialization includes design and technology development, manufacturing, field verification, and market development. Continuing technology advancements are realized through ongoing research and development at the Institute of Gas Technnology (IGT). This paper describes MCP's progress in stack and component scaleup and stack testing. The seven subscale (1 ft^2) stacks ranging in size from three to 70 cells which have been tested have demonstrated continuing improvements in power density. Performance equivalent to state-of-the-art bench scale single cells has been achieved. The first commercial-scale MCP stack is in production.

BACKGROUND

MCP's MCFC commercialization program has been established to develop a market responsive, natural gas-fueled on-site fuel cell power plant using the MCP internally manifolded IMHEX® concept. Contingent orders for this unit can be accepted in 1994 - 1995, with deliveries starting in 1996 - 1997. The market entry MCFC unit will have a rating of 250 to 2000 kW and will be designed for by-product heat utilization to provide process and/or space conditioning needs.

The MCP team is executing a highly integrated and coordinated program comprising design and development, manufacturing, field verification, and market development. Significant program funding is being provided by the Gas Research Institute (GRI), the Electric Power Research Institute (EPRI) the U.S. Department of Energy (DOE), San Diego Gas and Electric (SDG&E), and the State of Illinois Department of Commerce and Community Affairs (DCCA). Union Oil of California

(UNOCAL), Southern California Gas Company (SoCalGas), and the South Coast Air Quality Management District (SCAQMD) are supporting a 250 kW demonstration plant funded primarily by DOE.

Technology advancements have been realized by IGT in the areas of reduced cathode dissolution, low-cost electrodes, and strengthened components. Stack testing at IGT includes successful 1 ft^2 stacks of 3, 5, 10, and 24 cells with demonstrated performance improvements in each stack. A 70-cell, 1 ft^2 stack test has recently been completed by MCP. Carbonate pumping does not occur.

The first commercial-area (10 ft^2) IMHEX® stack is scheduled for operation at MCP in mid-1991. Separator plates are being fabricated and electrode fabrication has begun. Sintering trials of full-area electrodes were successful. Additional full-area stacks of 20 and 50 kW are planned prior to our first 250 kW demonstration in 1993.

A system study and mechanical design of a 250 kW, gas-fueled, MCFC facility (with waste heat recovered for cogeneration) has been undertaken. Analysis indicates a capital cost of mature technology at less than $1,200/kW and a cost of electricity of under 75 mills/kWh (at a gas price of $3.50/MM Btu) before credit for by-product heat. It is therefore probable that the final design will be competitive with grid power in most commercial/light industrial markets.

INTRODUCTION

The IMHEX® MCFC stack design concept is internally manifolded and patterned after plate and frame heat exchanger technology. In contrast with externally manifolded stacks, which have manifolds strapped to the sides, the internally manifolded IMHEX® concept has manifolds which are contiguous with the separator plates. Therefore, a number of advantages are realized over externally manifolded (EM) MCFC stacks.

First, EM stacks use gaskets to seal the gas feed manifolds against the stack. The gaskets wick electrolyte from the cells. Because of the potential gradient of the stack, the electrolyte is "pumped" from one end of the stack to the other. This results in flooding of the cells on one end of the stack and drying of the cells on the other end with a consequent decrease in performance of the end cells. Because the IMHEX® stacks have no gaskets, electrolyte pumping does not occur.

Second, the EM stacks require compression in three directions to seal the manifolds against the four sides of the stack and to maintain compression for cell-to-cell contact and sealing. The IMHEX® stack requires compression only in the Z-direction to maintain contact and sealing.

Finally, the IMHEX® stack intrinsically accommodates changes in stack height during commissioning and steady-state operation. The manifolds attached to the sides of the EM stacks do not easily adjust to changes in stack height.

STACK TESTING

Summary

A summary of subscale stack testing at IGT and MCP is presented in Table 1 and Figure 1. A steady progression of performance improvements, duration, and stack size is evident. Post-test chemical analysis of the IGT stacks and post-test visual inspection of the MCP stacks shows that carbonate pumping does not occur in IMHEX® stacks. Post-test chemical analysis of the MCP stacks is in progress.

Stack MCP-1

Stack MCP-1 (20 cells) was the first test of active components manufactured on MCP's prototype commercial-scale manufacturing equipment. The bubble barrier was eliminated by use of a reinforced matrix. This stack also utilized advanced aluminum and nickel corrosion protection on the wet seal rails and the anode side of the separator plate, respectively.

The performance of Stack MCP-1 was still increasing at the end of the test at 1350 hours as shown in Figure 2. Post-test analysis is in progress and there is no visual evidence of electrolyte migration. Successful operation of Stack MCP-1 demonstrates the successful transfer and scale-up of active component technology and fabrication from IGT's laboratory to MCP's factory.

Stack MCP-2

Stack MCP-2 is the tallest MCFC stack fabricated and tested in the world to date. Figure 3 shows the stable performance of Stack MCP-2. The stack was terminated after about 1250 hours of stable performance and an additional 330 hours of diagnostic testing. After more than 1500 hours of operation, there was no evidence of degradation of the end cells that would be an indication of electrolyte migration. The high performance for Stack MCP-2, presented in Figure 1 and Table 1, confirms the ongoing improvements that have been achieved in the IMHEX® stack-testing program. Figure 1 also shows that the average cell performance of Stack MCP-2 is equivalent to the IGT bench scale reference performance. The stack has not been disassembled but visual inspection indicates a total lack of carbonate pumping. The available data clearly show that the IMHEX® MCFC stack concept does not suffer from electrolyte pumping.

Stack MCP-3

Stack MCP-3 is the first commercial-scale IMHEX® stack to be fabricated, assembled, and tested at MCP. Startup is scheduled for the third quarter of 1991. The separator plates have been welded and the aluminum corrosion protection will be applied to the wet seals in the near future. The anodes have been tape cast and sintered and characterization is in progress. Cathode taping is about 50% complete and sintering trials will begin shortly. Matrix taping is in progress.

ACKNOWLEDGEMENTS

The authors and their companies, IGT and MCP, wish to thank DOE, EPRI, GRI, the State of Illinois DCCA, San Diego Gas and Electric Company, Southern California Gas Company, Union Oil of California, and the South Coast Air Quality Management District for their critical technical and financial support.

FIGURE 1: POWER DENSITY W/FT ^{2}

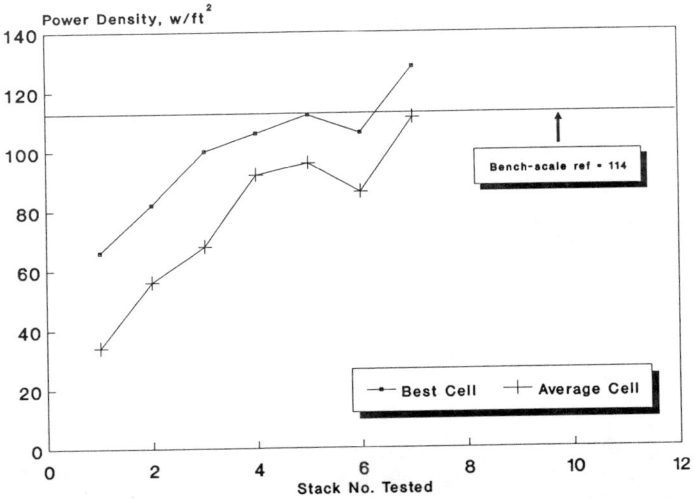

TABLE 1: SUBSCALE STACK TESTING HISTORY

Stack	No. of Cells	Power kW	Power Watts/cell	Hours of Oper	Start Date
IGT-1	10	0.4	40	430	08/88
IGT-2	10	0.6	60	850	01/89
IGT-3	3	0.2	67	580	05/89
IGT-4	5	0.5	100	340	11/89
IGT-5	24	2.5	104	1225	04/90
MCP-1	20	1.9	95	1350	12/90
MCP-2	70	8.7	115	1580	04/91

FIGURE 2: PERFORMANCE OF STACK MCP-1
Fuel - RNG

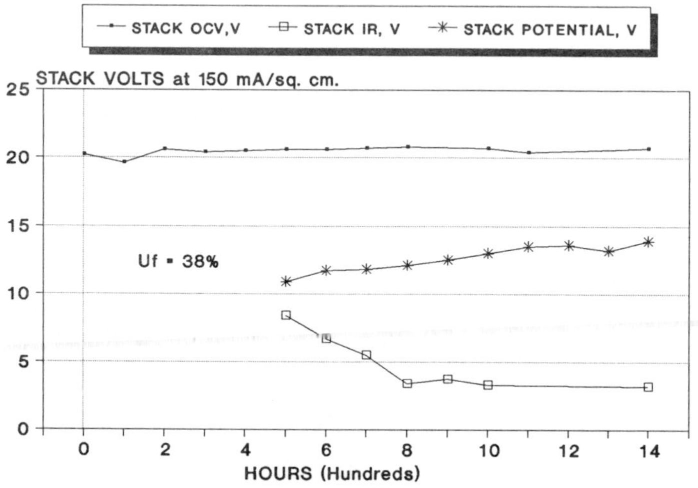

FIGURE 3: STACK MCP-2/70
75 PERCENT FUEL UTILIZATION

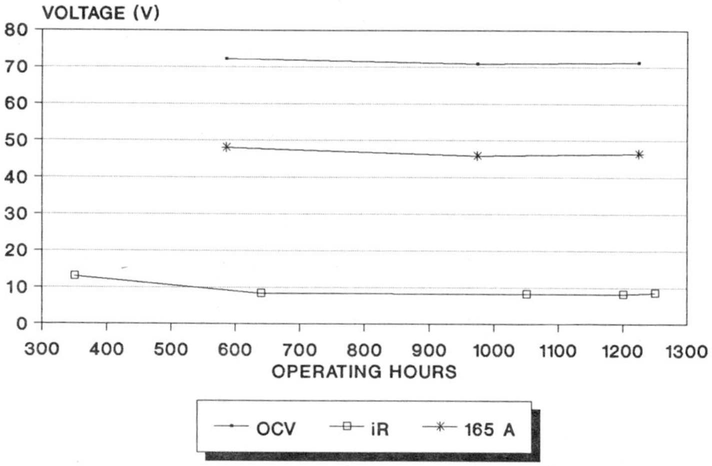

THE ENERGY CRISIS : FUEL CELL TECHNOLOGY VIEWPOINT

GINALYN P. REGIS
Fuels and Energy Division (FED)
Industrial Technology Development Institute (ITDI)

ABSTRACT

Our present pattern of energy utilization will someday be substantially altered. Impetus for change will rise from our inability to sustain continued increases in oil imports to prices, or to mounting concerns about the greenhouse effect, acid rain, or urban pollution.

This paper explores the role of electrochemistry for Fuel Cell technology in replacing oil with other energy sources at a minimum capital cost. This has a profound role in developing a new energy economy in which energy utilization is optimized by storage. The effects of electrochemistry experimented with chosen catalyst on non-traditional energy sources and on chemical processes are considered in some detail.

Developing a replacement of energy sources (conventional or non-conventional) will suffice in support to upturning our technology and economic activity for a developing country like ours.

A. FUEL CELL APPLICATIONS - AN EMERGING TECHNOLOGY

I. Historical Background:

Fuel cell is not a new one. The first fuel cell was made by Sir William Grove in 1839 when he observed that when hydrogen and oxygen are supplied separately to two platinum electrodes immersed in sulfuric acid, a current is produced by in the external circuit connecting the two platinum electrodes through a load. Since then, about 1958, a wide variety of fuel cell research programs has been reported and the most successful systems are the Gemini and the Apollo Spaceflights. The power-output of these units is of the order 1 to 2 kilowatt (KW). The efficiencies of these systems are in the regions of 50 to 70% (1). The most recent space science mission is the aborted space shuttle "Atlantis" in April 28,1989 (2).

In the United States interest in fuel cells for terrestial applications increased rapidly followed by the energy crisis in 1973. The U.S. government's thrust was to develop electrical power plants. In mid-1980, the Phosphoric Acid Fuel Cells (PAFC) was developed. Other programs were initiated, the U.S. Army 1.5 KW program for mobile electrical power and the 4.5 MW power installed at the United Technologies Corporation (UTC) (3).

Parallelly, in Europe, research and development on fuel cells is being carried out. The Commission of the European Communities (CEC) started fuel cell R&D programme in March 1985 with a total fund of 10 million European Currency Unit (MECU)* (4).

At the Industrial Technology Development Institute (ITDI) located in Taguig ,Bicutan,Manila, a research and development programme for fuel cells is being undertaken. Its objective is to develop fuel cell our indigenous sources.

II. Technical Background:

II-1 Definition:

Fuel cell is a device that produces electrical energy from the controlled electrochemical oxidation of fuel (5). Fuel cellsmaybe characterized as modular, polllution free and highly efficient device that offer applications over a wide range of power levels from watts to megawatts (6).

II-2 Basic Components of Fuel Cell: (Figure 1)

The basic components of a fuel cell are the electrodes

* MECU = 1.210 $

(anode and cathode) separated by an electrolyte which maybe either acidic or basic. The anode is the electrode in which

oxidation takes place. The electrode which is usually platinum provides sites for the electrochemical reaction of the fuel and oxidant and also act as conductors for electron flow to the external circuit. The electrical charge is transferred through the electrolyte within the cell by migration of hydroxyl ions from cathode to anode.

II-3 Operation of Fuel Cell Power Plant: (Figure 2)

For a fuel cell power plants, the major components are the: fuel processor, fuel cell power section, and the power conditioner.

Fuel Processor - is used to convert available fuels to hydrogen. The fuels used in fuel cells may either be hydrogen or other types of conventional fuels. Hydrogen can easily produced by four methods: (1) the direct thermal method; (2) the thermochemical method; (3) electrolysis; (4) photolysis. Water electrolysis has a long history in hydrogen production especially in obtaining high purity gases.

Fuel Cell Power Section - is the heart of the power plant wherein the elctrochemical conversion takes place. It consists of one or more fuel stacks - each stack contains many individual fuel cells usually connected in series to produce a stack output ranging from a flow to several hundred volts (direct current). The material which was customarily used was C-fiber (graphite) but it will be "burned" or oxidized during the use, and hence mears out and decreases its dimension (size). An excellent material which is now being used is the SPE (Solid Polymer Electrolyte) membrane. The diagram is shown in Figure 3.

The Power Conditioner - is used to convert the direct current produced by the fuel cell to alternating current.

III. The Classification of Fuel Cells

Just like the nuclear power plant, fuel cell maybe classified according to the following characteristics:

1. Type of Fuel (direct or indirect)

2. Type of Electrolyte (acid or base)

3. Temperature (low - very high)

4. Type of Catalyst

Table I

Classification of Practical Fuel Cells

Application	Fuel	Oxidant	Electrolyte	Temperature
Remote:				
Space	Direct H_2	Liquid O_2	Aqueous Alkaline Solid Polymer	Low, Intermediate Low
Undersea	Direct H_2 Direct Hydrazine	Liquid O_2 Hydrogen Peroxide	Aqueous	Low
Military:				
Low power, 100 W	Indirect Hydride	Air	Aqueous Alkaline Solid Polymer	Low
High power, 500 W	Indirect Hydrocarbon	Air	Phosphoric Acid	Intermediate
Commercial power				
Dispersed (on-site)	Indirect carbon Indirect methanol, ethanol Direct coal gas	Air	Phosphoric Acid Molten Carbonate	Intermediate High
Central station	Indirect coal Direct coal gas	Air	Phosphoric Acid Molten carbonate Solid Oxide	Intermediate High Vey high

In the future, fuel cells could be integrated with coal gasifiers to provide large, central-station, base load power plants that utilize coal directly. The capacity of such plants would range from 150 to 1000 MW. A plant of this kind is projected to be more than 45% efficient, on the basis of the heating value of the coal consumed.

Molten-Carbonate Fuel Cells (MCFC) Power Plants (Figure 4)

This system is integrated with a coal-gasifier. Interest in the system stems from its high efficiency, excellent environmental character, potential for competitive costs and modularity.

Cell Structure (Figure 5)

1. A separator and current collector plate that separates the fuel gas from the air of the adjacent cell in a multicell stack and also provides the electrical connection between cells.

2. An anode and current collector that conducts the electrons from the anode to the separator plate. This current collector must also provide passage for fuel flow.

3. An anode that consists of a porous nickel treated with a refractory oxide to reduce sintering. At the 650 centigrade temperature, no other catalyst is required.

4. An electrolyte system comprising a mixture of lithium-potassium carbonate and irent powder (presently a lithium aluminate). This mixture forms a paste when molten and freezes to form a "tile" when cooled.

5. A cathode that is similar to the anode except that it uses nickel oxide (doped with lithium to impart electronic conductivity).

6. A cathode current collector that has similar requirement and configurational options as the anode current collector. Since nickel is thermodynamically unstable material, options include lithium-doped nickel oxide and corrosion-resistant stainless steel.

Phosphoric Acid Fuel Cell (PAFC) Power Plants: (Figure 6)

Utility PAFC power plants will derive hydrogen by the conversion of a wide variety of primary fuels such as methane (natural gas), petroleum products (naptha), coal liquids (methanol) or coal gases. Despite the increasing number of developers and the seeming wide range of applications (kilowatts to megawatts), the basic PAFC system is virtually the same in all cases.

Cell Structure of PAFC system: (Figure 5)

1. A carbon or graphite separator - current collector plate that separates hydrogen from the air of the adjacent cell (in a multicell stack) and also provides the electrical series connection between cells. This plate must be impermeable to hydrogen and oxygen, a good electronic conductor, and stable to both fuel and air environments in the presence of concentrated phosphoric acid at 200 degrees centigrade.

2. Anode current collector ribs that conduct the electrons from the anode to the separator plate. The ribbed configuration provides gas passages for hydrogen distribution to the anode. These carbon or graphite ribs are made either a part of separator current collector plate or of the anode. In either event, they must have good electronic conductivity and be stable to the fuel environment.

3. An anode that consists of a porous graphitic substrate with the surface adjacent to the electrolyte treated with a platinum of platinum alloy catalyst.

4. An electrolyte matrix that retains that concentrated phosphoric acid. The matrices are fabricated from silicon crbide powder and a teflon binder. The electrolyte matrix has minimum ionic resistance, while separating the fuel and oxidant gas streams.

5. A cathode that is similar to the anode but uses a modified noble-metal catalyst and an increased catalyst loading to enhance the oxygen reduction kinetics. The cathode also is made hydrophobic by increasing the teflon content to prevent flooding of the electrode and provide for adequate oxygen diffusion to the active sites.

6. Cathode current collector ribs that are also virtually identical to the anode ribs. The only differences are in the height and width of the ribs which control the mass transport characteristics in the gas passage and the contact resistance between components. In a stack, the cathode ribs are perpendicular to the anode ribs to enable that simple location of external manifolds.

These single cells are stacked to produce the desired output power and voltage. For instance, the 4.5 MW power plant installed in New York, U.S.A., contains 20 stacks,each having nearly 500 cells of 3400 square centimeter area each.

Solid-Oxide Fuel Cell (SOFC) Power Plant

Several interest on SOFC started in 1960's because of its several advantages:

a.) Low inherent cost

b.) High power density

c.) Fuel flexibility

d.) lack of liquid phase

e.) very high power plant efficiency

But the enthusiasm was lost due to the problems of material stablility at 1000 degrees centigrade. In 1976, they developed SOFC using a Zirconia electrolyte. Recent SOFC cell and stacks are still under studies for its longer lifetimes.

Cell Structure of SOFC Fuel Cell System (Figure 7)

SOFC fuel cells employ a "tubular" stacks consists of: a porous support tube, a fuel electrode, a solid electrolyte, an air electrode, and an electronically conducting interconnection.

1. The porous tube provieds a strong mechnical support for the other components. It allows access of the fuel gas to the fuel electrode.

2. The fuel electrode must be electronically conductive, not crack otherwise lose conductivity after a thermal cycle, allow fuel gas to reach the electrolyte interface, and catalyze the fuel oxidation reaction.

3. The solid electrolyte must be an absulute gas barrier, provides good ionic conductivity but an electronic insulation. It is a combination of 10% Yttria-stabilized zirconia.

4. The air electrode must conduct electrons, withstand the oxidizing environment, allow oxygen to reach the electrolyte interface, adhere to the electrolyte thermal cycle without cracking and catalyzing the reduction of oxygen to oxide ions.

5. The interconnection must have high electron conductivity, low volatility, a thermal expansion match with other components and chemical compatibility with both oxidizing and reducing environments.

IV. Advantages/Disadvantages of Fuel Cell Power Plants:

ADVANTAGES:

Very efficient : In small as well as in large sizes

Modular : Can be sited where needed

A good neighbor : Quiet
low emissions, non-pollutant
Operates without external water

Waste heat recovery

Fuel flexibility

Durable reliable

Attractive for diverse applications

DISADVANTAGES:

High capital cost

CONCLUSIONS:

The World energy situation is tremendously changing. Aside from nuclear and fuel cell discoveries, there are other new emerging technologies that are coming out. Among which are the Ocean Thermal Energy Conversion Systems (OTEC) , Tidal and Wave Energy.

The latest is the controversial cold fusion reported by Stanley Pons of the University of Utah and Martin Fleischmann of the University of Southhampton in the United Kingdom. But these are not yet clearly apparent. If their claims are proven, this is a good contibutor to declining energy sources. Some books reveal that the fossil fuels will decline its production by the year 2010 and coal by the year 2030.

The most reliable and efficient power plant to be given emphasis is the fuel cell power plant. Because of its high efficiency, our hydrocarbon reserves can be utilize effectively. Its greenhouse effect is not dangerous because it produces only minimal amount of Carbon Dioxide gas. It is not a problem to urban pollution because during the operation it does not emit smoke.

Fuel cell application when given priority has a greatest potential energy sources compared to nuclear power plant. One requirement is an in-depth knowledge and expertise in the oxidation-reduction kinetics of fuels. The precious electrocatalyst (platinum is presently used) is quiet expensive, but there are no other studies for its replacement such as palladium and óther metal alloys. In the near future, the problem on its high cost will be solved.

REFERENCES:

1. Bockris, J. O'M & Srinivasan, S, Fuel Cells: Their Electrochemistry, Mc Graw Hill Book Co., p. 21

2. Wåldrop, M. M., Science, "Space Science on the Rebound" Research News, 05 May 1989, Vol. 244, pp. 525-526.

3. Lemons, Ross A., United States Fuel Cell Research and Development, Los Alamos National Laboratory, Los Alamos, N.M. 87544, U.S.A.

4. Zegers, P., CEC Fuel Cell R & D, Commission of the European Communities.

5. Loftness, "Energy Handbook", 1978

6. Srinivasan, S.; J. Electroanal. Chem., 118 (1981) 57-69, "Fuel Cells for Electric Utility & Transportation Applications"

7. Linden, David, Handbook of Batteries and Fuel Cells; "Utility Fuel Cell Power Plants", Mc Graw Hill Book Co., 1984, pp. 43-2 to 43-31.

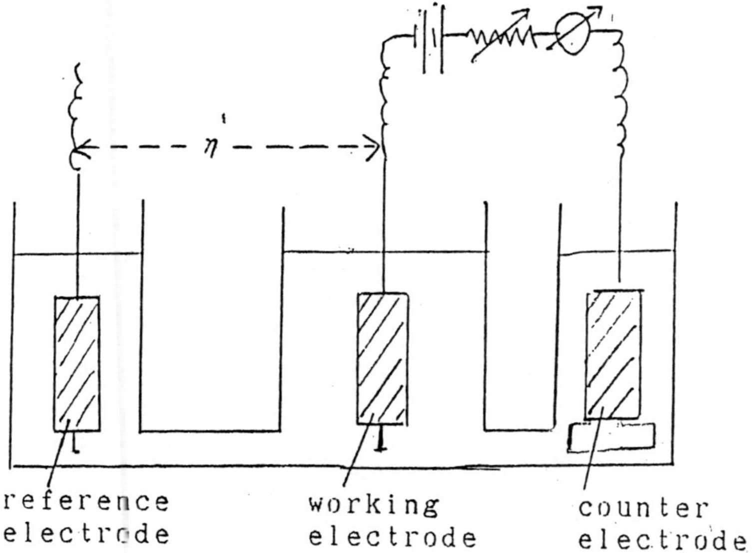
η
reference electrode
working electrode
counter electrode

FIG. 1

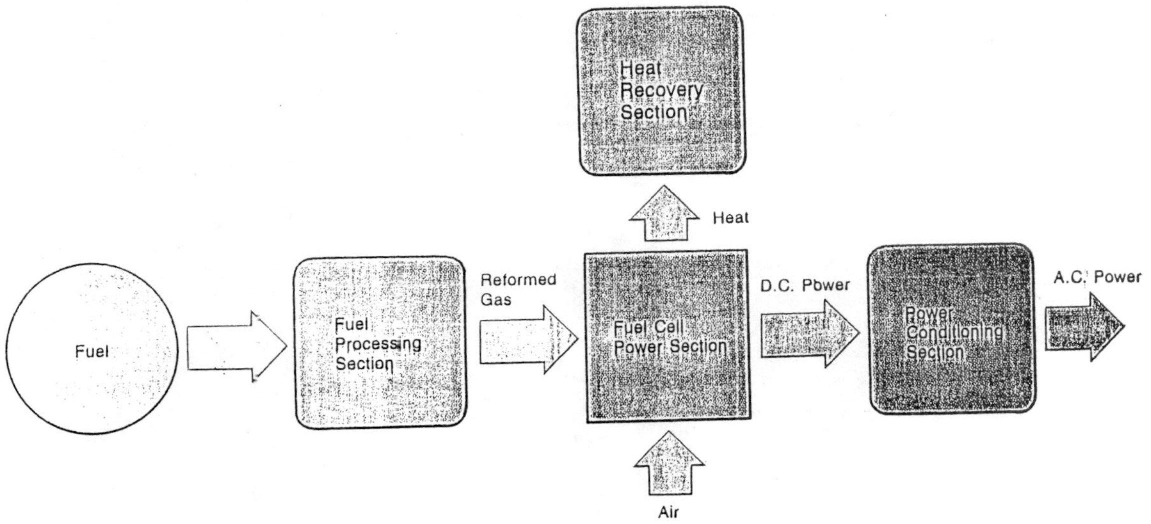
Heat
Recovery
Section
Heat
Fuel
Fuel
Processing
Section
Reformed
Gas
Fuel Cell
Power Section
D.C. Power
Power
Conditioning
Section
A.C. Power
Air

FIG. 2

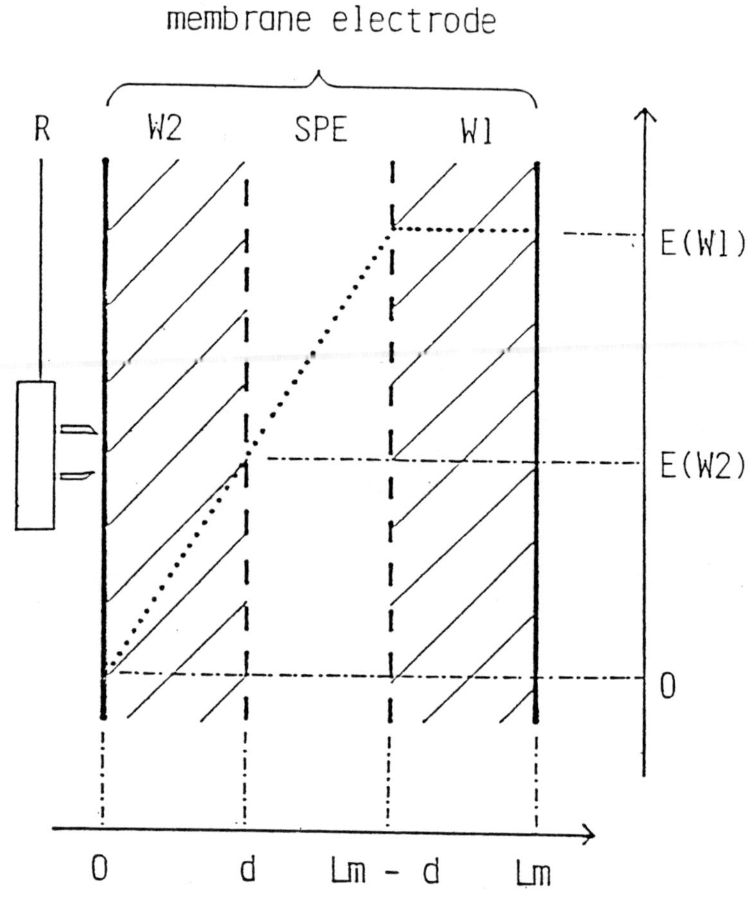

FIG. 3

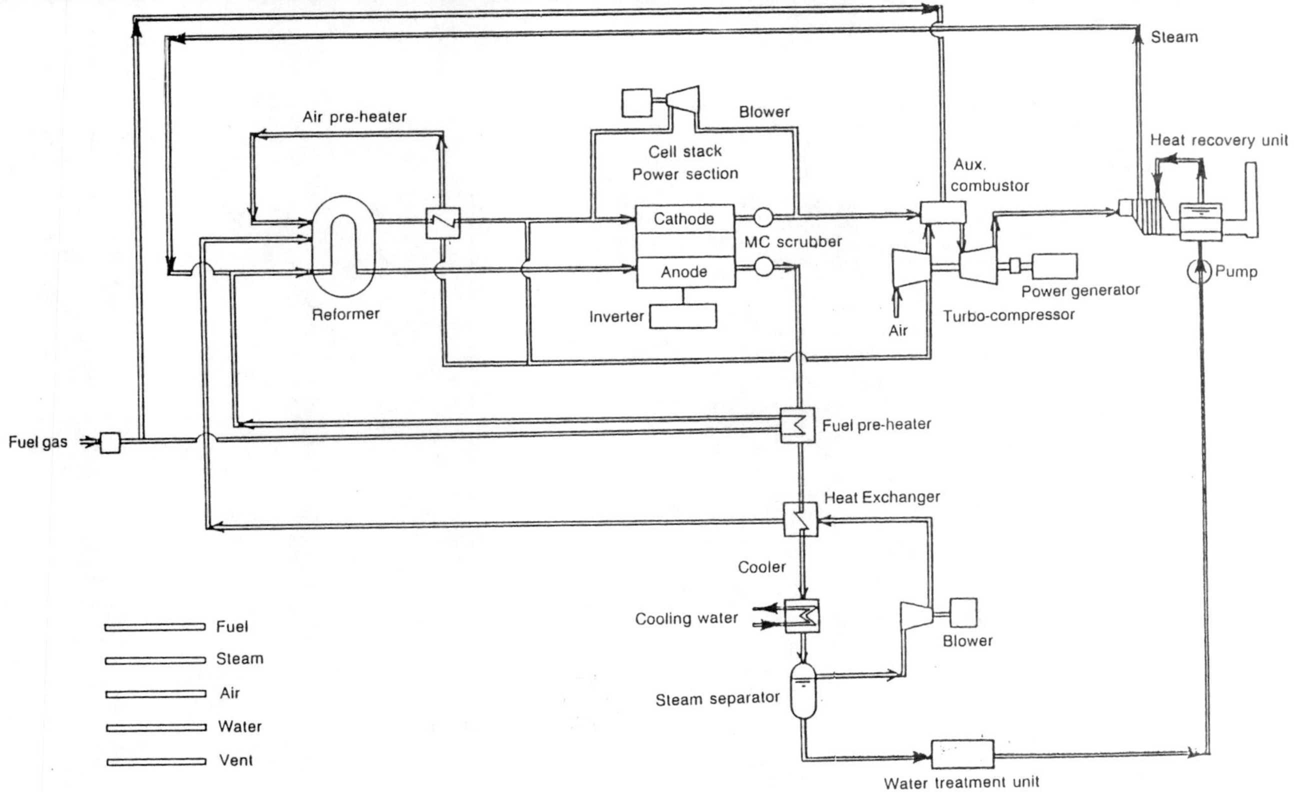

DIAGRAM OF 1 MW MOLTEN CARBONATE FUEL CELL POWER GENERATION SYSTEM

FIG. 4

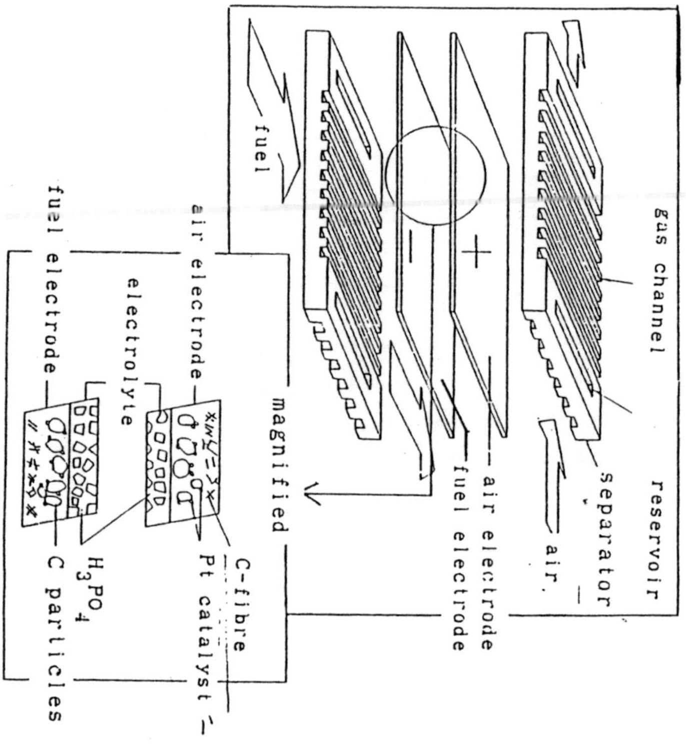
gas channel
reservoir
separator
air
air electrode
fuel electrode
fuel
magnified
air electrode
electrolyte
fuel electrode
C-fibre
Pt catalyst
H_3PO_4
C particles

FIG. 5

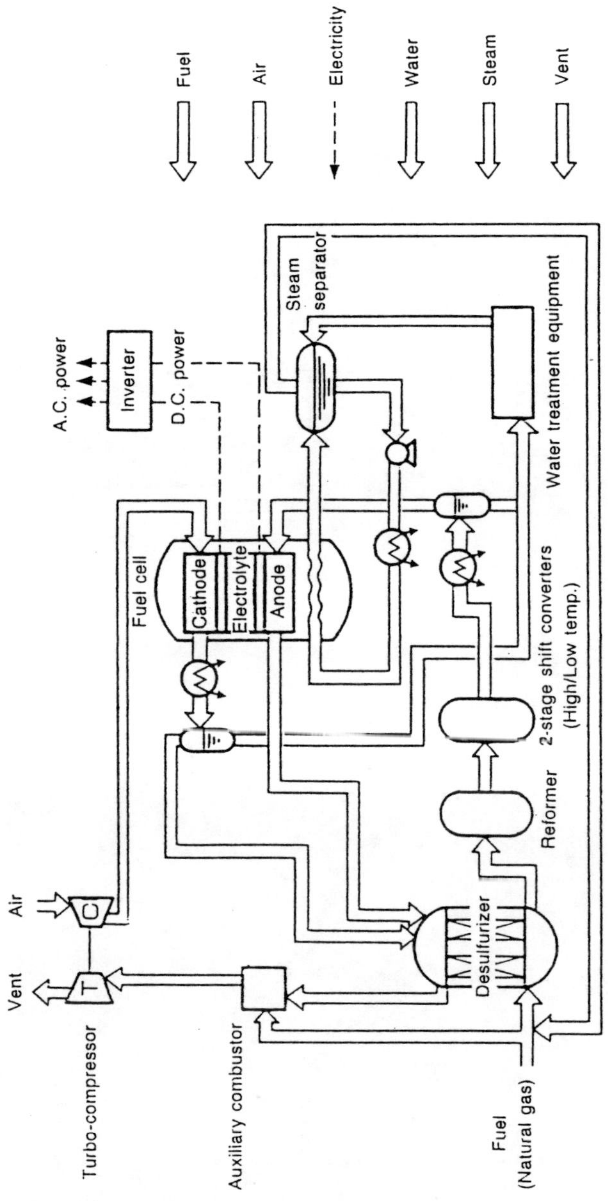
Turbo-compressor
Vent
T
C
Air
Auxiliary combustor
Fuel
(Natural gas)
Desulfurizer
Reformer
2-stage shift converters
(High/Low temp.)
Fuel cell
Cathode
Electrolyte
Anode
Inverter
A.C. power
D.C. power
Steam
separator
Water treatment equipment
Fuel
Air
Electricity
Water
Steam
Vent

FIG. 6

UNIT CELL STRUCTURES

Planar type

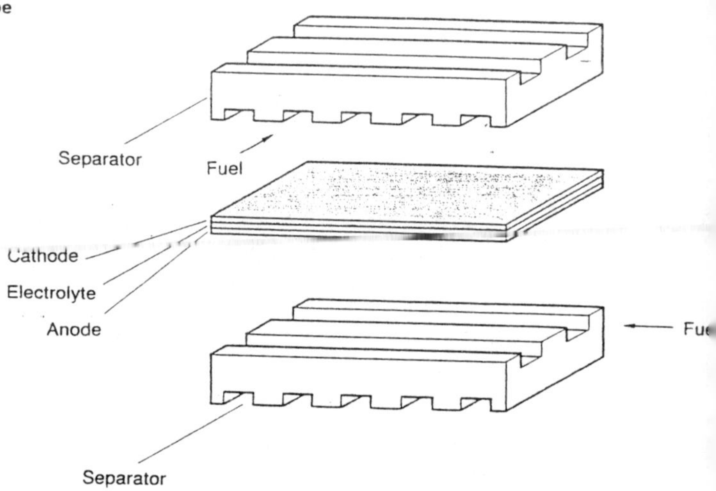

Tubular type

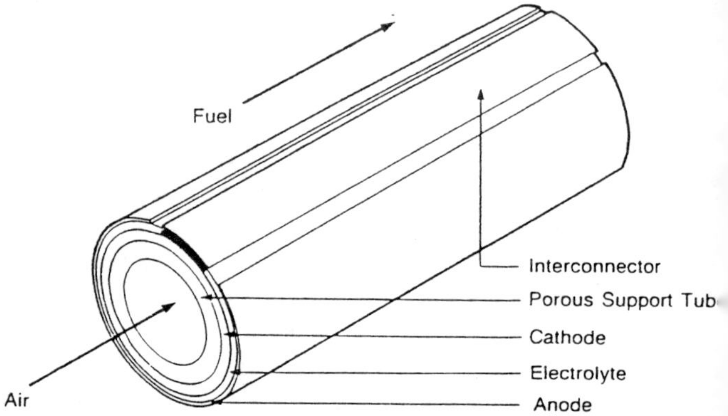

FIG. 7

INVESTIGATION ON SPE-ELECTRODE COMPOSITE SYSTEM FOR SOLID POLYMER ELECTROLYTE FUEL CELL

M.Swarnkar, M.K.Gupta, Ved Mitra* & S.K.Nema
Macromolecular Research Centre,
R.D.University,
Jabalpur, INDIA- 482 001
*Department of Non-Conventional Energy Sources,
New Delhi, INDIA

ABSTRACT

The development of fuel cell power generation system considered as a clean power generator has recently become a focus of interest. In the present study, we have communicated the development of SPE-electrode composite system by using perfluorosulphonate ionomeric membrane and Graphite/polyimide - IMIDCARB-8 electrode. Fine dispersion of Electrocatalytically active Platinum (0.05 mg/cm^2) is insitu generated within the polymeric matrix close to surface of SPE membrane. The incorporation of electrode on Platinum containing membrane facilitate locating electrocatalyst in three dimensional reaction zone between SPE and electrode. Thus the composite system so obtained comprises of layered configuration in three zones. The method of preparation of the SPE-electrocatalyst-electrode composite system has been described. Thermal, mechanical and chemical characteristics of the proposed system have also been presented. It may be concluded that this composite system can be potential candidate for emerging SPE fuel cell.

INTRODUCTION

The most efficient means by which the chemical energy of Hydrogen is converted directly into electrical energy to produce power, is offered by the electrochemical processes of fuel cells. The cell divides the overall reaction of the Hydrogen and Oxygen to form water into (i) the anodic oxidation of Hydrogen and (ii) the cathodic reduction of Oxygen. As a result of these reactions, an electric current flows through the external load. To develop the different types of fuel cell technologies, considerable research efforts have been made worldwide [1]. These developments include Phosphoric Acid fuel cell, Molten Carbonate fuel cell, Solid Oxide fuel cell and Solid Polymer Electrolyte fuel cell. Annual budget of $ 50-100 millions is being allocated in USA itself for the development of fuel cells by various agencies and this reflects the significance of fuel cells in future energy scenario.

Among the various fuel cell technologies solid polymer electrolyte have offered major advantages over other fuel cells in the facts: (i) the temperature of operation is not very high (ii) the electrolyte is solid and hence the corrosion problems are minimised and (iii) the high power density is attainable. SPE fuel cell has become the centre of attraction for the scientists engaged in this field although the first SPE fuel cell was developed by General Electric Company [2] with current density of 100 mV/cm^2 at 0.5 V and many other research organizations developed SPE and fuel cell by using different types of electrodes and polymeric materials with minimum amount of platinum (upto 0.4 mg/cm^2). Research and Development at Los Alamos Laboratory and a team of A & M Taxas University [3] has succeeded attaining high power density cell using low Platinum loading (0.4 mg/cm^2) in the electrodes and by localization of Platinum (0.05 mg/cm^2) on the active surface of electrodes and electrode Kinetics of the Hydrogen oxidation and the oxygen reduction reactions was enhanced by elevating both the temperature and pressure.

As a part of multiinstitutional National Programme on development of Solid polymer electrolyte fuel cell -SPELL in India, Macromolecular Research Centre has taken up the development of required SPE and polymeric electrodes. Under this programme, a three layered composite material was developed in our laboratory, in which the Solid polymer electrolyte, the electrodes and an electrocatalyst are present in monolithic system. This type of system is amenable for the commercial production. The 'SPELL' fuel cell may find application in Indian aerospace and other general industries. Unique aspects of this composite SPE-electrode system is that (i) we have visualized the Platinumless electrodes and 0.05 mg/cm^2 platinum has only been incorporated at the interface of electrode and SPE and (ii) electrode material is Graphite/polyimide, as graphite blended polyimide has good porosity for the diffusion of gases and adequate conductivity. In this system, three dimensional reaction zone formation takes place by the interdiffusion of solid polymer electrolyte and electrode and the electrocatalyst is predominantly present in this three dimensional reaction zone.

EXPERIMENTAL

Materials

Nafion-117 obtained from du Pont, USA was used as solid polymer electrolyte. Super fine graphite powder obtained from Hindustan Graphite Ltd; Bhopal, India and Polyamic Acid as reaction product of Pyromellitic dianhydride and Diamino diphenyl ether (10% solution in Dimethyl Acetamide) as precursor to polypyromellitimide were used for electrode preparation. 2% solution of Hydrogen Hexachloroplatinate (iv) hydrate from Aldrich Chemical Company, USA was used for the preparation of electrocatalyst.

Pretreatment of Perfluoroionomeric Membrane

Firstly, Nafion membrane was pretreated by immersing it in boiling HNO_3-H_2O (1:1) for 30 min to remove impurities and then boiled it in distilled water for 1 hr to remove nitrates and introduce a reproducible amount of water into the membrane. Then the membrane was dried in vacuum oven at 100 ^oC for 2 hrs. This sample is designated as SPE- 01.

Incorporation of Electrocatalyst on SPE Surface

Calculated vol of 2% solution of Chloroplatinic Acid, diluted upto 50% by methanol, was coated on the surfaces of SPE-01. This membrane was heated at 200°C for 3 hrs under Nitrogen atmosphere, as reported earlier [3].This sample was designated as SPE-02. The conversion of Chloroplatinic Acid in to electrocatalytic form of Platinum has been determined by novel and simple method worked out in our laboratories [4].

Electrode Imporporation on Modified SPE

After modification of solid polymer electrolyte by incorporating an electrocatalyst, 50 μ thick film of Graphite/polyimide (80:20) has been obtained by coating the surfaces of the modified SPE by suitable quantities of Graphite/polyamic Acid Solution and heating the same at 150°C for 2 hrs for accomplishing imidization which is monitored by appearance of 1775 cm^{-1} band in FT-IR [5] spectrum (taken in KBr pellet). The Electrode having 80: 20 carbon/polyimide was designated as IMIDCARB-8 whereas SPE-02 when it was coated one side with IMIDCARB-8 is designated as SPE-03 and the one which is coated on either side is designated as SPE-04.

Characterization

SPE-electrode composites were subjected to following characterization studies:

Thermal Studies

ThermoGravimetric Analysis (TGA) of the composites were determined by using Perkin Elmer's TGA-7 at a heating rate of 20°C/min.

Burst Strength

This was measured in two compartment cell specially designed and got fabricated. The composite membrane was placed between two compartments and differential pressure from 1 to 5 Kg/cm^{2} was applied to assess the burst strength.

Ion-Exchange Capacity

The ion-exchange capacity of the samples were determined by the speical technique developed in our laboratory [6] wherein the pH of known quantity of solution is recorded for initial and after immersing the ionomeric film of calculated volume into 0.1N Sodium Chloride solution for overnight. Ion-exchange capacity was calculated as equivalent to difference in H^{+} ion concentration for a given sample of known volume.

Resistivity

Resistivity of the composites were measured by placing two parallel graphite electrodes 4 cm apart on to the SPE-02 (3 cm wide) and by measuring the corresponding resistance R at 1 KHz.

$$\rho = \frac{RS}{L}$$

Where, ρ = Specific resistance

R = Resistance (in Ω)

S = 3 x sample thickness (in cm)

L = 4 cm

Mechanical Strength

Film (1 cm x 15 cm size) of SPE-01 to SPE-04 of uniform thickness were subjected to mechanical strength analysis according to ASTM/D/882 standard method. Instron's universal testing machine of 1 KN capacity was used for the analysis.

Chemical Stability

The composites SPE-03 and SPE-04 were first boiled with water and then their mechanical strength and thermal stability were again measured to compare the values without these treatment.

RESULT AND DISCUSSION

From commercial considerations and for achieving higher performance in fuel cells, it is essential that SPE, electrocatalyst and electrode should have monolithic structure. This feature would provide adequate structural integrity for active surface areas. At the same time the proper distribution of electrocatalyst between the porous electrode and solid polymer electrolyte is essential for favourable electrokinetics and thereby imparting higher performance. In our experiment we have, therefore, dispersed active Platinum particulates over the surface of Nafion membrane (SPE). The heating of the Nafion membrane at 200°C under Nitrogen atmosphere with dilute Chloroplatinic Acid solution spread on surface results in the formation of electroactive Platinum. Since the membrane is heated above its glass transition temperature (Tg), there is fair likelyhood of the penetration of Platinum particulate within the matrix of membrane. It implies that Platinum particulate may not only be located at the surface but they may be dispersed to some extent towards the bulk of the membrane. This dispersion of the Platinum at the region between the surface and the bulk provides a reaction zone for the electrocatalyst which is in very close proximity with proton conducting moities.

As evident from the experimental section that the electrode is formed by taking viscous paste of graphite and polyamic Acid solution over the electrocatalyst containing Nafion membrane. The heating of this paste at 150°C (which is a little above glass transition temperature Tg of the Nafion) results in the formation of electrode (IMIDCARB-8) adhered with SPE-02. During the heating process the electrocatalyst may also diffuse into the bulk of electrode through its surfaces. This interdiffusion of electrocatalyst into electrode and solid polymer electrolyte may result in a three dimensional zone wherein the Platinum electrocatalyst will be present. Since Platinum concentration has been retained as 0.05 mg/cm^2 on each surface of the Nafion membrane when it is in the cold condition, the total Platinum in each surface of our composite i.e. SPE-03 and SPE-04 will remain to be 0.05 mg/cm^2. The configuration thus obtained has been illustrated as schematic diagram in Fig.1. The region A and C in Fig.1 represents solid polymer electrolyte and electrode i.e. IMIDCARB-8. The region B is the zone wherein interdiffusion of the Platinum takes place between SPE membrane and electrode. This only represents the region where the Platinum is predominantly present. SPE-electrode composite shown in the schematic diagram symbolizes the monolithic version of the unit cell of the fuel cell.

The method of preparation of SPE electrode composite involves heating and cooling several times. These thermal treatment may affect the thermal stability and mechanical properties of the individual materials and that of the final system has been determined by thermogravimetric Analysis (TGA). Fig 2 and 3 show thermogram and derivatogram individually for SPE-01, SPE-02, SPE-03 and IMIDCARB-8 respectively. It is evident from Fig.2 that the incorporation of electrocatalyst in Nafion membrane does not affect the thermal stability and decomposition temperature is approximately 50°C higher in the case of SPE-02 than SPE-01. This may be attributed to the thermal treatment which the membrane has been subjected to and this has been also observed previously [7]. The transformation from clustured to homogeneous morphology has also been reported [7], when the Nafion membranes are heated above 170°C but below its decomposition temperature. Nevertheless, no corelation between thermal stability and the morphology of the membrane has been reported. Fig.3 indicates that there is a two step decomposition in the case of SPE-02 and as well as in the case of SPE-03. The weight loss in the case of SPE-02 is 3.8% at 375°C whereas it is around 9.9% in the later case, which indicate that the Platinum electrocatalyst has been diffused into the electrode surface thereby causing the decomposition of electrode. This is also corroborated with the fact that there is no decomposition observed at 375°C in the case of IMIDCARB-8 prepared separately.

The result of the mechanical properties evaluation are given in Table.1. It is seen from Table.1 that there is substaintial improvement in the tensile strength of SPE-02 but whereas there is a little reduction in its elongation. These changes may be attributed to the thermal treatment resulting in the change of morphology from heterogeneous (clustered) to homogeneous one. The effect of the electrocatalyst incorporation is not quite apparent as there has not been much change in the properties of the Nafion when it is heated to 200°C without electrocatalyst impregnation. The tensile strength and elongation is reduced to almost 60% in the case of SPE-03 in comparision to SPE-01. However, the modulus of this sample has gone up by two times. As expected the mechanical properties have shown a downward trend which may be attributed to the poor mechanical property exhibited by IMIDCARB-8. As a matter of fact IMIDCARB-8 has very poor mechanical property and as such its mechanical property can not be determined. The incorporation of IMIDCARB-8 over SPE-02 and SPE-03 serves the intended purpose of imparting sufficient mechanical properties to the final composite material i.e.SPE-04. The mechanical property of the SPE-04 as given in Table.1 may be adequate for the fuel cell purpose. This is amply demonstrated by the burst strength results wherein all the four samples (SPE-01 to SPE-04) have withstood the differential pressure upto 5 Kg/cm^2. As this burst test is done only for the limited period of time it may be desirable that this experiment may be conducted for longer time to establish the endurance capability of the composite with regards to the pressure in both anolyte and catholyte compartments. The poor mechanical property may also be compensated by utilizing suitable polymeric or metallic reinforcement over the composite membranes.

Resistivity measurements have shown that the SPE-02 has shown higher resistivity which is presumably due to the loss of clustered morphology and lesser water absorption. The electronic conductivity of the IMIDCARB-8 needs further improvement and this may warrant the use of other polymeric systems. The ion exchange capability of SPE-02 is 0.0415 meq/cm^2 and this has been recorded to be higher than sample of SPE-01 (0.0305 meq/cm^2). The good part of enhancement in ion-exchange capability of SPE-02 has been accounted for 4 to 8% of unreacted chloroplatinic Acid or its thermally decomposed products left behind the membrane after heating at 200°C. This has been confirmed by the wide difference in ion-exchange capability found between the values of samples of pretreated Nafion heated at 200°C with and without chloroplatinic Acid. The ion-exchange capabilities of samples of SPE-02 and SPE-03 are almost same whereas the IEC of sample SPE-04 could not be determined with reasonable accuracy.

Since the SPE-electrode composite could not be tested for the life test in the actual fuel cell, the energy performance could not be studied. In order to evaluate the overall performance the composite membrane will be tested in the actual working conditions. The problem of non adherence of the electrode with the SPE membrane at elevated temperature in water may be overcome by switching over the polymer which are stable to hydrolytic dissociation. The incorporation of the reinforcing structure may also eliminate this problem.

CONCLUSION

From the above discussion, it can be concluded that this type of SPE electrode composite may be the promising candidate may be the promising candidate for SPE fuel cell. Although, it is the part of the development of the SPE fuel cell and it may also require modification in the composition of the electrode material and its preparation techniques. However, after testing in the state of the art unit fuel cell, more modifications can be done to make this composite suitable not only for fuel cell but for other applications also.

ACKNOWLEDGEMENTS

The authors express sincere thanks to Dr. U.D.N.Bajpai, Department of Chemistry, R.D.University, Jabalpur for helpful discussion and M/S Hindustan Graphite Ltd., Bhopal and Vikram Sarabhai Space Centre, Trivandrum for providing us the samples of fine graphite powder and polyamic Acid solution respectively for our studies. The secretarial assistance of M/S S.D.Namdev and Rajendra Dubey, MRC, Jabalpur is gratefully acknowledged.

REFERENCES

1. Graham Hagey, in Proc. of the 1st Int. Fuel cell workshop Sep. 1989, Takeda, Japan, P.3-8.

2. L.J.Nuttall, A.P.Fickett and W.A.Titterington, Hydrogen generation by Solid polymer electrolyte water electrolysis, In Hydrogen Energy (Ed. by T.N.Veziroglu) p. 441, Part A. Plenum Press, New York (1975).

3. Supramaniam Srinivasan, David J. Manko, Mohammad Enaytullah and A. John Appleby; in Proc. of the 1st Int. Fuel cell workshop, Sep. 1989, Takeda, Japan, p. 121-164.

4. The Estimation of conversion of Electrocatalytic grade Platinum by thermolytic dissociation of chloroplatinic Acid on surface of polymer films using pH measurements.
-Macromolecular Research Centre Technical Report No. MRC:TECH-REP: GEN-2:91.

5. Manoj Gupta, Sandeep Tiwari, U.D.N.Bajpai & S.K.Nema, in Polymer Science : contemporary themes, vol. 1, P. 87-92, 1991, Ed. S.Sivaram, Tata McGraw Hill Pub. Company, New Delhi.

6. The Novel technique for determination of Ion-exchange capabilities of Ionomeric polymer films by pH measurements.
-Macromolecular Research Centre Technical Report No. MRC:TECH-REP: PAWAS-1:91.

7. Sudhir Nema, S.K.Tiwari & S.K.Nema, communicated to J.Eup. Poly. Sci.

TABLE 1: MECHANICAL PROPERTIES OF THE SPE & ITS COMPOSITES

Sample Designation	Tensile Strength (Kg/cm^2)	Elongation (%)	Modulus (Kg/cm^2)
SPE-01	183	137	1670
SPE-02	253	117	2450
SPE-03	103	75	3610
SPE-04	45	9	1530

FIGURE 1 : SCHEMATIC DIAGRAM OF SPE-ELECTRODE COMPOSITE

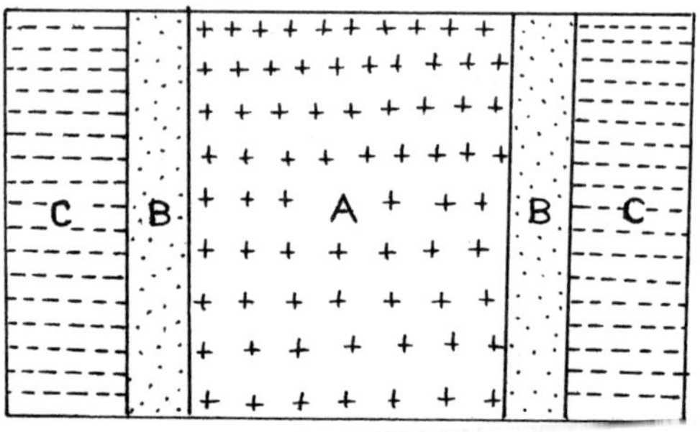

Wherein (A) represents SPE
(B) represent an electrocatalyst zone
(C) represent electrodes

FIGURE 2 : THERMOGRAM OF THE COMPOSITES

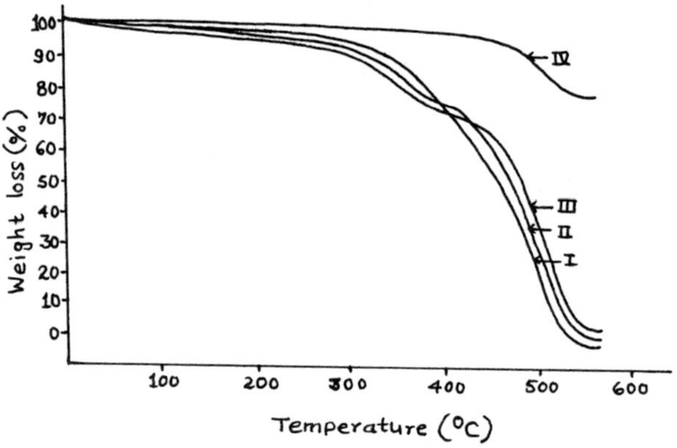

(I) SPE-01
(II) SPE-02
(III) SPE-03
(IV) IMIDCARB-8

FIGURE 3 : DERIVATOGRAM OF THE COMPOSITES

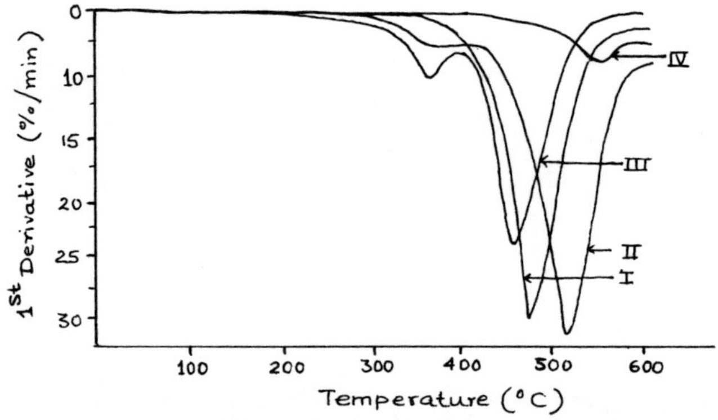

(I) SPE-01
(II) SPE-02
(III) SPE-03
(IV) IMIDCARB-8

CHAPTER FIVE

Hydrogen Fuel Cell Vehicles

DEVELOPMENT OF A PHOSPHORIC ACID FUEL CELL/BATTERY POWER SOURCE INTEGRATED IN A TRANSIT BUS

Arthur Kaufman
H-Power Corporation
21 Lackawanna Place
Bloomfield, New Jersey 07003, U.S.A.

ABSTRACT

This paper presents a summary of the major steps in the design and development of the fuel cell/battery power source for a transit bus. Preliminary specifications for the power plant and other bus components were first developed and then refined through a trade-off analysis which considered the performance, availability and cost of the major components: battery, fuel cell/reformer, power electronics, traction motor, and the master controller. Only current technology fuel cells and batteries were considered.

A half-scale brassboard system suitable for testing in a laboratory was developed and tested under standard bus operating conditions. The results have verified the projected performance and efficiencies of the fuel cell/battery powerplant and have provided the basis for proceeding to the next phase, that of assembling and testing a full-scale system in a small (27 ft) transit bus.

INTRODUCTION

The fuel cell produces no particulates, and produces extremely low emissions of oxides of nitrogen and carbon. The environmental benefits due to low emissions, the fuel saving potential due to high system efficiency, and the excellent acceleration performance of electric drive systems, make fuel cell/battery powered buses very attractive for urban transit applications.

BUS AND POWERPLANT SPECIFICATIONS

The fuel cell provides the average power requirements for the bus. A lead-acid battery is used to boost the power output for acceleration and grade climbing. During low traction power demand periods, the fuel cell charges the battery. During deceleration and downhill travel, the motor acts as a generator and also charges the battery. Since the fuel cell costs significantly more than a lead-acid battery, the fuel cell/battery bus should be designed with as small a fuel cell as possible. The overall power requirements should also take into account the accessory power requirements which may amount to 14 Kw or more.

Trade-offs between fuel cell power rating, battery type and size, and powertrain gear ratios were conducted to develop the final bus design and specifications. Major considerations were:

- Bus performance equivalent to standard diesel bus acceleration and grade performance,
- Ability to operate over standard DOT urban bus cycle as well as over specified GUTS Arlington Loop route,
- Ready availability of components, and
- Reasonable costs in relation to diesel bus costs.

Table 1 shows the specifications for the bus and its major components developed as a result of the trade-offs. These specifications resulted in a realistic layout and weight balance for the fuel cell/battery powered bus.

Detailed simulations of the bus performance were made for the standard DOT Transit Bus and GUTS Arlington Loop routes. The simulations provided the system performance and energy consumption data that were used for comparing with test data.

BRASSBOARD DESIGN

The specifications for the phosphoric acid fuel cell (PAFC), battery and power train were used to develop the specifications for a brassboard system suitable for testing in the laboratory. A 25 Kw fuel cell was specified, along with a half-size lead-acid battery.

The brassboard PAFC subsystem was designed and built by Fuji Electric Company. The fuel cell subsystem included a step-up chopper to raise the fuel cell output voltage as needed to enable battery charging and to meet other bus system requirements. Chrysler PEI integrated the fuel cell subsystem with the battery management system and the system controller. An Exide D31D battery was selected for the brassboard tests. A resistive load bank was used to provide the traction power demand.

BRASSBOARD TESTS

The complete brassboard test program covered three test phases:

Test Phase 1 - Acceptance
Test Phase 2 - Performance Characterization
Test Phase 3 - Proof-of-Feasibility.

Acceptance Tests - The purpose of these tests was to verify that the performance of the fuel cell subsystem and the battery modules complied with the specifications.

The brassboard fuel cell subsystem exceeded the power and efficiency specifications. The steady load power output of the fuel cell subsystem (at the chopper output) was measured to be 23.23 Kw, 3.2% above the 22.5 Kw specification, while the fuel consumption was 4.5% below the 10.79 kg/hr specification. As a result, the subsystem efficiency (including chopper losses) at full power was measured to be 38.8% without auxiliaries, and 36.5% with auxiliaries. The fuel cell auxiliaries (cooling pumps, air and fuel supply pumps, etc.) thus consumed only 2.3% of the gross output of the fuel cell.

The start-up time (33 minutes) and power-up time (7 minutes for power increase from 25 to 100% rated power) also were within specifications. The power-up time for the full size fuel cell is projected to be 4 minutes for power increase from 0 to 100%.

Performance Characterization - The purpose of these tests was to characterize the performance of the fuel cell and battery subsystems individually and then in the hybrid configuration. Steady-state operation was used for characterization. Varying loads were also applied to test and verify the communication between the fuel cell and the battery subsystems, automatic control of fuel cell power output, and proper operation of the system controller.

Good agreement was obtained between the fuel cell performance measured during these tests and that measured during the earlier acceptance tests.

The battery, which consisted of 17 Exide D31D 12 V modules, performed well. Its stiff voltage-current characteristic required some state-of-charge measurement software changes.

The power demand from the fuel cell was controlled by the battery and system controller, based on the battery voltage and traction power requirement. The fuel cell power output was controlled by a 0 to 10V analog signal supplied to the fuel cell internal controller. The control system worked well and excellent correlation was obtained between the load demand (the traction power) and the fuel cell/battery system power output.

Proof-of-Feasibility - The purpose of this final test phase was to verify the capability of the brassboard, after proper scaling corrections, to power the full size bus over standard routes. The measured fuel consumption and battery state-of-charge data were compared with the corresponding data projected by a computer simulation program. The analysis therefore also provided verification of the assumptions used in developing the simulation program.

The test procedure was carefully developed to ensure that the loads corresponded with the traction power requirements for the test routes. Test cycles included the DOT/UMTA Transit Coach Duty

Cycle (TCDC), the Indianapolis (INDY) route, and the Georgetow University Transportation Society (GUTS) Arlington Loop route. The measured fuel economies for the brassboard and the extrapolate and projected fuel economies for the baseline fuel cell/battery TB are shown in Table 2. The extrapolated data from the brassboar measurements are in excellent agreement with simulation results Good correlation was also obtained between the measured an projected system response, bus acceleration times, batter conditions during the runs, and the system power output under rout conditions.

These results provide the proof of feasibility of the fuel cell battery system as a potential power source for urban transit buses

ECONOMIC ANALYSIS

The cost-competitiveness of equivalent PAFC/battery, diesel, an methanol-ICE powered buses was analyzed and compared on a lif cycle cost basis. A sensitivity analysis was conducted to provid a range of possibilities related to cost and bus operational usage The bus technologies were assumed to comply with the environmenta standards for the 1995-2000 time frame, and current as well a energy-based fuel pricing structures were evaluated.

At the current diesel and methanol fuel price differential, PAF powered buses will cost 18.7% more to own and operate than diese buses, assuming that reliable particulate emission controls wil become feasible for diesel buses. Under similar assumptions methanol-ICE powered buses will cost about 22.5% more than diese buses. With energy-based fuel prices, the fuel cell bus economic are within the competitive range of the diesel and methanol powere bus. The fuel cell bus has an economic advantage over th methanol-ICE bus due to its higher operating efficiency and lowe maintenance costs. Table 3 shows the estimated life cycle costs o diesel, PAFC, and methanol engine buses for the two fuel pric scenarios.

TEST-BED BUS DESIGN

The PAFC, when coupled with a suitable surge battery, is capabl of providing the total power and energy requirements of urba transit buses. The results of the trade-off analysis an performance projections show that a current technology 50 Kw PAFC operating in parallel with a lead-acid battery, would be capable o supplying the needed power for acceleration and grade climbing The range of the bus would be determined by the size of the fue tank. It would be feasible to carry sufficient fuel in a 27 f transit bus for a range of over 150 miles without refueling A artist's concept for the PAFC/battery test-bed bus is shown i Figure 1.

The PAFC/battery bus system is capable of higher operatin

efficiencies than internal combustion engine powered diesel and methanol buses over the DOT/UMTA Transit Coach Operating Duty Cycle as well as the GUTS routes. The brassboard PAFC exhibited an efficiency (based on LHV of 4767 kcal/kg methanol) of 36.5% after accounting for all internal losses. PAFC system efficiencies of the order of 42% are achievable with currently available stack technology, and would be used with the test-bed bus. Unlike internal combustion engines, the fuel cell efficiency does not degrade significantly at part load.

The electric drivetrain provides superior traction capabilities to the bus. Performance simulations have shown that because of the high initial acceleration of electric drive systems, overall power requirements are lower than for equivalent diesel buses. The acceleration is smooth and controllable. Added energy saving benefits are offered by regenerative braking, which allows storing and reusing braking energy which is normally wasted by conventional power sources. Regenerative braking would also reduce brake wear, a major maintenance item for current transit buses.

ENVIRONMENTAL BENEFITS

The fuel cell bus will meet all present and future environmental standards without any performance or cost penalty. The by-products of the electrochemical reactions are water and carbon-dioxide, with some carbon-monoxide in quantities much lower than those emitted by internal combustion engines using hydrocarbon fuels. Nitrogen oxides are generally below measurable limits because of the low catalysis temperatures in the reformer and stack. Particulate emissions are absent. The operation of the fuel cell stack and reformer is silent. The only noise and vibration from the assembly will be from the coolant pump and various air intake blowers, estimated to be well below the noise and vibration levels of internal combustion engines.

DEVELOPMENT SCHEDULE

Detailed design of the test-bed bus is currently under way. Fabrication will begin after a design review in March 1992. The full-scale propulsion system will be assembled and tested by H-Power in late 1992. After testing the propulsion system on a dynamometer, it will be installed in the bus by Bus Manufacturing U.S.A. Inc. The bus fabrication is expected to be completed by July 1993.

A total of three test-bed buses will be fabricated and delivered to DOE for further evaluations. In addition, Transportation Manufacturing Corporation (TMC) will develop a design concept and economic analysis for a standard 40 ft fuel cell/battery transit bus. Successful completion of this phase will enable H-Power and TMC to make business decisions on the commercial viability of PAFC/battery transit buses.

FIGURE 1
Artist's Concept for the Phosphoric Acid
Fuel Cell/Battery Powered Test-Bed Bus

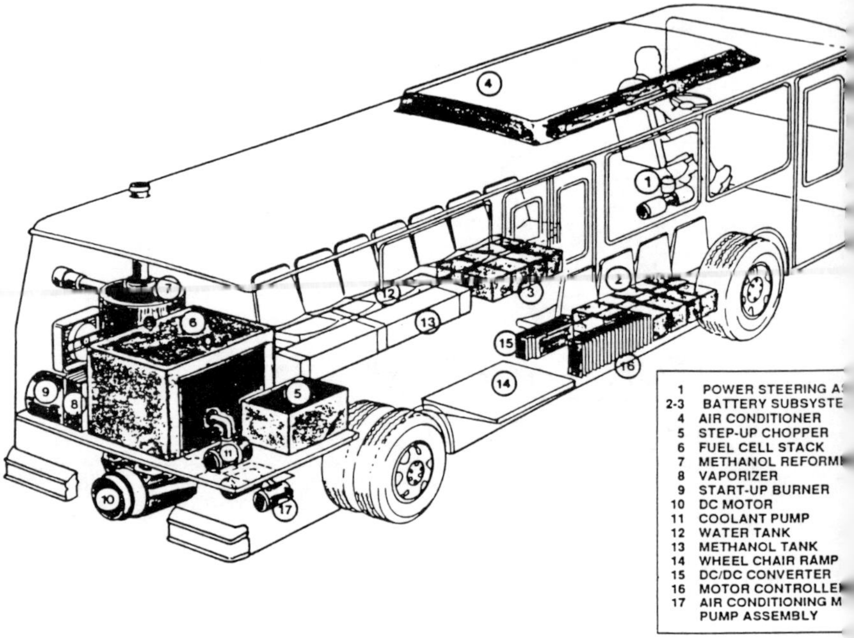

TABLE 1
Summary of Specifications for the
Fuel Cell/Battery Powered Bus

<u>Fuel Cell Subsystem</u>	Liquid-Cooled Phosphoric Acid
Nominal Power Rating	50 Kw
Fuel	Commercial Grade Methanol
Efficiency	42% (LHV) at Chopper Output Terminal
Single Cell Performance	0.66V @ 240 Ma/cm2
Number of Cells	160
Total Weight	2,598 lb (1,178 kg)
<u>Battery</u>	Lead-acid SLI Type
Nominal Voltage	204 V
Capacity (C/20)	76 Ah
Subsystem Weight	1,250 lb (567 kg)
<u>Motor</u>	DC Traction Motor
Base Speed/Max Speed	850 rpm/3,500 rpm
Weight	1,050 lb (476 kg)
<u>Controller</u>	Transistorized
Weight	120 lb (55 kg)
<u>Bus Body</u>	Standard Transit, forward control
Length	26.7 ft (8.14 m)
Width	90 in (2,286 mm)
Floor Height	14 in.
Seating	24 passengers
Weight, including axle	13,000 lb (5,900 kg)
<u>Accessories</u>	All standard bus accessories, including air conditioning
Total Bus Weight (GVW)	22,325 lb (10,125 kg)

TABLE 2
Comparison of Measured and Projected Fuel Economies.

Cycle	Brassboard Measured MPG	Test -Bed Bus Extrapolated* MPG	Test -Bed Bus Projected** MPG
TCDC	4.90	2.52	2.58
INDY	4.61	2.37	2.40
GUTS	4.48	2.30	2.23

* = Extrapolated from measured data
** = Projected by simulation model

TABLE 3
Comparison of Life Cycle Costs of Urban Transit Buses
(40 Ft Buses)

Bus Type	Life Cycle Cost (cents/mile) Current Fuel Prices	Life Cycle Cost (cents/mile) Energy-Based Prices
L-PAFC/Battery	55.9	54.1
Diesel	47.1	53.5
Methanol-ICE	57.7	55.0

THE BALLARD/BC BUS DEMONSTRATION PROGRAM

B. N. Gorbell and K. B. Prater
Ballard Power Systems
107 - 980 West 1st Street
North Vancouver, BC V7P3N4
Canada

ABSTRACT

In this paper, we present a brief background on fuel cells and specifically on the Solid Polymer Fuel Cell (also known generically as the Proton Exchange Membrane or PEM fuel cell). We then discuss the issues which must be addressed to achieve commercial penetration of these fuel cells for motive applications. Finally we describe a program underway in Vancouver, Canada to demonstrate the operation of a commercial transit bus fully powered by Ballard Solid Polymer Fuel Cells and fueled by hydrogen.

INTRODUCTION

The Industrial Revolution and the subsequent Transportation Revolution were powered by heat engines. The term "heat engine" includes the steam engine, the internal combustion engine, and the gas turbine engine. These devices rely on a multi-step process which begins with the combustion of a fuel to convert the energy in that fuel to heat. A small fraction of that heat is then converted into mechanical energy within the engine. The fraction of the heat which can possibly be converted into mechanical energy is limited by the amount of unused heat energy which is expelled to the environment in the exhaust.

The theoretical limitation on heat engine efficiency is determined by the Carnot Cycle. For a steam engine, operating at 250 °C, one cannot possibly convert more than 29% of the energy in the heated steam into mechanical energy. In a real system, thermal and frictional losses reduce the that percentage further. A typical automobile operating in the city expels about 85% of the energy in the fuel it consumes into the environment as useless heat.

The fuel cell, by contrast, is a device which converts a fuel and an oxidant directly into electricity by an electrochemical process. The fuel is not burned to produce heat and the efficiency of energy conversion is not limited by the Carnot Cycle limits placed upon heat engines. The fuel cell, like a battery, has no moving parts in the energy conversion device, and thus suffers no frictional losses. Realistic energy efficiencies in excess of 50% can be achieved with fuel cell systems.

Though invented by Sir William Grove 150 years ago, the fuel cell found its first real application in the US space program of the 1960s. In the last decade, there have been a number of demonstrations of stationary applications of fuel cells for the production of electricity and heat. Only in the last few years, with a breakthroughs in a fuel cell technology based upon a solid polymeric membrane as the electrolyte, has it become possible to seriously consider commercial vehicles powered by fuel cells.

THE FUEL CELL

A fuel cell, like a battery, is an electrochemical energy conversion device, but a fuel cell is continuously refuelable and, thus, is a source of continuous electrical power. It is energy efficient, non-polluting, silent, and reliable. In some embodiments, it is a low temperature device which provides power instantly on demand, and exhibits a long operating life with minimal maintenance. For a detailed discussion of fuel cells, see reference [1] and references cited therein.

A generalized fuel cell is shown in Figure 1. It consists of two porous electrodes at which the energy conversion process takes place. A fuel (hydrogen, in the case of the example) is supplied to one electrode while an oxidant, usually oxygen from the air, is supplied to the other electrode. At the fuel electrode (the anode) the fuel molecules give up electrons to the electrode and create an electric current in the wire leading from the fuel electrode through the electrical load to the oxidant electrode (the cathode). The remnants of the fuel molecules are now positively charged ions, hydrogen ions in the example shown in Figure 1.

The intrinsic tendency for the fuel and oxygen to react, as demonstrated by normal combustion, is manifested as a voltage between electrodes of the fuel cell. When the electrons, drawn by that voltage, arrive at the oxidant electrode, they are removed from the electrode by oxygen molecules, producing negatively charged oxygen ions. The build-up of positive ions at the anode and negative ions at the cathode would soon stop this process if there were no mechanism for these charges to be dissipated. There must, therefore, be an "electrolyte" between the two electrodes to allow the migration of ions from one electrode to the other.

In the example in Figure 1, the electrolyte allows the positively charged hydrogen ions to migrate from the anode to the cathode where they react with the negative oxygen ions to produce the product of this reaction, water. In this example, the electrolyte is an acid, completely analogous to the acid electrolyte in the automotive battery.

A liquid acid electrolyte is used in the most mature fuel cell technology--the Phosphoric Acid Fuel Cell (PAFC). The selection of a liquid electrolyte imposes several constraints on the design and operation of such a fuel cell. The pressures of the fuel and oxidant gases must be carefully regulated to assure that no substantial pressure differential exists across the electrolyte. Should a pressure differential develop, the electrolyte would be forced away from one electrode and into the other, terminating operation.

Product water, if not removed, would dissolve in the electrolyte, increasing its volume and flooding the electrodes. To prevent such flooding, the PAFC is operated at about 190 °C. At that temperature, product water is vaporized as it is formed and carried from the fuel cell with excess oxidant flow.

The hot, corrosive electrolyte must be contained to prevent its contact with other systems or with personnel. This places severe restrictions on the design and materials choices for the PAFC. In conjunction with the other system requirements, the containment requirement results in the PAFC system being rather large and heavy.

The Solid Polymer Fuel Cell eliminates the disadvantages associated with liquid electrolytes by replacing the liquid electrolyte with a thin polymeric membrane. This electrolyte replacement also significantly simplifies the requirements for supporting systems. Recent developments by Ballard [2], using an experimental electrolyte membrane developed by Dow Chemical, have substantially reduced the size of the fuel cell itself and made the application of fuel cells to motive power possible.

THE SOLID POLYMER FUEL CELL

The Solid Polymer Fuel Cell uses as its electrolyte one of several ion exchange polymer membranes. These membranes are electronic insulators, but excellent conductors of hydrogen ions. The materials used to date consist of a fluorocarbon polymer backbone, similar to Teflon, to which are attached sulfonic acid groups. The acid molecules are fixed to the polymer and cannot "leak" out, but the protons on these acid groups are free to migrate through the membrane.

The electrolyte membrane looks rather like a thick sheet of food wrap and can be handled easily and safely. The anode and cathode are prepared by applying a small amount of platinum black to one surface of a thin sheet of porous, graphitized paper which has previously been wet-proofed with Teflon. The electrolyte is then sandwiched between the anode and cathode, and the three components are sealed together under heat and pressure to produce a single "membrane/electrode assembly" (MEA). This assembly, which is the heart of the fuel cell, is less than a millimeter thick.

The anode and cathode are contacted on the back side by flow field plates made of graphite in which channels have been formed. The ridges between the channels make electrical contact with the backs of the electrodes and conduct the current to the external circuit. The channels supply fuel to the anode and oxidant to the cathode.

The electrode reactions in the SPFC are analogous to those in the PAFC. Hydrogen from the fuel gas stream is consumed at the anode, yielding electrons to the anode and producing hydrogen ions which enter the electrolyte. At the cathode, oxygen combines with electrons from the cathode and hydrogen ions from the electrolyte to produce water. The water does not dissolve in the electrolyte and is, instead, rejected from the back of the cathode into the oxidant gas stream. As the SPFC operates at about 80 °C, the water is produced as liquid water and is carried out of the fuel cell by excess oxidant flow.

Ballard has converted this technology into a fuel cell stack, delivering a nominal 5 kW of dc power using hydrogen as the fuel and air as the oxidant. Ballard has also developed a self-contained fuel cell system with a single on/off switch control. The development and sale of such stacks and systems are not sufficient, however, for the ultimate commercialization of this technology.

CHALLENGES FOR COMMERCIALIZATION

The potential advantages of fuel cell technology for transportation applications--substantial reductions in environmental impact and improvement in fuel efficiency--provide a significant opportunity for the commercialization of this technology. But several challenges must be met before major market penetration is achieved.

Appropriate products must be developed to meet the specific needs of the transportation industry. It must be recognized that these needs are rather different for the various components of that industry. The requirements for powering an automobile are not the same as those for a locomotive or even for a bus. The development of appropriate fuel cell power systems will require some further developments in the fuel cell stack. The issues here are largely those dealing with the trade-offs between performance and manufacturing cost. No performance breakthrough is required to produce Solid Polymer Fuel Cells for transportation applications.

The greatest challenges for the development of appropriate fuel cell power systems lie in the balance of the system. Such a system is illustrated in Figure 2. It includes the fuel storage sub-system, in the case of hydrogen, or the fuel processing sub-system, in the case of hydrocarbon fuels; the air supply sub-system; the heat/water management sub-systems; the power conditioning sub-system; and the control sub-system. Each of these sub-systems requires further development for the ultimate commercialization of fuel cells for transportation applications.

The cost of the complete power system must be reduced to levels acceptable for each transportation market. The acceptable cost varies dramatically from such commercially operated fleet vehicles as locomotives and buses to privately owned and operated automobiles.

An infrastructure must be developed to support fuel cell powered vehicles. This infrastructure includes a manufacturing/sales/distribution infrastructure, a maintenance infrastructure, a fuel infrastructure, and an appropriate regulatory infrastructure. Again, the infrastructure requirements are more easily satisfied for fleet vehicles than for privately owned automobiles.

But, in order to establish the required infrastructure and in order to achieve the production levels necessary to achieve competitive costs, one must first overcome the risk aversion associated with a new technology. The potential users, the potential manufacturers, the potential fuel suppliers, and the potential regulators must be convinced that the technology is appropriate, safe, and sufficiently close to commercialization to warrant the commitment of time and financial resources to the development of products and the establishment of infrastructures.

The most effective method for overcoming risk aversion is the demonstration of the technology in real applications in the commercial environment. Ballard has embarked upon such demonstrations in the motive and utility fields. The remainder of this paper will discuss the transit bus demonstration program presently under way in British Columbia, Canada.

THE BALLARD/BC BUS DEMONSTRATION PROGRAM

Ballard, in conjunction with the Province of British Columbia and the government of Canada, has begun the first phase of a program to develop a commercial transit bus powered by Ballard's Solid Polymer Fuel Cells. Phase I began in November of 1990 and is funded at the level of $4.84 million Canadian.

The objective of Phase I is the demonstration of a commercial transit bus fully powered by Ballard fuel cells which provides the same driver acceptance as the corresponding diesel powered bus. The specific operating goals are listed in the following table. The 120 kW power system is to be assembled for testing in October 1991 and the bus is to be operational in the Fall of 1992.

While the program is largely funded by the governments of British Columbia and Canada, the project, including allocations of funds, is controlled by a steering committee headed by a representative of the local transit authority, which will test the bus. The committee, with representation from the US and the UK as well as Canada, includes potential users, academics, and representatives of regulatory agencies. Ballard's objective in establishing such a steering committee was to create a program which would be driven by the demands of the market, rather than technical targets.

As a result of input from the committee, the Ballard/BC Bus program differs from other such programs in three significant ways: 1) it is based upon the use of a commercial bus as the demonstration platform; 2) it will use fuel cells to provide all the motive power for the bus, rather than relying upon a substantial battery pack for acceleration; and 3) it will use gaseous hydrogen as the on-board fuel.

Ballard, as prime contractor, and SAIC as the sub-contractor responsible for system integration, have selected a commercially available 32-foot bus from National Coach Corporation as the demonstration test bed. These buses are presently in use in transit systems. The commercial bus is delivered as a diesel-powered vehicle and its performance is fully documented in that configuration. The bus itself will be received by the program in November 1991.

While the ultimate commercial fuel cell bus may incorporate batteries to permit regenerative breaking to be incorporated into the bus, this program seeks to demonstrate the capability of the fuel cell and to avoid any doubts that the fuel cell is the primary source of motive power on the bus. The 120 kW fuel cell system will be based upon Ballard's present 5 kW stacks configured as three parallel strings of eight cells operating at a nominal voltage of 200 VDC.

Gaseous hydrogen was initially chosen as the on-board fuel for system simplicity. Further discussions with potential users, however, have confirmed that choice as a user preference. BC Transit is already testing natural gas fueled buses and is satisfied with the safety of gaseous fueled buses. It would be their strong preference to have a hydrogen fueled bus rather than to have the weight, capital cost, and maintenance issues associated with a reformer on-board each bus. They also recognize the environmental advantages, if one is to use a hydrocarbon fuel, of having a single reformer at the depot from which CO_2 can be collected, rather than a fleet of mobile reformers from which CO_2 could not realistically be collected.

The fuel system will be based upon DOT approved glass-wound composite cylinders operating at 3000 psi. With this fuel system, an operating range of 150 km at full power is projected for the Phase I demonstrator.

The configuration of the bus and its component systems is illustrated in Figure 3.

On the assumption of a successful Phase I demonstration, Phase II will involve improvements in range, based upon improved hydrogen storage technology, the use of more compact fuel cells and ancillaries, and the increase in the size of the bus to 40 feet.

SUMMARY

The Solid Polymer Fuel Cell as a technology has now reached the point where is can be demonstrated as the power system for a commercial bus. This demonstration is seen as an important step on the path to the ultimate commercialization of this technology.

REFERENCES

1. A. J. Appleby and F. R. Foukes, Fuel Cell Handbook, Van Norstrand Reinhold, New York, 1989.

2. K. B. Prater, J. Power Sources, 29 (1990) 239-250.

FIGURE 1: GENERALIZED FUEL CELL

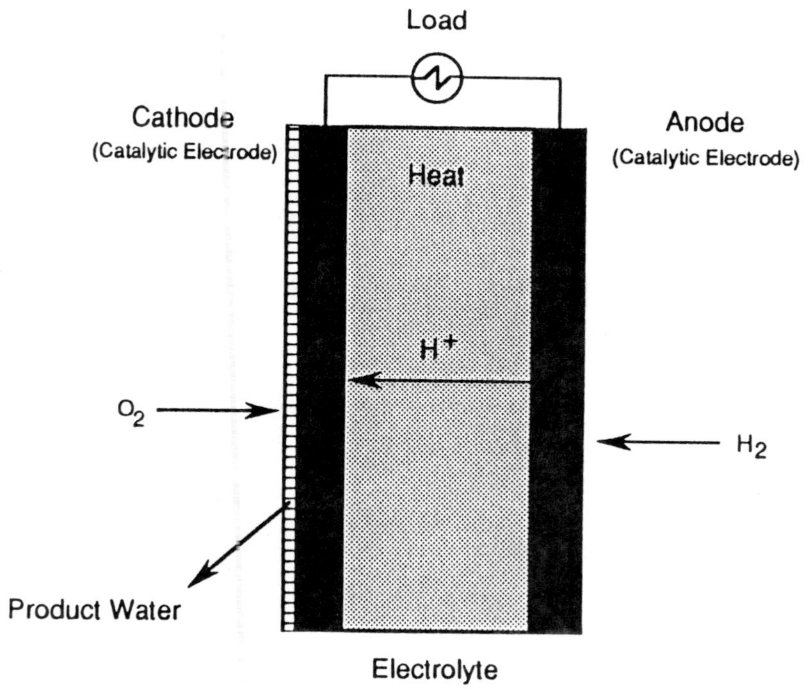

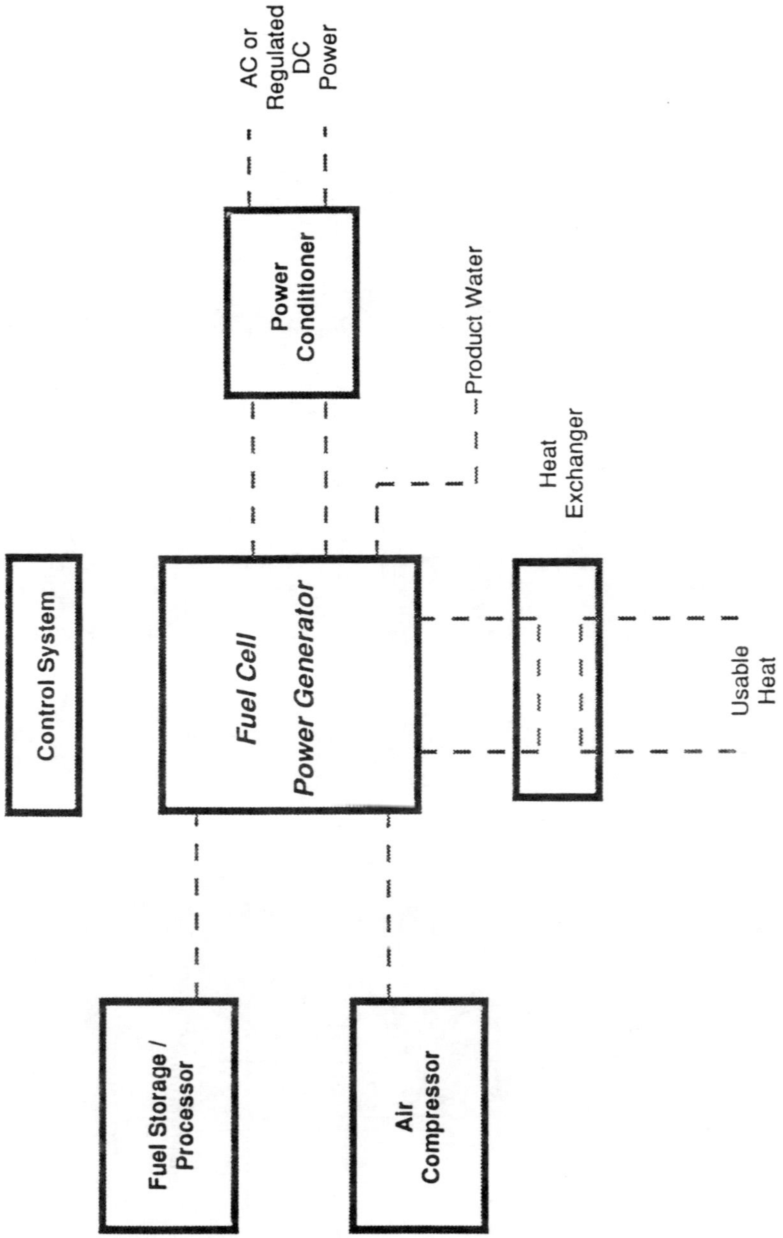
Control System
Fuel Cell
Power Generator
Fuel Storage /
Processor
Air
Compressor
Power
Conditioner
AC or
Regulated
DC
Power
Product Water
Heat
Exchanger
Usable
Heat

FIGURE 3: FUEL CELL POWERED TRANSIT BUS

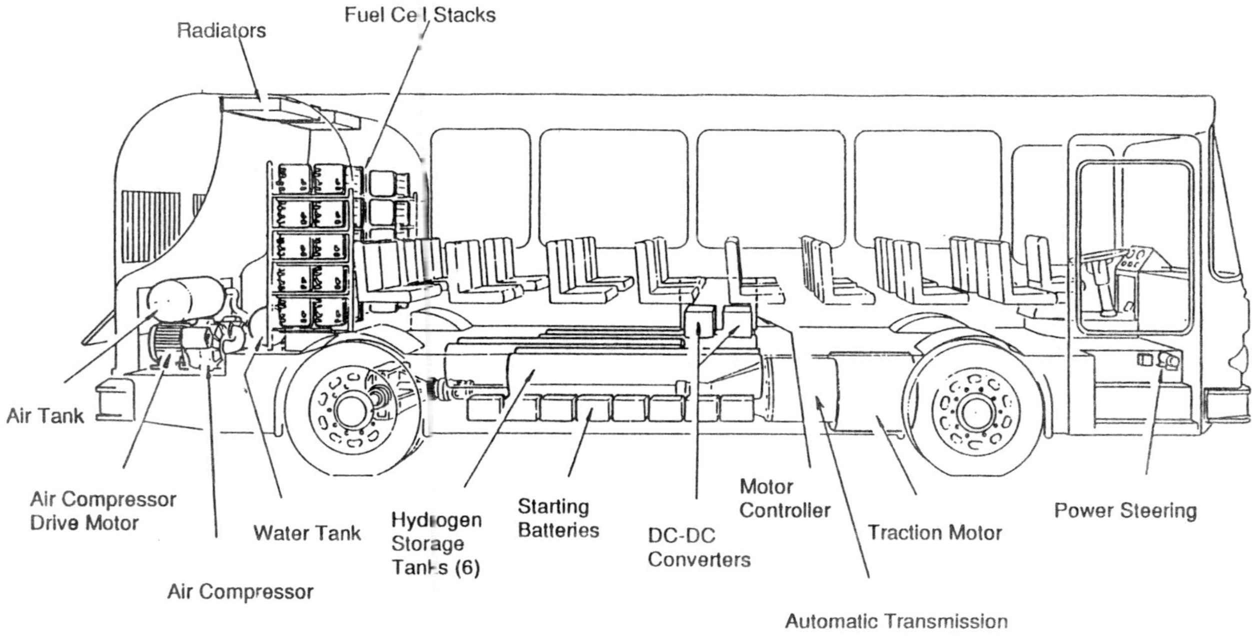

LaserCel1 Prototype Vehicle

Dr. Roger E. Billings, Dr. Maria Sanchez, Dr. Paul Cherry, D. Bradley Eyre
American Academy of Science
TechCenter, Suite 1000
26900 East Pink Hill Road
Independence, MO 64057-3284

Abstract

A solid polymer electrolyte fuel cell has been installed in a prototype vehicle to assess the potential of fuel cells in vehicular applications. The fuel cell is based on an advanced technology developed at the American Academy of Science known as the LaserCel. Hydrogen is stored on the vehicle in a safe, metal hydride powder form. Hydrogen is released from the hydride by heat supplied from the exhaust of the fuel cell. Inside the cell, hydrogen is reacted with the oxygen from air to form water. Energy from this reaction is released from the cell in the form of electricity which is used to power an electric motor which is coupled to the wheels through a standard transmission.

Background

For several years, the advantages of utilizing hydrogen as a fuel, have been demonstrated in vehicle prototypes. Hydrogen can be synthesized from water and most of the renewable energy sources. It can be safely stored in its powdered, metal hydride form, and the only resulting air pollution is water vapor and very small amounts of nitric oxide when it is combusted.

Unfortunately, there are several drawbacks which have limited the widespread shift to a "hydrogen energy economy". The most serious disadvantages include:

1. Hydrogen is more expensive to produce than the conventional hydrocarbon fuels in extensive use today.

2. The safest method of storing hydrogen is in the metal hydride form. Though safe, this form of hydrogen storage is very heavy, and therefore has limited the range of hydrogen automobiles to 250 kilometers (150 miles).

3. The existing gasoline infrastructure is not compatible with gaseous hydrogen distribution. In order to begin the implementation of this new fuel, it would be necessary to build an entirely new distribution system.

Recent advances in solid polymer fuel cells, have made possible the application of this technology to vehicle transportation systems. Since fuel cells are two to three times as efficient as the internal combustion engine, and since fuel cells can be operated in reverse as an electrolyzer, their application in an automobile largely overcomes each of the major impediments to widespread application of hydrogen vehicles.

1. Although hydrogen is more expensive to produce than the conventional fossil fuels, the higher utilization efficiency of the fuel cell more than compensates for this difference making the cost of operating a vehicle on hydrogen less than operating on gasoline.

2. Since hydrogen is utilized two to three times as efficiently in the fuel cell as in an internal combustion engine, vehicle range is extended to 500 kilometers (300 miles) or more without increasing the weight of the hydride storage vessel or the quantity of the hydrogen stored.

3. By operating the fuel cell in reverse as an electrolyzer, it is possible to produce hydrogen for storage in the metal hydride vessel by connecting the vehicle to a source of electric power and water. Since both are readily available, the problem of an incompatible infrastructure is overcome.

For these reasons, hydrogen fuel cells are the "key" to the commercial implementation of hydrogen as a vehicular fuel.

LaserCel1 Prototype

The U.S. Postal Service donated a Ford Fiesta which had been retrofitted for operation as an electric vehicle by JET Industries, and had been used for five years to deliver the mail in El Paso, Texas. Upon receipt, the vehicle was in poor condition, and, after three years of storage, was nonoperational.

In postal service, the vehicle was heated with a gasoline-fueled furnace. As the first step in preparing the prototype for the fuel cell conversion, the gasoline heater, and fuel tank were removed. Also, the special equipment installed on the vehicle for mail delivery, such as a large hood-mounted mirror and a mail shelf, were removed from the vehicle. New seats were installed but no effort to switch the steering column to the left side was attempted.

Modifications to the vehicle body styling were conceived and implemented by Design Factory of Kansas City. Included in the design changes were the installation of European-style headlights, sport mirrors, high performance tires, and body skirting on all four sides to lower the vehicle profile.

The electrical system of the car was redesigned to incorporate a new controller and various monitoring and regulating equipment were installed for added safety.

The hydrogen system consisted of a metal hydride storage vessel mounted in the rear of the vehicle, a heat exchanger, gas compressor, control system, and the hydrogen fuel cell. A schematic diagram of the hydrogen system is presented as Figure 1.

FIGURE 1. FUEL CELL VEHICLE BLOCK DIAGRAM

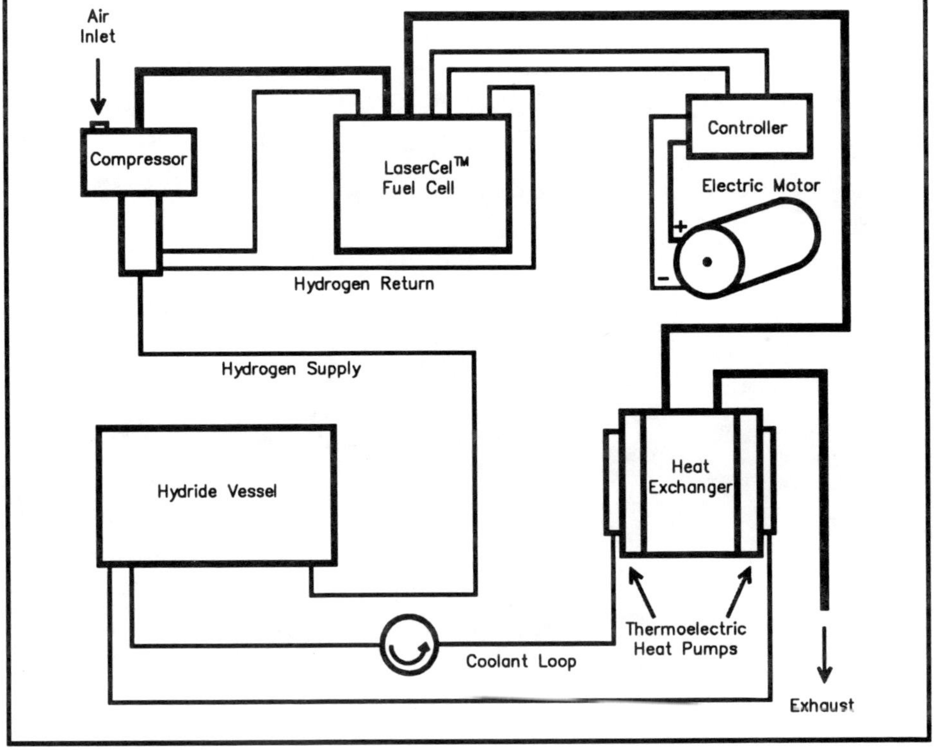

Hydrogen Fuel Cell

The fuel cell is located in the forward "engine compartment". It is based on a cell geometry developed at the American Academy of Science, referred to as the LaserCel. This name is applied to the geometry because of the utilization of a high-powered YAG laser in the fabrication of the anodes and cathodes used in the cell.

The cell utilizes a solid polymer electrolyte produced by the Asahi Chemical Company of Japan. In operation, hydrogen is applied to the anode side of the cell at a pressure of approximately two atmospheres (absolute). The hydrogen molecule is broken down on the surface of a noble metal catalyst (platinum) forming hydrogen ions or protons and releasing electrons to the anode. A proprietary method of catalyst application provides satisfactory results with loading densities of less than 1.0 milligrams per square centimeter. The hydrogen

FIGURE 2. SPE FUEL CELL

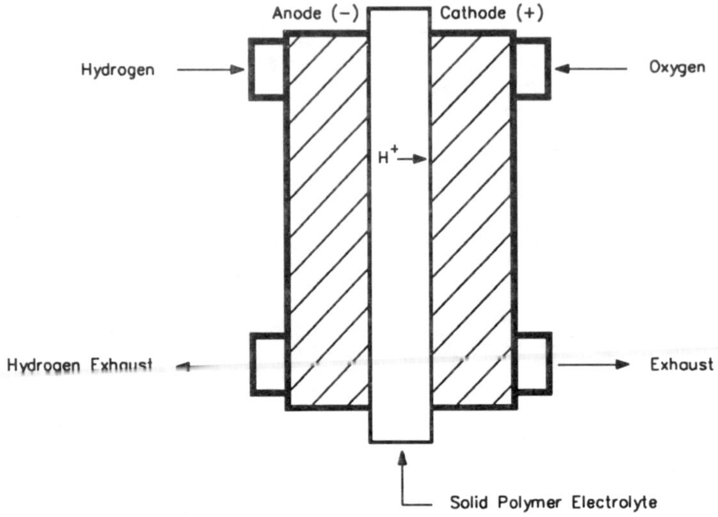

FIGURE 3. SPE FUEL CELL POWER CURVE

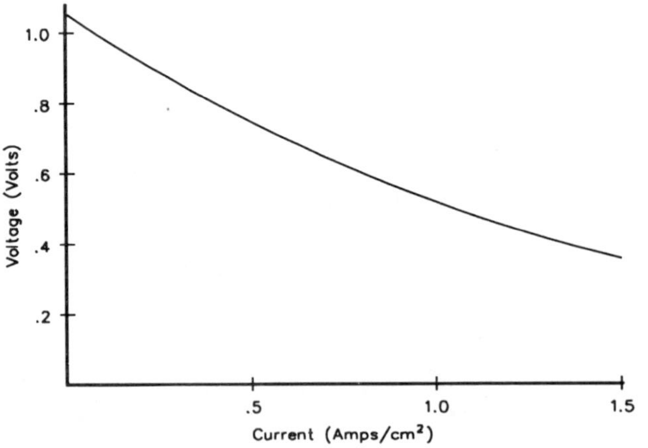

ions are transported through the membrane to the cathode side where they react with oxygen to form water. See Figure 2. The reactions involved in the fuel cell are as follows:

Anode Reaction	$H_2 \text{--->} 2H^+ + 2e^-$
Cathode Reaction	$2H^+ + 2e^- + 1/2\ O_2 \text{--->} H_2O$
Overall Reaction	$H_2 + 1/2\ O_2 \text{--->} H_2O$

In order for the electrons to travel from the anode to the cathode side of the cell, they must flow through an external load. In the case of the fuel cell vehicle, they pass through the electric motor providing the source of propulsion power.

In the LaserCel1 prototype, the fuel cell consists of two stacks of cells inside of a single enclosure. The small stack consists of 16 cells, and produces 12 volts to power the vehicle's accessories. The large stack, consisting of 135 cells, outputs 100 volts to power the drive motor. It is not necessary to balance the output of the two stacks because they are electrically isolated.

The cell operates at a current density of about 0.5 amps per square centimeter at 0.75 volts per cell. The cell voltage as a function of load is presented in Figure 3. The dimensions and mass of the LaserCel fuel cell shown in Figure 4 are:

Height	0.23 meters
Length	0.25 meters
Width	0.16 meters
Mass	34.0 kilograms

FIGURE 4. HYDROGEN FUEL CELL AND COMPRESSOR

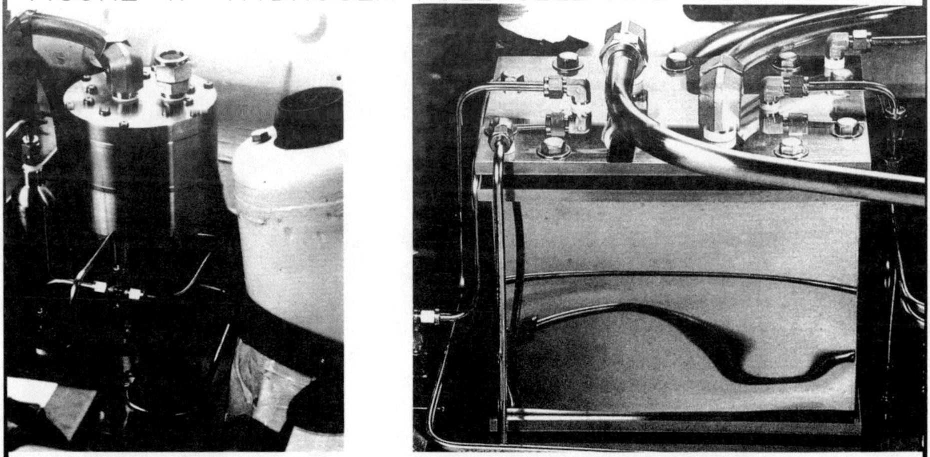

Gas Compressor

Hydrogen Fuel Cell

FIGURE 5. GAS COMPRESSOR DESIGN

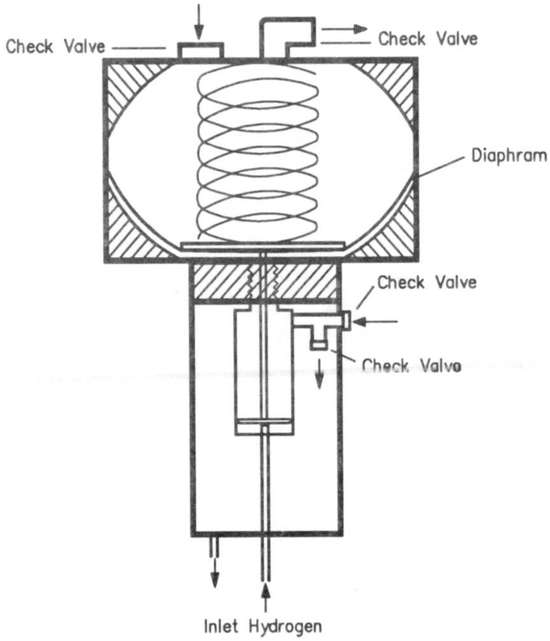

In order to achieve adequate cell performance, reaction kinetics dictate that hydrogen flow through the cell must be 0.5 times greater than actual cell consumption. Surplus hydrogen is captured upon exiting the cell and is recirculated. Oxygen is supplied to the cell at a rate of 2.5 stoichiometry, and at a pressure of 1.15 atmospheres (absolute).

Gas Compressor

The hydrogen gas pumping needed for recirculation and the air compression required for the fuel cell are provided by the proprietary gas compressor shown in Figure 4. The compressor utilizes hydrogen pressure from the metal hydride storage vessel to activate a gas piston. The piston actuates a diaphragm air compressor and circulates hydrogen through the cell. The design of the gas compressor is depicted in Figure 5 and Figure 6 shows the system gas recirculation. The gas compressor is controlled electronically, as detailed in Figure 7, by a time delay circuit and a gas pressure switch. As hydrogen is consumed by the fuel cell, gas pressure drops. The drop in pressure is detected by the switch which initiates a new cycle. In this way, the gas compressor automatically is load following.

FIGURE 6. GAS RECIRCULATION SYSTEM

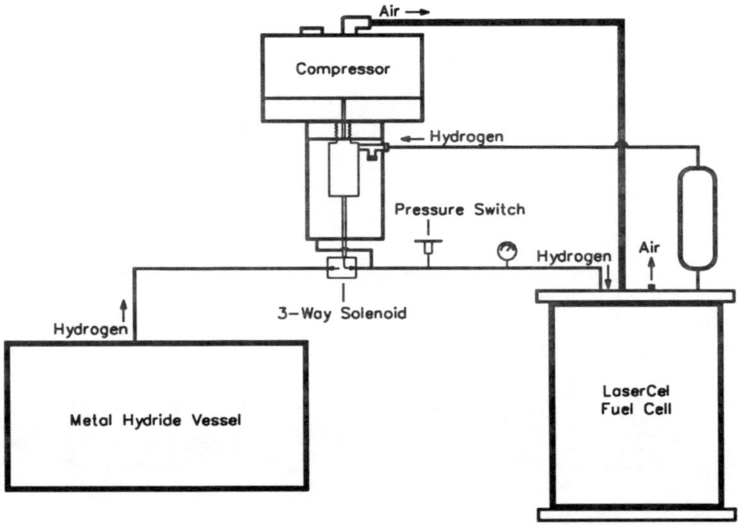

FIGURE 7. COMPRESSOR CONTROL SCHEMATIC

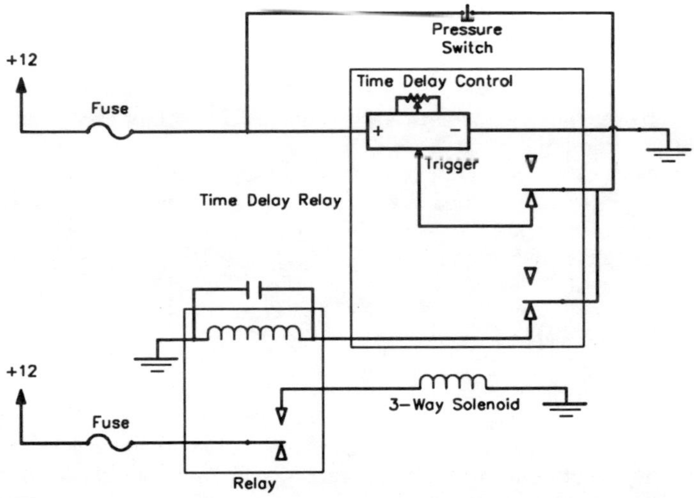

FIGURE 8. HEAT EXCHANGER SYSTEM

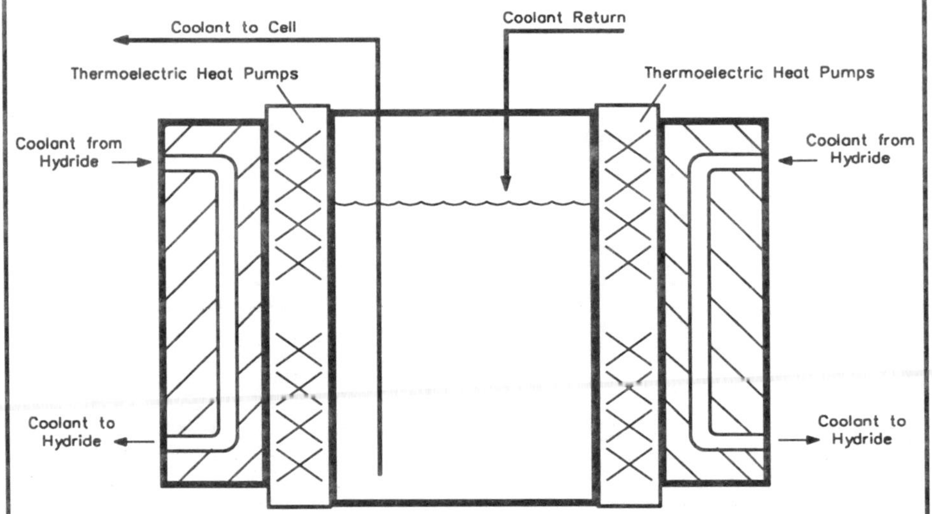

Cooling System

The fuel cell has heat plates interspersed every three cells to provide temperature control for the system. The heat plates are cooled by water which is circulated to the system heat exchanger. The heated water is utilized to provide the hydrogen gas humidification required by the cell through a technique which is proprietary and therefore not disclosed in this discussion.

In the system heat exchanger, heat from the heat plates as well as from the cell exhaust gases is transferred into an ethylene glycol water solution which is used to heat the vehicle and to supply heat of dissociation for the metal hydride storage vessel. The heat exchanger is equipped with eight thermoelectric heat pumps which are automatically energized should the fuel cell overheat. In actual operation, these heat pumps have not been required to date. The heat exchanger design is outlined in Figure 8.

Metal Hydride Storage Vessel

The hydrogen for the vehicle is stored in a metal hydride storage vessel located near the rear of the automobile. See Figure 9. This hydride vessel stores 2.3 kilograms of hydrogen and has a total mass of 136 kilograms. With this quantity of hydrogen, the vehicle has a range of 300 kilometers. (By modernizing the electrical propulsion system of the car, it is estimated that the range could be extended to 400 kilometers.)

FIGURE 9. METAL HYDRIDE STORAGE VESSEL

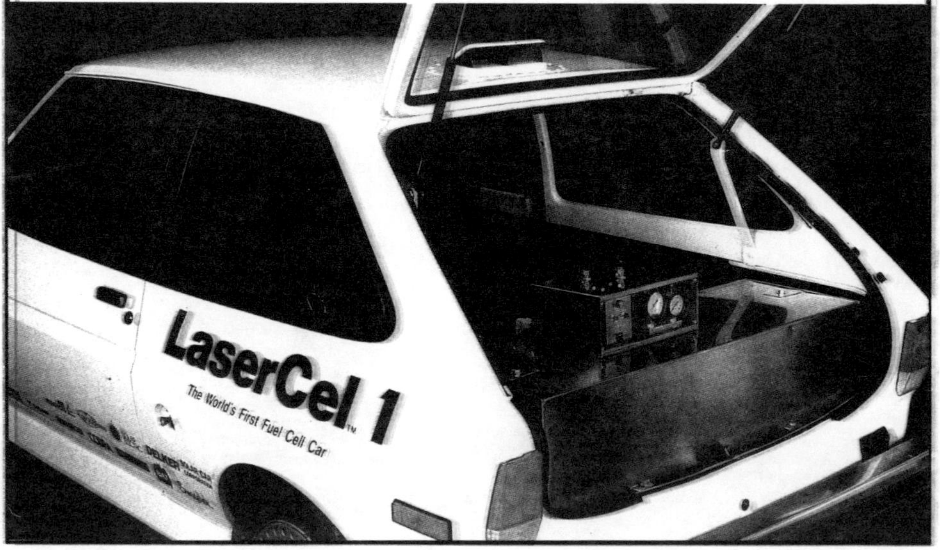

The hydride alloy utilized is $Fe_{44}Ti_{51}Mn_5$. This alloy was selected for its ease of activation, and tendency to inhibit decrepitation. The alloy has a useable weight percent of 1.9 and operates at temperatures ranging from approximately 10° to 90° C.

A recharge "quick" connect is installed on the left side of the vehicle. To recharge the hydride, hydrogen is supplied at a pressure of 17 to 20 atmospheres. Heat must be removed from the hydride bed to achieve a quick recharge. This is accomplished by connecting an external heat exchanger to the vehicle through the twin "quick" connects installed on the rear of the vehicle to the right of the hydride storage vessel. The external heat exchanger consists of a finned tubing coil through which cooling fluid is pumped as air is forced over the fins to dissipate the heat. The design of the external heat exchanger is shown in Figure 10.

Accelerator Battery

To minimize the fuel cell capacity required for the prototype, an accelerator battery was installed in the vehicle. The accelerator battery is utilized to provide part of the power during periods of high acceleration, and then is recharged during cruise. The vehicle accelerator battery is lead-acid, and has a total capacity of 25 horsepower-hours. The 13, 6-volt batteries were provided for the project by the Exide Corporation.

FIGURE 10. EXTERNAL HEAT EXCHANGER

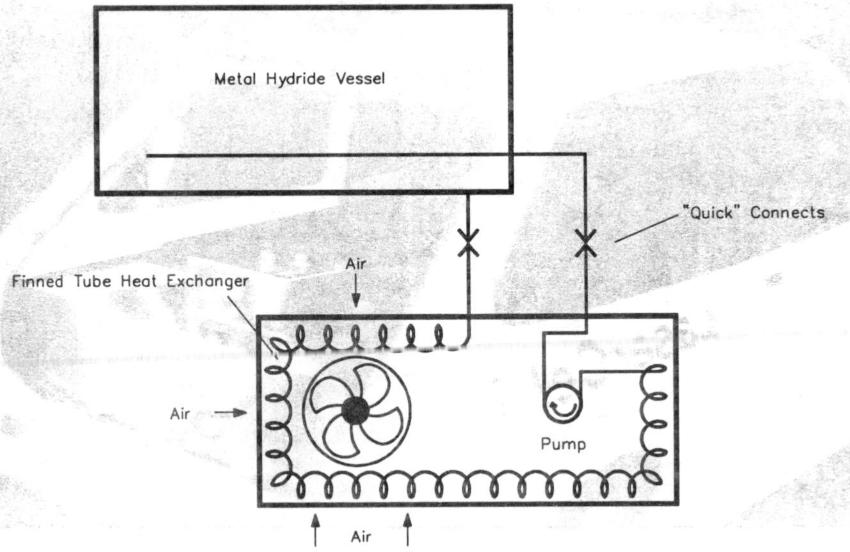

A safety analysis of the prototype design and construction was performed by Air Products of Allentown, Pennsylvania. The Pennsylvania State Energy Office provided part of the funding for the construction of the prototype. The LaserCel1 prototype will undergo extensive testing during the next 18 months.

Conclusions

The hydrogen fuel cell prototype vehicle demonstrates that an SPE-type fuel cell is technically capable of powering a standard automobile. Before such systems become a commercial reality, several design improvements should be incorporated into the design. An intelligent power controller should be developed to manage charging and discharging of the accelerator battery. This would greatly reduce the accelerator battery capacity required by the vehicle. A reduction in vehicle operating cost could also be realized if reliable operation could be achieved at a current density of 2 amps per square centimeter.

In general, the boost to the advantages of hydrogen energy resulting from the application of a fuel cell to a vehicular system are exciting and warrant considerable attention.

CHAPTER SIX

Hydrogen Applications

THE ROLE OF THE ENTREPRENEUR IN BRINGING TO PASS OUR SUSTAINABLE HYDROGEN ENERGY FUTURE

G. L. Benson, Ph.D.
Irvin L. Young Entrepreneurship Chairholder
and Wisconsin Distinguished Professor
College of Business and Economics
University of Wisconsin -- Whitewater
USA

It is a pleasure for me to be here. When Dr. Billings called and asked if I would be willing to speak at this Conference, I was a little reluctant to do so because I do not have a scientific, engineering, or technical background. Rather, I am a business school professor whose area of specialization is entrepreneurship.

Dr. Billings and I go way back to our childhood days when we grew up and went to school together in the same community and school system. After completing high school, our paths diverged as he became a scientist while I became a University professor.

Over the years, I have followed his pioneering work in the field of hydrogen energy and technology with great interest and admiration because he is truly one of the world's leaders in hydrogen technology and one of the most brilliant scientific minds of our generation, in my opinion.

So when he called and asked me to participate in the Conference as a speaker I was honored, but not at all sure exactly what I might be able to contribute.

Having given this Conference and this speech a great deal of thought over the past several weeks, I have concluded that my message to you is that the key to our nation and the world developing a sustainable hydrogen energy future lies in putting hydrogen technologies in the hands of entrepreneurs, not in reliance on government or large energy, automotive or other similar types of companies or organizations, because entrepreneurs and entrepreneurial companies understand how to get new technologies from the garage, the basement or the laboratory to the market place.

I would like to begin by quoting from a paper written and presented by an associate of mine, Professor Thomas DiLorenzo, at Washington University in St. Louis, Missouri as a part of their Contemporary Issues Series in August 1991.

The paper is entitled, "Distrust of Energy Markets Leads to Policy Failures" and I would like to quote from this paper or article at some length, as follows:

> "The collapse of the iron curtain has brought home the dismal failure of economic central planning. It has also emphasized, in sharp contrast, the power of the marketplace to produce economic growth and improve standards of living, including environmental gains. Why, then, do we see fresh proposals to centrally orchestrate a U.S. energy policy?...

One explanation for this apparent paradox is the lack of understanding of the shortcomings of previous government attempts to displace energy markets. This report focuses on the ineffectiveness of government control of petroleum and natural-gas resources...

Reviewing the regulatory history of energy markets bolsters confidence in free markets and private property and recommends a healthy skepticism of interventionist energy policies. Letting the market work is a far more efficient way of meeting the energy needs of American businesses and consumers....

Although economists have developed theories of the optimal rate of extraction of exhaustible resources, no critical resource has ever been exhausted....

In 1980, the U.S. Geological Survey (USGS) of the Department of Interior reported:

> It is clear that considering the minimum quantity of petroleum in the United States as $15,000,000,000 barrels, and continuing the present rate of increase in production, the supply would be exhausted by about 1985."

Apocalyptic forecasts of impending shortages of natural resources and the economic crises that would ensue provide the rationale for government intervention in natural-resource markets...

An ambitious program dubbed "Project Independence" was begun in 1974. The program's objective was to give the nation "the capacity for self-sufficiency" in energy by 1980. The 800-page Project Independence report was prepared by a phalanx of 750 energy "specialists" in government and in private consulting firms. It compiled the results of ten hearings conducted around the country that produced testimony from almost a thousand witnesses...

The Wall Street Journal summarized the study's options for "curing" the nation's energy problems:

> Subsidize new synthetic-fuel industries. Uncle Sam would purchase their high-cost output, or support their prices, and stand ready to bail them out if they fail...
>
> Erect an oil-tariff wall around the U.S. to protect new high-cost domestic energy industries from being undercut in the future by cheaper foreign oil...even though it is the sort of trade protectionism that the U.S. has criticized Europe's Common Market for practicing.

Store huge quantities of crude oil against a future cutoff of foreign supplies.

Concentrate on long-range conservation measures. High investments in mass transit and renewal of inner cities would "balance the energy budget" by cutting demand instead of boosting supply...

These four "options" stressed governmental solutions to energy problems and gave the private sector no significant role. In this way, the nation's energy difficulties, which were exacerbated rather than eased by government policy, were used as a rationale for massive government spending and pork barrel projects....

The energy crises began to ease during the late 1970's, not because of the success of the policies developed by the politicians and bureaucrats, but because of the operation of the laws of supply and demand.. Higher world prices for crude oil encouraged exploration and development as well as conservation; consequently, supplies increased and demand diminished....

Yet, the Department of Energy is still active in issuing energy warnings. The Energy Information Administration (EIA) has managed to have Congress declare October "Energy Awareness Month". It is interesting that the EIA is firmly convinced that the nation's citizens are not aware of the importance of energy, despite the fact that consumers are constantly reminded of the cost and importance of energy every time they pay a utility bill or fill up their automobiles at a gasoline station.

Despite the overwhelming evidence that markets are vastly more efficient at allocating resources than are government bureaucracies, some within the federal government continue to berate the market and to declare an "urgent need" for "national energy policy," i.e., to replace private market decision making with bureaucratic controls and mandates.

The history of government intervention in petroleum and natural-gas markets is a tale of one bureaucratic fiasco after another. Price and allocation controls created shortages, inequitable distribution schemes, and higher prices.

The Persian Gulf War also provided a valuable energy policy lesson. When the petroleum market responded rationally to the invasion of Kuwait in August 1990, with price increases that reflected the increased uncertainty of future supplies, there were calls by members of Congress and others for price controls, antitrust suits, and even the nationalization of the oil industry. Thankfully, these wild reactions were ignored, and by January 1991, when it became apparent that the gulf crises would soon end, oil and gas prices were at or below their preinvasion levels. The market handled an international crisis in the petroleum industry with no shortages, no inequitable government allocation schemes, and only a temporary increase in prices.

Unfortunately, such lessons were all but lost on many members of Congress and other policymakers. When President Bush announced his energy policy initiatives during his January 29, 1991, state-of-the-union address, he was harshly criticized by a number of legislators, bureaucrats, and environmental activists who advocate a government-controlled energy policy. The basis of this criticism was that the president's policy purportedly relied too heavily on market forces and too little on governmental mandates and controls.

Advocates of greater governmental control of petroleum and natural-gas markets can be expected to continue to spread anti-market propaganda and to offer more bureaucracy as the "solution" to volatile energy markets. Let the voters beware of these "bureaucrophiliacs," who remain stubbornly uninformed by the lessons of history and immune to economic common sense."

I believe that Professor DiLorenzo is right when he says that markets are vastly more efficient in allocating resources than are government bureaucracies and I further believe that the key to getting hydrogen technologies into the market place in a commercially and economically feasible manner as the most reasonable and cost effective way of helping to solve both our energy and our environmental problems is to put that technology in the hands of entrepreneurs.. While many people are sitting around waiting for the evolution and development of a coherent federal government energy policy in the U.S. and complaining about the lack thereof, entrepreneurs such as Dr. Billings are already at work solving these problems, and they have been so engaged for years.

What is an entrepreneur, you might ask. An entrepreneur is one who assumes the risk, financial and otherwise, of undertaking a new business venture. Within the context and for the purposes of this presentation, I would like to suggest that there are two basic types of entrepreneurs -- those who operate independently within the market system and those who work for other companies and organizations -- such corporate or organizational entrepreneurs are often referred to as intrapreneurs.

Such entrepreneurial types are visionary, intuitive, creative, innovative, risk taking, hard working, persistent, determined people who understand, intuitively and otherwise, how to get new technologies into the market place. They understand and they have had experience with new startups, buyouts, franchising, acquisitions and mergers, joint ventures and strategic alliances, licensing agreements, financing, marketing and sales strategies and the myriad of other issues, problems and strategies involved in bringing technologies to the market place. The key then is to get hydrogen and other types of technologies into the hands of these entrepreneurs.

One such idea for doing that is the Science Information System (SIS) currently being proposed by the American Academy of Science.

Shown below (and on the overhead) is a schematic of this proposed SIS System.
The purpose of the SIS System is to provide a low-cost, easy-to-use and effective method of transferring technology from its source to industry.

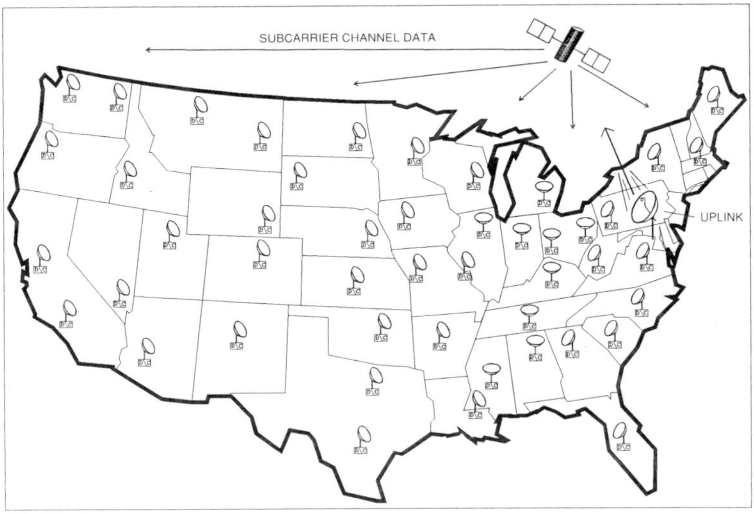

In the Science Information System, up-date information will be broadcast daily from the main up-link station via a single, narrow channel, over a satellite radio communication system, to each of the SIS Data Center Locations. Through the use of Functionally Structured Distribution, thousands of Data Center Locations could operate in the System, giving access to technology information to literally millions.

I have suggested to Dr. Billings and others at the American Academy of Science on several occasions that the down link SIS Data Center Locations should be located at Universities where Entrepreneurship Programs are found, for example, as one way of placing such technologies at the fingertips of entrepreneurs.

The Science Information System, when fully implemented, will be a satellite-based computer network based upon Functionally Structured Distribution (FSD). The plan calls for the installation of 500 Data Centers.

Other suggestions that I have, aside from implementing such a system as an SIS System, for placing such technologies at the fingertips and at the disposal for entrepreneurs include the following:

(1) Capital formation dedicated to alternative energies and ecopreneurship needs to be encouraged amongst venture capital firms and other financing and funding schemes and programs.

(2) National and International Business Plan Writing Competitions need to be encouraged in areas such as hydrogen technologies, energy and the environment.

(3) Organizations such as the Association of Collegiate Entrepreneurs (ACE), Students In Free Enterprise (SIFE), the Association for Private Enterprise Education (APEE) and others need to sponsor awards, projects and competitions etc. targeted at solving our global energy and environmental problems.

(4) The International Association for Hydrogen Technology should establish an Entrepreneurship Division.

(5) A Journal, perhaps entitled New Hydrogen Technologies, should be published perhaps by the International Association for Hydrogen Technology.

(6) Spin-off applications for hydrogen technologies need to be aggressively sought and pursued.

(7) Entrepreneurship Programs all across the U.S. and around the world need to be refocused away from pet rocks, fast food, fads, poster and T-shirt type entrepreneurial ventures and toward entrepreneurship focus on solving societal problems such as energy and the environment and the same thing is true of organizations such as ACE, SIFE, APEE, USASEE, ICSR, and others.

(8) Public policy and government intervention strategies need to be refocused on enabling technology based entrepreneurs and technology transfer in areas such as hydrogen energy technologies.

(9) More cross-disciplinary and inter-disciplinary training and education is needed at colleges and universities and elsewhere between engineering, the sciences and business -- especially in the form of technology based entrepreneurship.

(10) Large corporations and organizations need to become far more entrepreneurial, innovative and creative with regard to finding applications for hydrogen technologies and far less concerned with protecting their existing technologies.

These then are a few of my thoughts with regard to "The Role of the Entrepreneur in Bringing to Pass Our Sustainable Hydrogen Energy Future" and with regard to how that might most effectively, efficiently and judiciously be accomplished.

I notice that my time is up. I could spend all day speaking on this topic because it is a topic of great interest to me and of great importance to the world that we live in. But I must stop now and allow a few minutes for a question and answer period.

Before I open this session up to the floor for questions, however, I would again like to thank Dr. Billings for inviting me here to speak. My only hope is that the thoughts that I have shared here will be in some way helpful in advancing the use and application of hydrogen and hydrogen technologies in helping to solve our global energy and environmental problems.

I would be glad to answer any questions that you might have on my presentation or entrepreneurship in general for the next five or ten minutes.

Thank you.

HYDROGEN PROJECTS RUNNING IN EUROPE 1991 -- APPLICATION REPORT

OLOF TEGSTROM
P.L. 1575
S-870 15 UTANSJO
Sweden
Tel. 0611-64150

Application research is one of the most important steps towards an introduction to society. As many countries in Europe have realized the damage of the increase of carbon dioxide (CO_2) in the atmosphere, they have introduced taxes on CO_2. That makes a balance in the market economy between the real fossil fuel bill and clean fuels like hydrogen. We have to realize the fact that the CO_2 today increases 2 ppm/year, the way we live. That means that in 20 years we have increased another 40 ppm. Another fact is that the time-constant in changing the market is at least 20 years. So it was when we introduced electricity, refrigerators, washing machines, etc. It is therefore important that we, as soon as possible, introduce the carbon dioxide-free fuels. In Europe the bill for fossil fuel damage has passed the price of hydrogen production from non-fossil fuels, which I will show at the end of my speech.

Another very important step in economy of hydrogen has been taken by Professor Michael Gratzel, Lausanne Switzerland, which has developed the new solar cell, based upon Titandioxide (cheap colour material) instead of the 50 times more expensive silicon material. Germany and Switzerland give already up to 70% state support to the introduction of these non-fossil generators. That means that the market increases rapidly in these countries. That is also why we can report seven interesting application projects which will give us all statistic material we need to change the energy market.

P. Number 1 to report is a Swiss project running for three years. Here we produce hydrogen from 60m² solar cells every day by electrolyzing normal rainwater. The solar cells are placed on the roof of the family house. Markus Friedli is the owner in Zollbruck and the family uses the hydrogen regularly for washing machine, house cooking, heating and the family car (a converted Toyota gasoline car).

His results showed the way to a new society module, called Module 2000 FH (F= family size and H= hydrogen domination). His experience has lead to a competence in society modules in the range of architecture - energy. Markus Friedli AG.

Q. Number 2 in my presentation is "Bayernwerk" Germany, which is 50 times bigger in hydrogen production. It creates the module for smaller villages (called Module 2000 VH). That means that 50 families in villas or bigger houses can live in highest living standard with hydrogen. Here we demonstrate the current production by fuel cells, heaters and liquid hydrogen for cars.

S. Number 3 is "Fachochscule-Wiesbadan" Germany that since five years produces hydrogen from wind. The experience from this type of source shows that an optimum is reached when we combine solar and wind, depending on where on earth we live. For coastal places it could be 50% of wind energy and 50% of solar energy. Here we find most every kind of application like hydrogen for lamps, current production by fuel cells, heaters, current production by motor-generators which eliminates the normal diesel-generators.

R. Number 4 in my series is from Triesdorf in Germany, where solar panels are giving current for electrolysis of water. Here we find the use of the 2 elements in water: Oxygen, which is pumped into a water culture system (fish production, etc.) and the hydrogen for cooking and heating.

T, U, and V are parallel projects, which complete the important statistical material when we introduce new society modules. They also complete the economy curve for hydrogen production (X). It shows that, already when we replace 10.000 m^3 oil/gasoline/diesel by hydrogen from wind and solar, we achieve a balance in the direct society economy of today. Within 10 years time we will find the hydrogen from sun and wind to be the "best oil-quelle" we ever had on this earth.

PHOTOVOLTAIC CELL

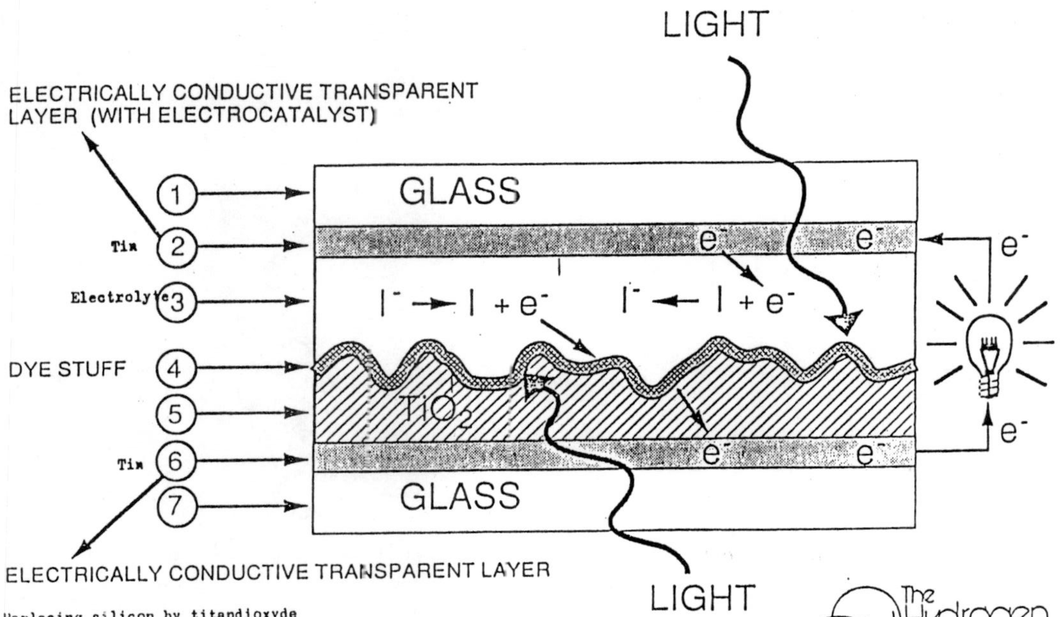

Replacing silicon by titandioxyde is the secret behind the price-reduction factor 50.

M. Graetzel 1989 (EPFL)

The Hydrogen Organization

Historic review:

Population - carbon dioxide (CO_2) - world energy consumption.

TEBETRON

Pl. 1576
S-870 16 UTANSJÖ Sweden

880830/0Tm

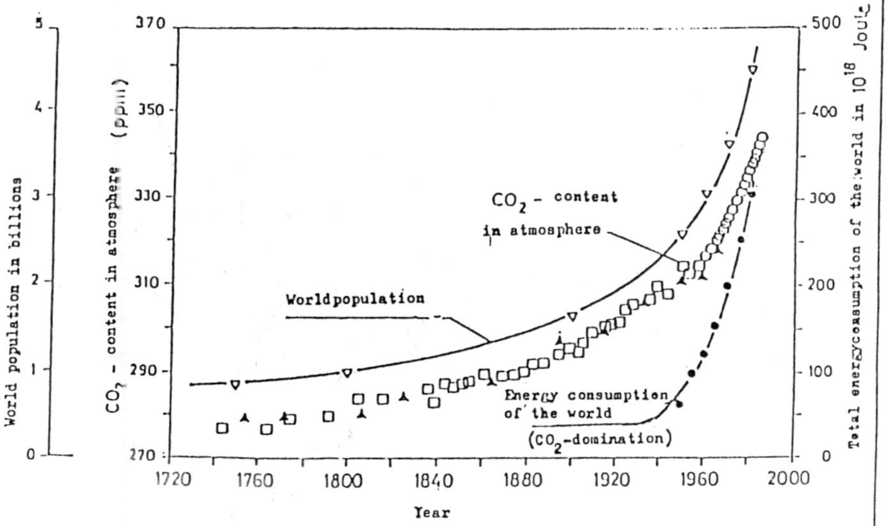

○ Measured CO_2- content at Mauna Loa, Hawai.

□ ▲ Measured CO_2-content confirmed by The University of Bern (drilling hole in the ice of Antarktis)

▽ Worldpopulation

● world energy consumption

} World energy conference Munic 1980.

4.

5.

% damaged or dead woods in Europe 1988

Summary (in English) of the TACTH Report 1984:2,

S Olof Gunnarsson
Ingemar Leksell

APPROACH TO AN ESTIMATION OF THE SOCIAL COSTS FOR HEALTH AND ENVIRONMENTAL IMPACTS, CAUSED BY MOTOR VEHICLE EXHAUST EMISSIONS IN SWEDISH URBAN AREAS

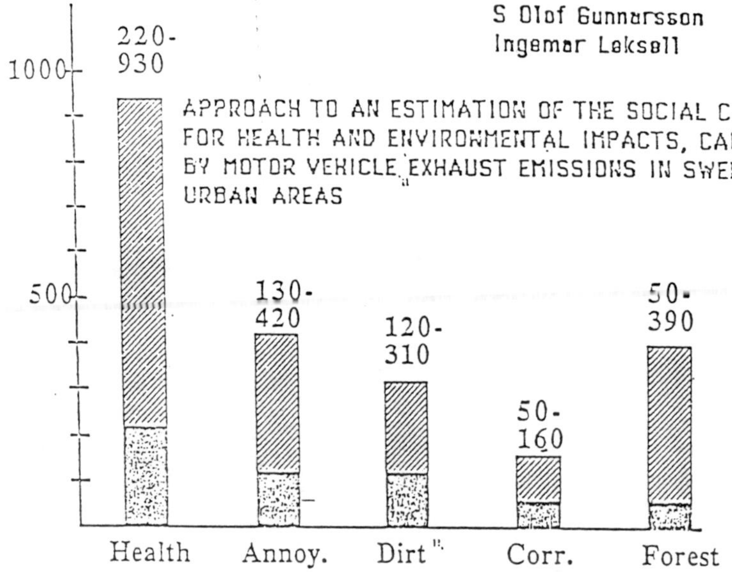

Fig. 1. Motor vehicle emissions in Swedish urban areas. Estimated costs 1982/83. Mill. SEK.

*) diesel-powered

Fig. 2. Motor vehicle emissions in Swedish urban areas 1982/83.
Marginal costs in SEK per 10 km.

Study TACTH 1987:6
All prices have doubled last 4 years.
1989 – 6 SEK/lit. Diesel
– 10%/year over inflation

"EUROPE MODULE 2000."

IN HYDROGEN SOCIETY

Markus Friedli AG
Energiesysteme

Längmattstrasse 32
CH-3436 Zollbrück

P

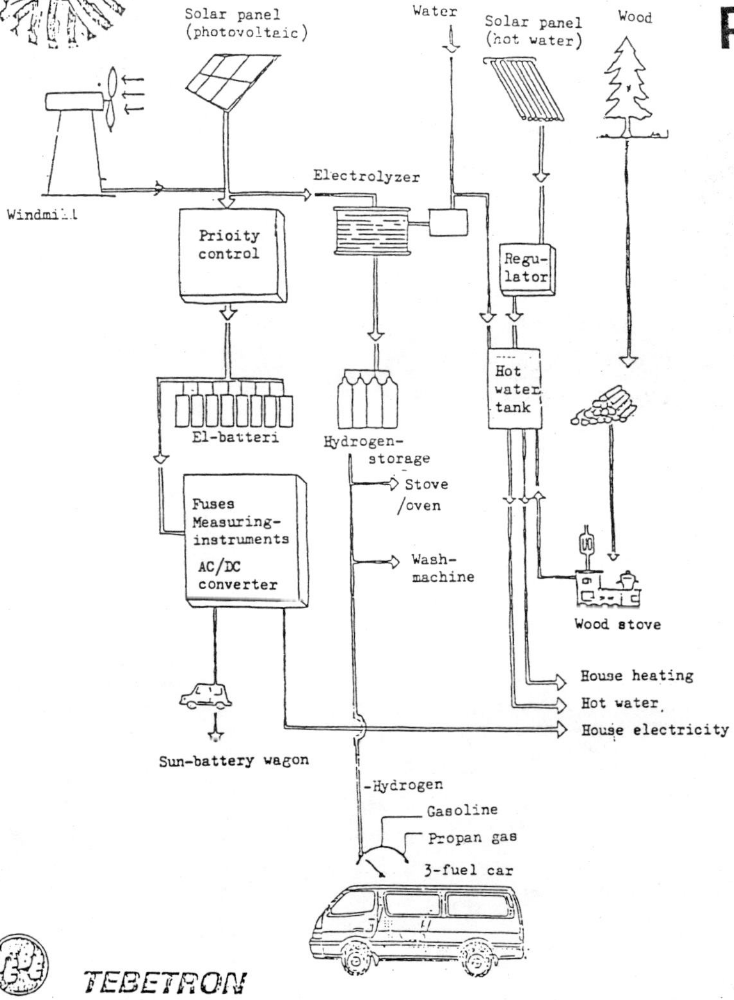

TEBETRON

P.L. 1575 S-870 15 UTANSJÖ Sweden Tel. 0611-641 50

910516

WORLDS FIRST HYBRIDE CAR: 3-fuel system HYDROGEN-GASOLINE-PROPAN

Markus Friedli AG
Energiesysteme
CH-3436 Zollbrück

P

Hydrogen car - TOYOTA

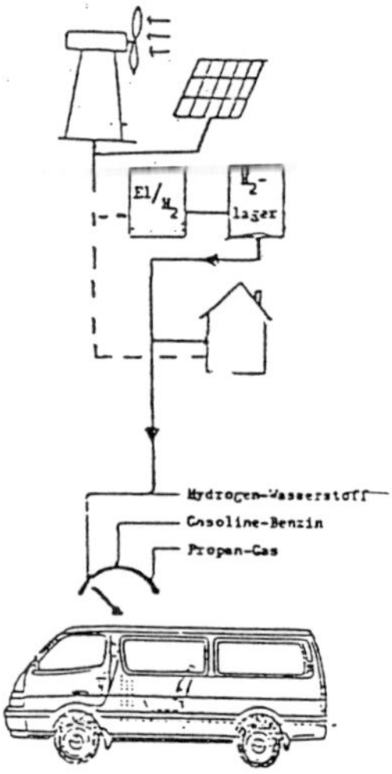

3-fuel system
Switch control.

Type	:	Toyota HIACE Kombi 4x4
Reg. Nr.	:	BE 352088
Chassie Nr	:	JT121UHB900000285

KFZ 2 x 80 kg Hydride V-Fe-Ti Mannesmann.
Loadpressure: 24 hours 30 Bar without extra cooling. (ca. 50% tank)
Hydride heating from ordinary motorcooling water circuit.
Startcondition: 30 Bar 26° C and ending 0 Bar 83° C
Injection of water .6 liters/10 km

Waterinjection curve:

Flat way drive : 0,6 lit./10 km
Mixed drive : 1,0 lit./10 km
Hard drive : 1,5 lit./10 km
Waterinjector: ordinary gasolinepump and gasolinecarburator.

Storage system	2 low-temperature hydride storage units, fluid-heated 280 kg. petrol equivalent 12 litres
Operating range	255 km at constant speed of 50 km/h 120 km at constant speed of 120 km/h 55 km in urban cycle

Hydride Storage Tank (by Mannesman

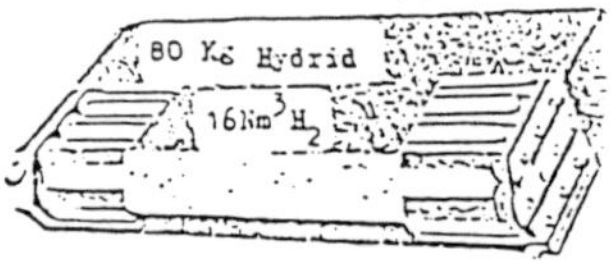

The tank capable of starting minus 29° C. (H > 2 Bar)

De flesta bensinbilar kan byggas
om enligt denna princip till VÄTGAS!

TEBETRON
Pl 1575
S-870 15 UTANSJÖ
SWEDEN

Tel.
Nat. 0611-641 50
Int. + 46 — 61164150

Telex:
71080 HAERNOS
Att. TEBE

Telefax:
61115900
Att. TEBE

SOLAR-WASSERSTOFF-BAYERN GmbH

SWB

8000 München 80
Rosenheimer Straße 143b
Telefon: (089) 41 86 98-0
Fax: (089) 41 86 98-23

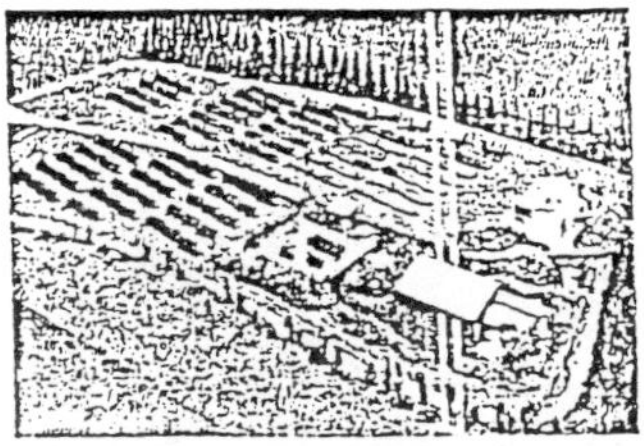

Model of the solar hydrogen plant in Neunburg vorm Wald, FR Germany

Courtesy, Solar-Wasserstoff-Bayern

Generating Solar Hydrogen:

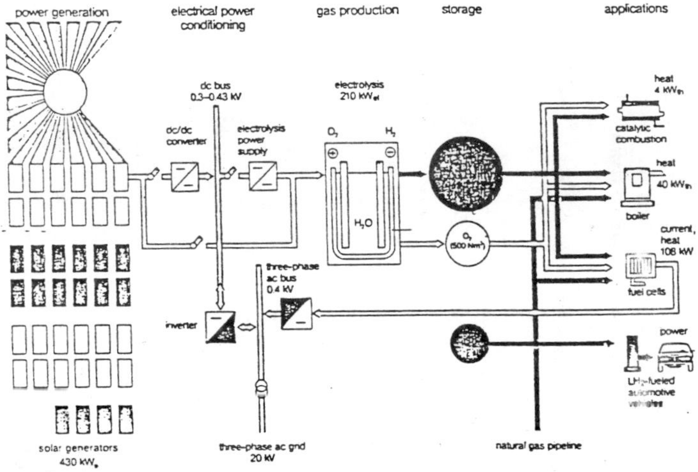

Diagram of the solar hydrogen plant and possible applications

The objective of this long-term project is to discover reliable parameters for evaluating the chances for a solar hydrogen energy economy under the solar conditions of central Europe. Phase 1 of the project runs until 1991. It is planned to start in the first half of 1990.

The design details: 430 kW_p photovoltaic generators, monocrystalline, polycrystalline and in the future amorphous silicon cell technology, in most cases non-tracking. Diaphragm cell electrolysis with solid electrolyte, advanced alkaline electrolysis; total output 210 kW_{el}. Hydrogen/oxygen pressurized storage (30 bar, 5,000 m^3 for hydrogen, 500 m^3 for oxygen). Four fuel cells (two alkaline, two phosphoric acid) with 108 kW_e total output. Two catalytic burners (total 4 kW_{th}), two boilers (total 40 kW_{th}). One liquid hydrogen filling station. The transition phase to a hydrogen energy economy will be investigated assuming the parallel use of natural gas and hydrogen.

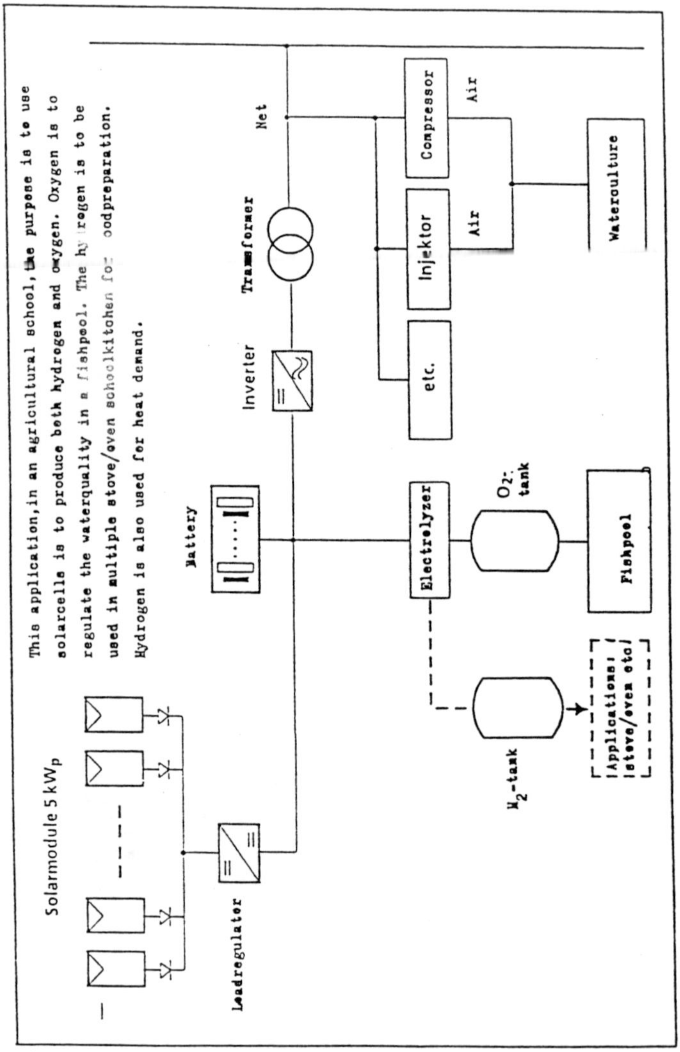
This application, in an agricultural school, the purpose is to use solarcells is to produce both hydrogen and oxygen. Oxygen is to regulate the waterquality in a fishpool. The hydrogen is to be used in multiple stove/oven schoolkitchen for foodpreparation. Hydrogen is also used for heat demand.
Solarmodule 5 kWp
Leadregulator
Battery
Inverter
Transformer
Net
Compressor
Injektor
etc.
Air
Air
Waterculture
Electrolyzer
O_2-tank
Fishpool
H_2-tank
Applications: stove/oven etc.

The Hydrogen in Europe today-open society applications.

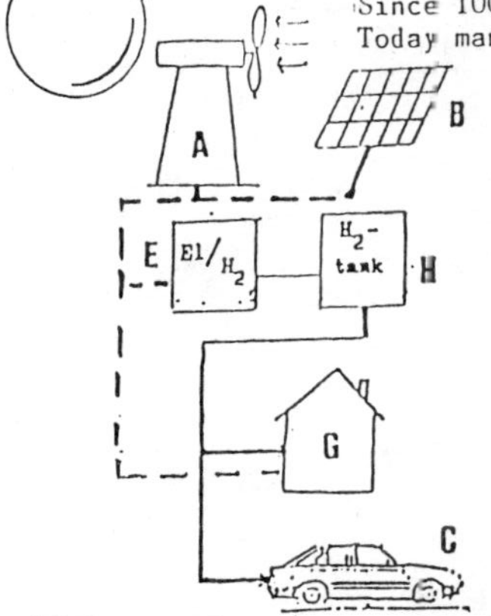

Since 100 years we have had HYDROGEN in society, but we called it citygas(60% H). Today many society moduls are running in open society. YOU CAN LOOK FOR YOURSELF !

Here you will find the addresses and the experience for houses and cars, which prove the ecological solution for the world. New solarcells, producing electricity for less than 3 cent/KWh.
Many countries in Europe have introduced enviromental taxes on CARBONDIOXYDE-, NOX-exhaustto force the fossil-fuel consumers to pay their bill to the nature.
Reports 1990 in Sweden,USA and Germany show that the fossil fuel bill in city-areas is >1 US$/liter diesel and gasoline and the increase >7%/year above inflation.

Result: Hydrogen from solar and wind is cheaper than gasoline+ pollution bill.

OLOF TEGSTRÖM
PL 1575
S-870 15 UTANSJÖ
Sweden

A Windmill
B Solarcells
E Electrolyzer
H Hydrogen storage
-1 hydride tank(Nm^3/Bar)
-2 Liquid H_2-tank (Nm^3)
-3 pressure tank(Nm^3/Bar)

G Applications:
-1 Stove/oven
-2 Heater(KW)
-3 Gas lamps
-4 Fuel Cell (KW_{el})
-5 Motor-generator KVA

C Car system:
-1 Hydryde tank(Nm^3/Bar)
-2 Liquid H_2-tank(Nm^3)
-3 Pressure tank(Nm^3/Bar)

MODUL-INDEX	Start year	A øm/KW	B m^2	E Nm^3/h	H (Nm^3/Bar) -1	H -3	G (H_2-use) -1	G -2	G -3	G -4	G -5	C (motorappl.) -1	C -2	C -3	remarks	For project investors: Contact person, Address, Tel. Fax.
P	1990	-	60	2	19/30		X	4				32/30			Toyota	Markus Friedli, Längmattstrasse, CH 3436 ZOLLBRÜCK, Schweiz. F-413568160
Q	1990	-	3000	45		5000/30		4+40		86			X		BMW	Axel Szyscka, SOLAR-WASSERSTOFF-BAYERN D-8000 München 80 FRG, F-4941869823
R	1991	-	50	1		10/30	X	1+1								Holm Schupert Klempnauer, BOLKOW-STIFTUNG, TRIFSDORF, Daimler Str.15 D-8012 OTTOBRUNN, FRG Tel.4989609703 1
S	1988	12/20	20	4		175/30	X	4+6	X	2	8,5+4				Opel Briggs&S	Prof.S.Schullen, Fachhochschule Wiesbaden, Heinrich-Lerch-Str.1 D-6090. Rüsselsheim FRG, Tel.49614268606.
T	1990	-	3500	52		Capacity test										Hartmut Steeb, HYSOLAR PROJ. DFVLR, Pfaffenwaldring 38-40 D-7000 STUTTGART 80 FRG, Tel.49 7116862477.
U	1988	-	100	2		Capacity test										
V	1986	9/5	58	0,2		30 /120				1						Anne Maan, Univ.of Oldenburg, Box 2503, D-2900 OLDENBURG FRG. Fax.49 4417983201. Tel 441 7983541

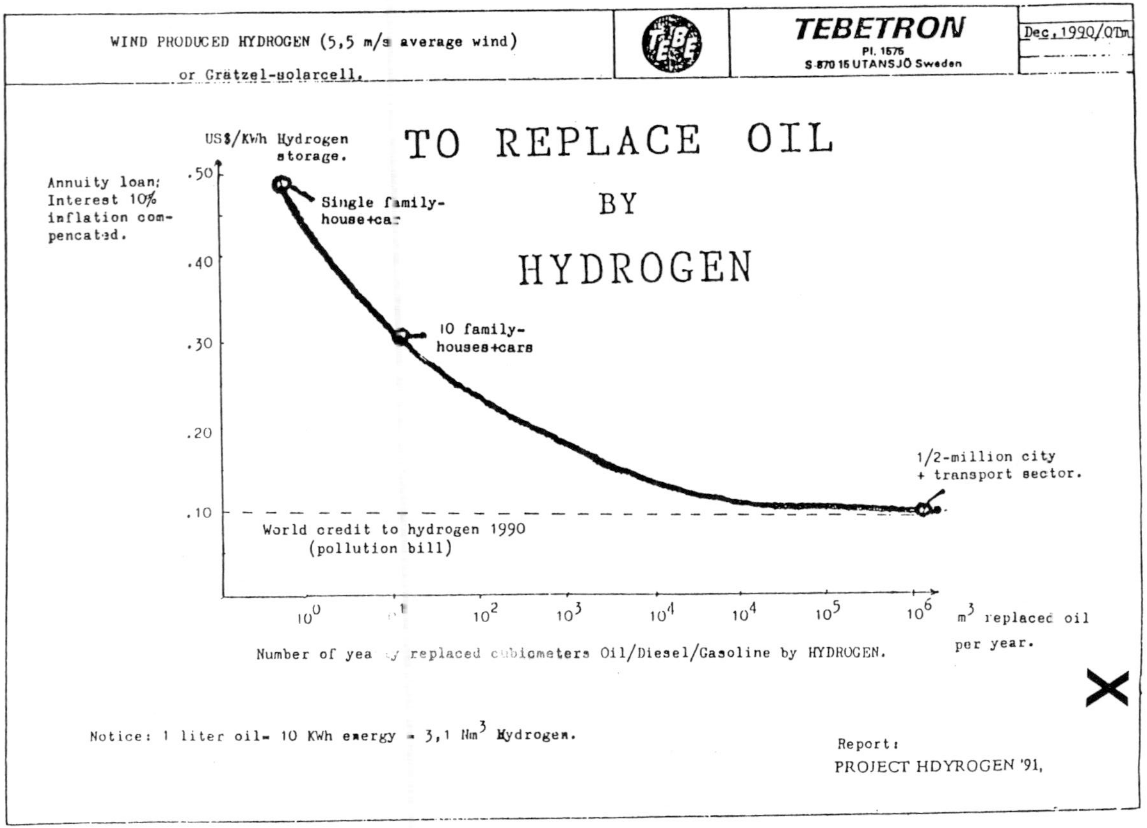
WIND PRODUCED HYDROGEN (5,5 m/s average wind)
or Grätzel-solarcell,
TEBETRON
Pl. 1576
S-870 15 UTANSJÖ Sweden
Dec. 1990/QTm
TO REPLACE OIL
BY
HYDROGEN
US$/KWh Hydrogen storage.
Annuity loan; Interest 10% inflation compencated.
.50
.40
.30
.20
.10
Single family-house+car
10 family-houses+cars
1/2-million city + transport sector.
World credit to hydrogen 1990 (pollution bill)
10^0
10^2
10^3
10^4
10^4
10^5
10^6
m^3 replaced oil per year.
Number of yea y replaced cubicmeters Oil/Diesel/Gasoline by HYDROGEN.
Notice: 1 liter oil= 10 KWh energy = 3,1 Nm^3 Hydrogen.
Report:
PROJECT HDYROGEN '91,

HYDROGEN AS AN ENERGY CARRIER WITH FOCUS ON ELECTRICITY STORAGE

Lars Henrik Nielsen and Lotte Schleisner
Risø National Laboratory, Systems Analysis Department
DK-4000 Roskilde, Denmark

ABSTRACT

In Denmark the utilization of hydrogen as an energy carrier has been considered especially regarding hydrogen as an electricity storage facility. The purpose has been to investigate the efficiency of an electrical storage system today and in the future based on hydrogen.

The energy flows of an electricity storage system based on hydrogen are described as expected today and in the future. Based on these efficiencies and on the expected development in price for each of the components in the electricity storage system the price per kWh electricity passing the store is estimated.

INTRODUCTION

Today there exists a well-known need for energy storage in large scale in the electrical energy system. This need is in many countries met by the use of pumped hydroelectric stores. However, an optimal utilization of energy stores requires development and implementation of new and more broadly applicable storage technologies.

The variation of load throughout the day, week or year imposes a demand for storage especially with the increase in use of large power plants designed to operate at maximum efficiency on base load. Also in connection with the increase in utilization of variable energy sources such as solar, wind, wave energy etc. and in the case of combined heat and power production there is a need for energy stores.

Hydrogen is often mentioned as the coming energy carrier and storage medium for the next century. Hydrogen may be extracted from water by use of any kind of primary source of energy, and may be converted to water again in a closed chemical cycle without emissions to the environment.

THE DANISH PROJECT CONCERNING HYDROGEN AS AN ENERGY CARRIER

The project titled "Hydrogen as an energy carrier" started in January 1991. The project is financed by the Danish Ministry of Energy and by the Danish public utilities ELSAM and ELKRAFT. The purpose of the project is to gather and effect the research which has been going on around the world until now concerning hydrogen, as well as estimating the perspectives for production of hydrogen and utilization of hydrogen as an energy carrier in the Danish energy system.

A special interest for the Danish public utilities has been the possible utilization of hydrogen as a storage medium for electricity, as in Denmark there exist very poor conditions for electricity stores in the shape of pumped hydroelectric stores. For that reason a first part of the project has been a study of the utilization of hydrogen as an electricity store and an investigation of the energetical and economical consequences. The study has been based on literature surveys and, in some cases, on personal communication. The description of the different technology processes for production, storage, transportation and utilization of hydrogen refers to already demonstrated technology. However, in some cases the technology has not yet been proven, and the efficiencies should be viewed as theoretical estimates.

THE SYSTEM ANALYZED FOR ELECTRICITY STORAGE

The system analyzed for electricity storage based on hydrogen consists of an electrolyzer, a hydrogen store and a fuel cell as shown in Figure 1. The electrolyzer produces hydrogen (and oxygen) based on electricity. Hereafter the hydrogen is compressed, cooled and led to the hydrogen store.

The store in this example is assumed to be utilized only as a seasonal store. As the amount of hydrogen stored in this case will be large the store chosen for hydrogen storage is a cavern. Caverns in saltdomes are the best way of storing large amounts of pressurized hydrogen as well technically as economically.

After the period of storage hydrogen is recovered from the store, expanded and converted to electricity and heat in fuel cells, whereby the storage cycle is closed.

In the case analyzed it has been assumed that all the different phases of the system are in progress at the same locality, and transmission of hydrogen and electricity has therefore not been included.

Although the store has been dimensioned for one year, the store can be used for smaller cycles too, for instance as a day and night store.

ENERGY FLOWS

The efficiencies for each of the energy processes in the system are given in Table 1. The table shows the efficiencies for the technologies today and the efficiencies which can be expected in the future. As well during the electrolysis process as during the conversion from hydrogen to electricity in the fuel cell system an amount of heat at a temperature of about 100°C will appear, which may be utilized for district heating.

Based on the efficiencies given in Table 1 the total electricity efficiency by passing an amount of electricity through the system (electrolyzer-compression-storage-expansion-fuel cell) can be estimated.

Energy Conversion Today

The overall conversion efficiency of the total system today for electricity is calculated to 53%. The link with the largest loss of electricity in the energy chain is the fuel cell with a loss of 35% of the total electrical input to the system.

It is assumed that 90% of the heat loss during electrolysis and during conversion of hydrogen in the fuel cell will be a potential heat source for district heating. Hereby the heat effciency of the system is calculated to 40%.

The total efficiency (electricity and heat) of the storage system based on the technologies today is 93%.

Energy Conversion in the Future

The energy flows through the system in the future are expected to be as shown in Figure 2.

The amount of incoming electricity to the system is set to 100% at the upper part of the energy tree. The dark flow going downward at the figure shows the relative amounts of electricity or hydrogen passing through the different processes (hydrogen production, compression, storage, expansion and electricity production) in the system.

The flows to the right show the energy losses by each process. The amounts of heat losses which may be at a temperature level usable for heating are shown as incoming flows from the left to the energy tree. The accumulated amounts of heat are shown as an energy flow going downward in the tree like the flow of electricity/hydrogen.

During recovery of hydrogen from the store the hydrogen is expanded to the pressure level of the fuel cell. By expansion of hydrogen in a turbine it seems possible to recover about one third of the compressor work as electricity. This is shown at the figure as a negative loss of electricity.

The efficiency of the system working only as an electricity store is in the future expected to be about 59%. Of the incoming amount of electricity to the system 35% is expected usable for heating purposes. By utilization of as well electricity as heat from the system the total efficiency of the system is 94%.

By placing the electrolyzer and the fuel cell close to a district heating system the loss of energy seems very small by utilization of hydrogen as an energy carrier.

The efficiencies mentioned above are reached by utilization of alkaline water electrolysis, saltdomes in caverns as storage medium, and solid oxide fuel cells as the main components in combination with an existing power plant to utilize high temperature heat losses from the fuel cell. It seems possible to reach better electrical efficiencies by use of other kinds of components. The use of high temperature electrolysis, for instance, may increase the electrical efficiency. However, this requires the presence of an amount of surplus heat at a high temperature level.

ECONOMIC ESTIMATE FOR A FUTURE ELECTRICITY STORAGE SYSTEM

An economic estimate is given below illustrating the order of magnitude of extra costs related to seasonal storage of electricity in a future 10-15 years ahead.

Efficiencies and relative energy flows in the system where hydrogen is the energy carrier are assumed as shown on Figure 2. The estimated extra costs shown on Figure 3 are expressed in $/kWh electricity entering the storage system. The costs include investments, operation and maintenance, but do not include cost of input electricity to the system. Thus, zero costs of energy inputs are assumed in Figure 3.

A cost estimate based on Figure 3 for electricity drawn from the store, therefore must be added to the cost of the energy input. Furthermore the value of the related heat and oxygen produced should be subtracted.

Based on Figure 2 and 3 the additional costs by drawing 1 kWh electricity from the seasonal store at times of the year with a large demand of electricity appear to be about 0.10 $/kWh electricity. About 28% of the expenses derive from hydrogen production and production of oxygen and heat in the electrolyzer based on electricity, 15% of the expenses derive from storage and compression etc. of hydrogen, while 57% of the expenses are connected to production of electricity and heat in fuel cells.

Assumptions on the operation of the storage system

It is assumed that the store performs one charging/discharging cycle in one year. The capacity of the hydrogen energy store is accordingly designed to contain all hydrogen produced in one year. Thus a very large volume of the compressed hydrogen gas store is required.

The economic calculation is based on an installed capacity of 100 MW on the input side of the storage system. Other capacities in the system chain are based on the assumption that full load hours for the electrolyzer and the fuel cell part each cover 1/3 of the one year cycle period. In practice, however, the sum of charging and discharging periods on partial load may cover the total cycle period.

The relative capacities of the system parts - the input/output capacities of the hydrogen store and the installed fuel cell capacity - in this case therefore correspond to the energy flows shown on Figure 2.

Economic assumptions and data

Investments, operation and maintenance costs used are shown in Table 2.

The economic and technical data for the technologies in the system are based on different references listed in [1]. Some of the references of special importance for this paper are listed in the reference list. The composition of economic and technical data have been based on own estimates.

For the electrolyzer a lifetime of 20 years has been assumed. The operating costs, that is utilities, operation and maintenance, have been set to 10,3% of the investments, which is a rather large amount compared with other technical plants.

The hydrogen store investment is expressed per m^3 storage volume and this estimated overall investment applies for solution-mined caverns in a saltdome. In Denmark caverns in saltdomes are used as stores for natural gas.This underground store is assumed to be a dry compressed gas variable pressure store (without brine-compensation).

To stabilize the cavern basegas is required to ensure the pressure in the cavern staying above a lower pressure limit. The amount of initially lost hydrogen basegas is of the same order of magnitude as the working volume of the store. This initial energy investment is not included in Figur 3. Costs of basegas in the economic calculations can roughly be included by adding about 5% to the cost of electricity during charging of the storage system for seasonal storage.

The costs of the hydrogen store are based on the costs of storing natural gas in caverns in Denmark. The lifetime of the hydrogen store has been set to 30 years based on the technical equipment. However, the saltdomes in caverns itself in all probability will have a lifetime above 50 years for storing hydrogen.

It is assumed here that the future investment cost of a highly efficient fuel cell - where hydrogen is the fuel - is about 770 $/kW installed. Relatively high yearly costs for operation and maintenance are expected and the lifetime of the total fuel cell system is put to 10 years. The major part of the initial investment covers well known technology having a substantial longer lifetime. For the fuel cell stack itself a relatively short lifetime has been chosen.

Calculated levelized costs per year for the system components are based on an interest rate of 5% p.a. and the assumed economic/technical lifetime of each component as shown in Table 2. The levelized costs are expressed in $/kWh electricity input to the system, and the accumulated costs through the technology chain of the system are presented in Figure 3.

To interprete Figure 3 assume as an example a price of electricity at low load of say 1 c/kWh in average during the charging period of the storage system. Including this cost of electricity input to the system the total costs add to about 7 c/kWh (electricity input). Seen on Figure 2, the electricity output when storing 1 kWh is 0.59 kWh, and therefore the cost of electricity drawn from the store is about 12 c/kWh (electricity output) minus the potential value of the heat and oxygen generated in the processes.

Such resulting costs of electricity can be compared with the marginal production cost of electricity in future energy systems in periods of high load of electricity.

The possible advantages in the total CHP system which may appear by combining an electricity storage facility based on hydrogen to the system have not been considered. E.g. it is expected that an electricity storage system can substitute power installed elsewhere in the system.

FINAL REMARKS

Considerable uncertainty is connected to the estimated technical and economic future development of the technologies that constitute our system. Furthermore, strongly simplified operation conditions and system integration has been assumed. Results therefore only reflect a first order assessment.

The operation conditions for e.g. the fuel cell part of the system, where pure hydrogen is the fuel, raise expectations for very high efficiencies. Furthermore, much lower cost figures and longer lifetime estimates for future fuel cells are seen in the literature that will project costs for the fuel cell part of the system about half of the costs presented in Figure 3.

It should be emphasized that this analysis does not include benefits related to the possibility of simultaneous utilization of a seasonal store for load management on cycle periods less than one year e.q. on a day/night basis. A more thorough system analysis is required to reveal overall effects of including an electricity storage facility in the energy system.

REFERENCES

1. SCHLEISNER, Lotte, NIELSEN, Lars Henrik, PEDERSEN, Allan Schrøder, KJØLLER, John (1991). Hydrogen as an energy carrier (focusing on electricity storage). Progress report (in Danish). Risø-M-2940. Risø National Laboratory, Denmark. 127 pp.

2. KASKE, G., SCHMIDT, P., KANNGIESSER, K.W. (1991). Comparison between high-voltage direct-current transmission and hydrogen transport. Marl, F.R.G. Int. J. Hydrogen Energy, Vol. 16, No. 2. pp. 105-114.

3. MINISTRY OF ENERGY (1990). Energy 2000: A plan of action for a sustainable development. Danish Ministry of Energy, Copenhagen. 115 pp.

4. FINNSTRØM, Bengt (1989). Hydrogen - a study of efficiencies. Swedish National Board for Technical Development, Sweden. Paper enclosed IEA Annual Report.

5. STEINBERG, Meyer, CHENG, Hsing C. (1989). Modern and prospective technologies for hydrogen production from fossil fuels. Upton, USA. Int. J. Hydrogen Energy, Vol. 14, No. 11, pp. 797-820.

6. WINTHER, C.J., KLAISS, H., NITSCH, J. (1989). Hydrogen as an energy carrier: What is known? What do we need to learn? Stuttgart, F.R.G. Int. J. Hydrogen Energy, Vol. 15, No. 2, pp. 79-91.

7. FUEKI, K. Professor Emeritus, the University of Tokyo, Chairman, WEC fuel cell study committee (1988). Fuel cell-past trends and future prospects.

TABLES

TABLE 1: ELECTRICITY EFFICIENCIES FOR THE TECHNOLOGIES IN THE ELECTRICAL STORAGE SYSTEM

	Electrical efficiencies	
	Today	Future
Electrolyzer (100 C, 10 bar)	91%	94%
Compression	96%	97%
Storage in Caverns	99%	99%
Expansion	101%	101%
Fuel Cell (1000 C)	60%	65%

TABLE 2: ESTIMATED FUTURE SPECIFIC INVESTMENTS AND YEARLY COSTS OF OPERATION AND MAINTENANCE

	Electrolysis	Solution-mined cavern in salt	Fuel cell
Investment	290 \$/kW el	43 \$/m^3	770 \$/kW el
Lifetime	20 years	30 years	10 years
Operation and maintenance % of investment per year	10.3	1.6	7.5

Price level: about 1990

FIGURES

FIGURE 1: THE SYSTEM ANALYZED BASED ON HYDROGEN

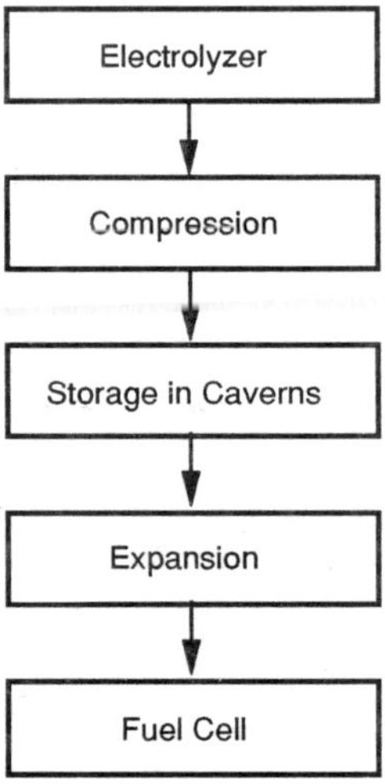

FIGURE 2: THE ENERGY CONVERSION EXPECTED IN THE FUTURE
IN AN ELECTRICITY STORAGE SYSTEM
BASED ON HYDROGEN TECHNOLOGY

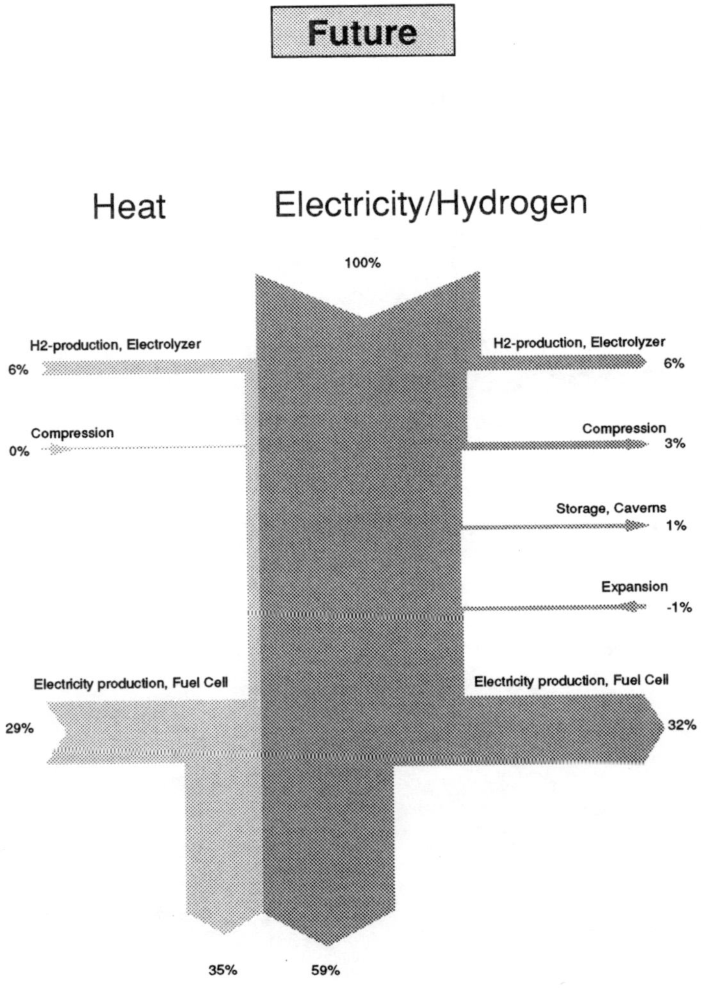

FIGURE 3: ACCUMULATED COSTS PER kWh ELECTRICITY DELIVERED TO THE STORAGE SYSTEM

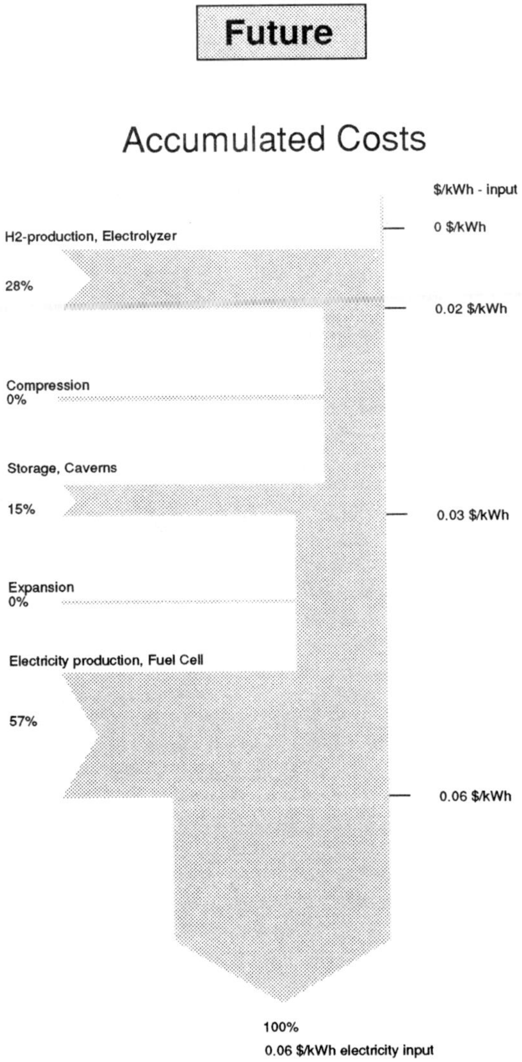

PERFORMANCE OPTIMIZATION OF SCRAMJET ENGINE WITH ACTIVE COOLING

Y. Tsujikawa, T. Onoda and S. Fujii
Department of Aeronautical Engineering
University of Osaka Prefecture
Sakai 591, JAPAN

ABSTRACT

The SCRAMJET engine can be regarded as a most promising one to propel a vehicle at hypersonic speeds to orbit. It simply consists of the external compression inlet, combustor and nozzle. In the analysis, there is most severe difficulties in the combustor due to the problem such as turbulent diffusion and mixing at supersonic speed range. For the combustor, the two dimensional flow model with coaxial injection of hydrogen parallel to the air flow is considered. In addition, the effect of the active cooling with slush hydrogen, which is a key aspect, has been analyzed.

INTRODUCTION

In WHEC# 8, numerous articles have described the goals, management structure and technical accomplishments of the NASP program [1]. Authors also have optimized scramjet engine geometrically [2,3]. It is regarded as a system which consists of a number of subsystems (components). The operating state is expressed by the parameters which govern such subsystems. The optimization of criteria, for example, specific impulse and thrust coefficient can be treated as non-linear programming problem. Using the multiplier method, the optimum values of parameters can be calculated in a short time.

The numerical calculation of the turbulent mixing and chemical kinetics of combustion reaction in the supersonic flow through a scramjet combustor is a computational problem of considerable interests. The introduction of detailed chemical reaction mechanisms into theoretical analyses of complex turbulent mixing flows often results in excessive computer run times and storage requirement. In the above study, therefore, some simplifications have been employed. A related concept involves the assumption that the fuel and oxidizer react instantaneously and form reaction products in chemical equilibrium. It was also shown that the completion of combustion is affected significantly by velocity difference between air stream and hydrogen jet and initial hydrogen temperature [4].

Hypersonic vehicle fly at trajectories that will subject the airframe and engine structures to extreme thermal, aeropressure and acoustic environment[5]. While traditional structural design approaches, such as external insulation and hot structure, apply to the future hypersonic vehicles the combination of both high peak heat fluxes and integrated heat load will require hydrogen active cooling for significant regions of the airframe and all of the engine flow path. The hypersonic vehicle thermal environment includes the most severe aerodynamic heating ever encountered by a flight vehicle, the shock interacting heating on the engine leading edges. Hypersonic engine leading edges may experience heat fluxes up to three orders of magnitude higher than the peak heat flux on the space shuttle nose during reentry. Other components that will require active cooling will be the entire engine structure and portions of the airframe forebody and aftbody.

As a fuel for aerospace application, the slush hydrogen is most promising because of its capability for active cooling. Figure 1 shows the statepoint of NBP LH_2 and of 50 percent solid content SH_2[6]. Relative to the heat of vaporization at one atmosphere we can have an enthalpy gain of 11.5 percent of sensible heat by going to the triple point(TP) and an enthalpy increase of as much as 18.1 percent of sensible plus latent heat by continuing to a 50 percent solid-liquid mixture. With regard to density, an increase of about 9 percent can be achieved by lowering the temperature of the hydrogen to the TP, or an increase of about 16 percent can be achieved by going all the way to the 50 percent quality statepoint in the solid-liquid region. Using the more dense SLH_2 rather than NBP liquid as the cryogenic working fluid will yield reduced vehicle size with a corresponding decrease in vehicle weight. The benefit has been quantified as being between a 13 and 26 percent saving in the take-off gross weight.

In the present study, the geometric dimension of the scramjet combustor is obtained, by including the effect of hydrogen active cooling with slush hydrogen represented by hydrogen injection temperature.

SCRAMJET ENGINE

The design of efficient scramjet engines, where the primary mode of combustion occurs in a supersonic rather than subsonic air streams, requires special consideration in a number of areas. The schematic presentation of the scramjet engine used in this study is shown in Fig. 2, and is a generic derivation of some recent scramjet powered vehicle concepts. For simplicity, and to be consistent with some of the engine configurations that have been proposed, it is also assumed that the scramjet inlet is located beneath the main body and sufficiently far aft that the effects of forebody compression are negligible. The external compression inlet is used. For simplicity, the air capture ratio is fixed to unity in the present analysis. Multi–oblique shock external compression inlet is adopted, which is also optimized by multiplier method using pressure recovery factor as objective function.

Combustor

The combustion chamber must be designed to match the mixing process which is gradual and stable because the flow is supersonic. The main requirements of a scramjet of this type is that the reaction rates which control the chemical reactions be very fast so that the mixing of fuel and oxidizer can be a much slower process and, therefore, be a rate-determining process. The turbulent mixing at supersonic speed range is affected by the method of fuel injection and accompanied with the complicated shock structure and intense interferences by air. The analytical study is very difficult, so that the data obtained from experiments must be fed back into the analysis. The two-dimensional (2D) axisymmetric flow model is made up. Solutions of parabolic partial differential equations for transport of momentum, energy and mass are accomplished by the finite-difference technique of Patankar and Spalding[7].

ACTIVE COOLING

Design concepts for actively cooled structures include forced convection, film and transpiration cooling techniques [8]. Early rocket design used grooved copper wafers to transpiration cool their combustor areas. In addition, film cooling was used in several technology development programs during the sixties. The early hypersonic programs and predecessors to the space shuttle engine development investigated forced convection concepts.

In the case of film and transpiration cooling, the coolant (hydrogen) is injected into outer flow. Therefore, in the present study, only the convective cooling are analyzed. The basic equations are obtained as:

$$\dot{m}_{H2} C_p (T_{out} - T_{in}) = \frac{\phi}{34.07} \dot{m}_{air} C_p (T_{out} - T_{in}) = \dot{Q}_T \tag{1}$$

$$h(T_W - T_{fluid}) = \dot{q}_{max} \tag{2}$$

$$\Delta P = (4f) \frac{L}{D_h} \frac{1}{2\rho} \left[\frac{\dot{m}_{H2}}{A} \right]^2 = P_{supply} - P_{exit} \tag{3}$$

In above equations, hydrogen temperature $T_{out}(= T_f)$ and the equivalence ratio ϕ are included into analysis.

RESULTS AND DISCUSSION

The combustion reaction is assumed to proceed entirely in supersonic speed range. Figure 3 shows the development of the velocity and temperature profiles along the duct. The jet reaches the pipe wall (RDW) about 59 jet diameters from the start and by reaching there the velocity profile looks generally like that in pipe flow. However the temperature is rather non-uniform there, more than 100 jet diameters combustion is virtually complete.

The completion of combustion is rather difficult to be decided. In the present study, it is determined as the axial distance where temperature profile has non–uniformity (difference between peak temperature and that of centerline) within 10 percent. The combustion efficiency is equivalent to a value of unburned hydrogen. The relation between two ways of estimation is shown in Fig. 4. Ninety percent uniformity is equivalent to unburned hydrogen of 2 percent. This location is regarded as completion of combustion (CC).

Figure 5 shows an influence of the equivalence ratio on the axial change of unburned hydrogen. A faster spread is expected for non–stoichiometric conditions. For large distance from the start, however, hydrogen remains at the vicinity of centerline, and is scarcely diffused.

The velocity ratio between hydrogen and air flow strongly affects the axial distances, such as EPC (end of potential core), RDW and CC. It is shown in Fig. 6. CC is found to decrease drastically with U_{air}/U_f.

The specific impulse versus combustor inlet Mach number is shown in Fig. 7. The optimum specific impulse is realized at $M_1 \approx 2.5$. It depends on the equivalence ratio.

Figure 8 shows the combustor length determined by the temperature uniformity of 90 percent. It increases keeping with combustor inlet Mach number. The completion of combustion is affected significantly by initial hydrogen temperature. For fixed equivalence ratio, hydrogen temperature is raised with increase of jet velocity. In the discussion below, the influence of equivalence ratio, velocity ratio and hydrogen injection temperature are estimated. They are connected tightly and shown by the following equation.

$$r = \sqrt{\frac{2T_f U_{air}}{5T_{air} U_f + 2T_f U_{air}}}$$

where r is fuel nozzle diameter. The influence of velocity difference between air stream and hydrogen jet on thrust and specific impulse are depicted in Fig. 9 and 10, respectively. Because of the change of nozzle diameter the specific impulse increases with velocity ratio, U_f/U_{air}.

Figure 11 shows the specific impulse versus hydrogen injection temperature with varying the equivalence ratio. From this figure, the hydrogen temperature is found to affect the specific impulse slightly.

CONCLUDING REMARKS

The performance of scramjet engine with active cooling has been estimated. The axisymmetric(2D) flow model is adopted to analyze the characteristics of combustor. The conclusions of this research can be summarized as follows.

(1) Supersonic flow in the combustor have been analyzed by adopting 2D axisymmetric flow model. The completion of combustion is examined by uniformity of temperature profile and/or unburned hydrogen.

(2) Equivalence ratio and temperature of hydrogen jet affect significantly to the heat release, therefore, specific impulse.

(3) The thrust and specific impulse strongly depend on the velocity ratio U_f/U_{air}. However, because of variable hydrogen nozzle diameter, the hydrogen temperature is found to affect the specific impulse slightly.

REFERENCES

[1] R. R. Barthelemy, " The National Aero-Space Plane Program," *National Aero-Space Plane and Space Applications, WHEC# 8*, 1 (1990).

[2] Y. Tsujikawa and M. Nagaoka, " Determination of Cycle Configuration of Gas Turbines and Aircraft Engines," *ASME J. Eng. GT & Power*, **113**, 100 (1991).
[3] Y. Tsujikawa, Y. Tsukamoto and S. Fujii, "Performance Analysis of Scramjet Engine with Quasi-One-Dimensional Flow Model," *Int. J. Hydrogen Energy*, **16**, 135 (1991).
[4] Y. Tsujikawa, T. Onoda and S. Fujii, "Performance Optimization of Scramjet Engine," *National Aero-Space Plane and Space Applications, WHEC# 8*, 171 (1990).
[5] D.B. Paul, C.L. Clay and M.P. Camden, " Hydrogen Active Cooling of Hypersonic Structures," *National Aero-Space Plane and Space Applications, WHEC# 8*, 157 (1990).
[6] R.L. DeWitt, T.L. Hardy, M.V. Whalen, G.P. Richter and T.M. Tomsik, " Background, Current Status and Prognosis of the Ongoing Slush Hydrogen," *National Aero-Space Plane and Space Applications, WHEC# 8*, 69 (1990).
[7] D. B. Spalding, B. E. Launder, A. P. Morse and G. Maple, " Combustion of Hydrogen–Air Jets in Local Chemical Equilibrium (A Guide to the CHARNAL Computer Program)," NASA CR-2407 (1974).
[8] R.B. Moore, G.E. Kinard and D. Nahmias, " Slush Hydrogen Production for the National Aero-space Plane Programs," *National Aero-Space Plane and Space Applications, WHEC# 8*, 107 (1990).

FIGURES

FIGURE 1: TEMPERATURE-ENTROPY DIAGRAM OF CRYOGENIC HYDROGEN

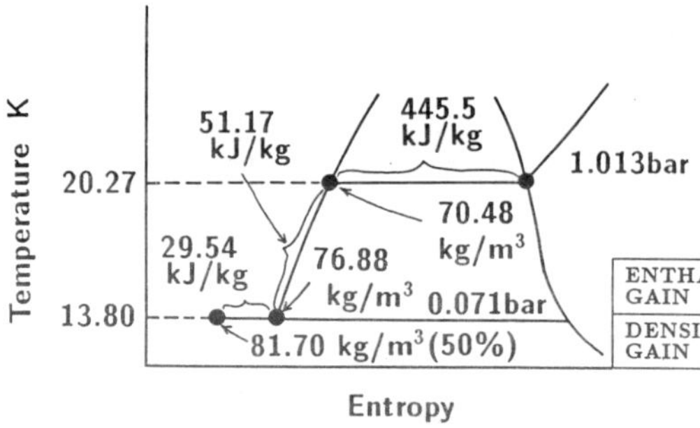

	TO T.P.	TO 50% SLH_2
ENTHALPY GAIN	11.5%	18.1%
DENSITY GAIN	9.1%	15.9%

FIGURE 2: SCHEMATICS OF SCRAMJET ENGINE

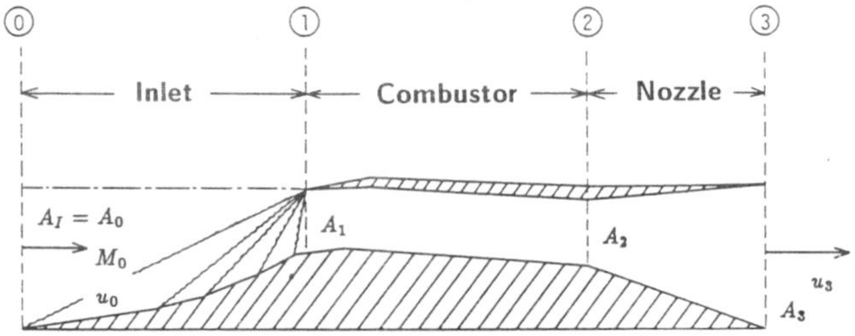

FIGURE 3: VELOCITY AND TEMPERATURE PROFILE

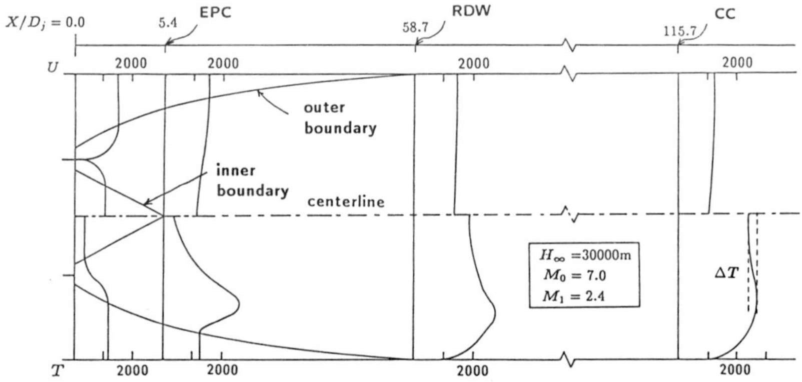

FIGURE 4: RELATION BETWEEN UNBURNED HYDROGEN AND UNIFORMITY OF TEMPERATURE

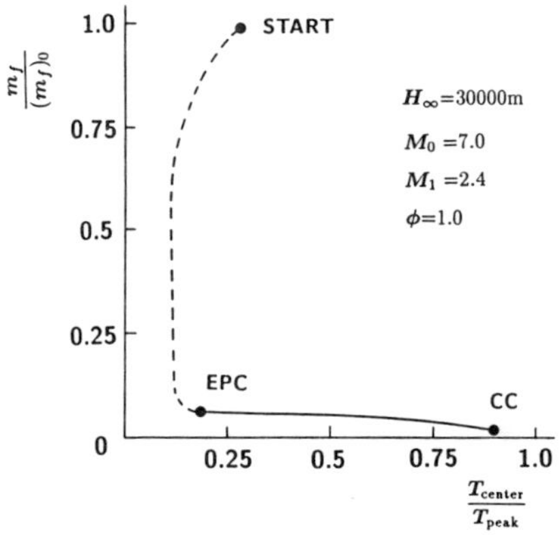

FIGURE 5: INFLUENCE OF EQUIVALENCE RATIO ON MIXING

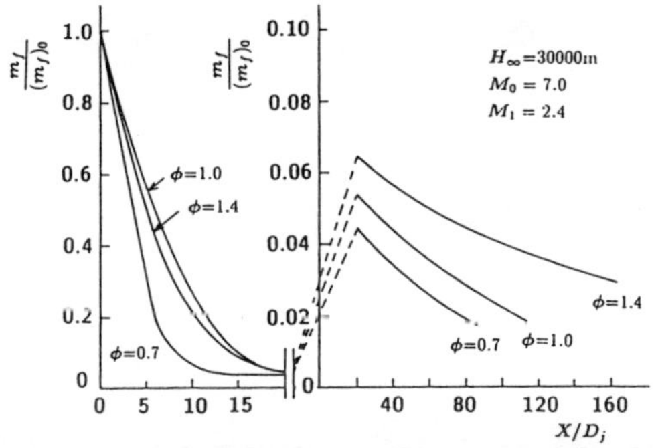

FIGURE 6: INFLUENCE OF VELOCITY RATIO ON COMBUSTOR LENGTH

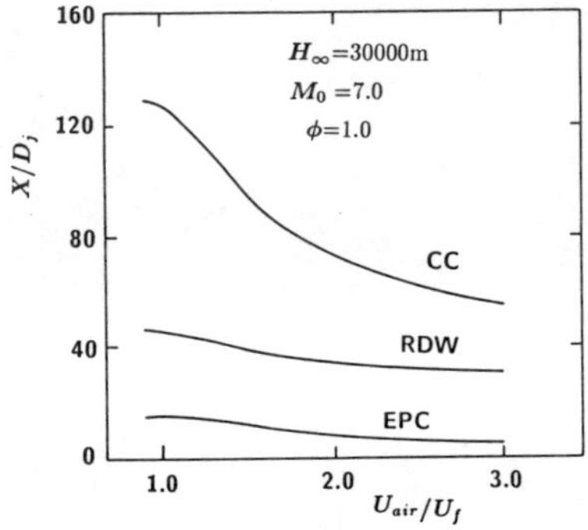

FIGURE 7: SPECIFIC IMPULSE VERSUS COMBUSTOR INLET MACH NUMBER

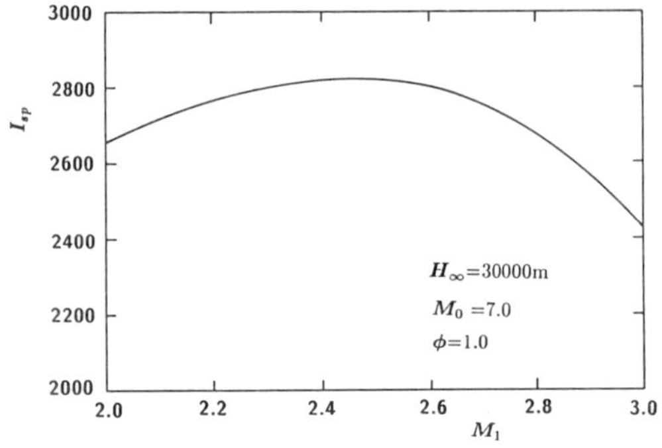

FIGURE 8: COMBUSTOR LENGTH VERSUS INLET MACH NUMBER

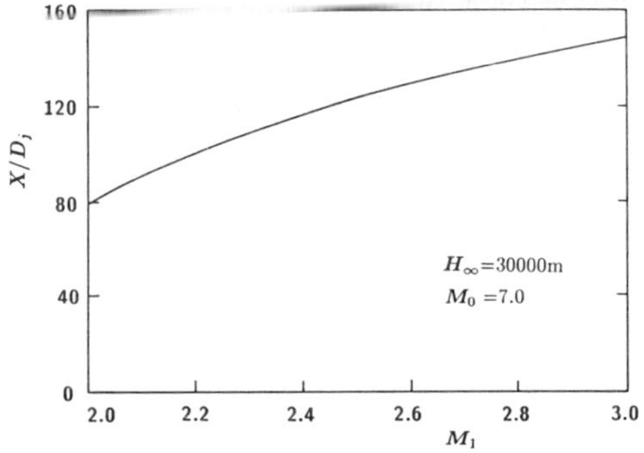

FIGURE 9: THRUST VERSUS HYDROGEN INJECTION TEMPERATURE

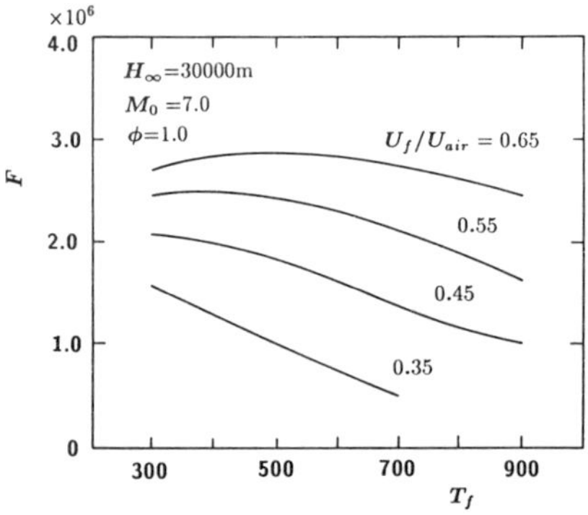

FIGURE 10: SPECIFIC IMPULSE VERSUS HYDROGEN INJECTION TEMPERATURE

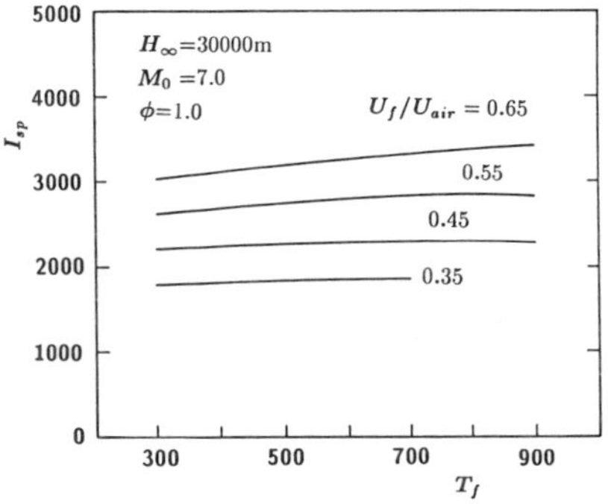

FIGURE 11: INFLUENCE OF EQUIVALENCE RATIO ON SPECIFIC IMPULSE

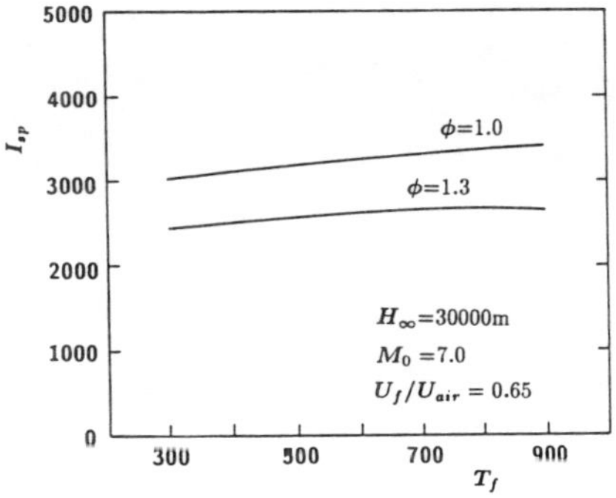

AN ELEMENTARY APPROACH TO DESIGNING METAL-HYDRIDES FOR PRACTICAL APPLICATIONS

K. Sapru
Technology Innovation Products & Services (TIPS), Inc.
1924 Smallbrook, Troy, MI 48098, U.S.A.

ABSTRACT

The study of metal-hydrides is a fascinating area of materials science and technology due to their inherent simplicity and the ability to be tailor-made for a number of applications. In the past these materials have been studied mostly for their use as storage medium for automotive applications. Recently, in addition to other applications they have also been gaining attention as anode materials for rechargeable batteries. This is because of their potential for superior performance and their environmentally benign nature. A discussion is presented on how to use the periodic table of elements to design new and unique alloys for both gas-phase and electrochemical applications.

INTRODUCTION

Hydrogen as an energy carrier and its role in the ongoing worldwide efforts to develop alternatives to fossil fuels is no longer just a curiosity, but a real near and medium-term possibility. The missing link in the chain of production-storage-use of hydrogen has been a safe and economical way of storing it. The relatively new field of metal-hydride technology has not only solved this problem, but has also opened up possibilities for many hydride-based products, that could play a very important and positive role in our future energy/environment scenario. These products include hydrides for transportation, high energy density rechargeable batteries, heating and cooling systems, and temperature controllers. A recent review on the practical applications of metal hydrides is given by F.E. Lynch [1]. A "hydride tree" (figure 1) is drawn to show one basic materials technology leading to numerous products.

The key factor in realizing the commercial potential of hydrides lies in our ability to design new hydride-formers (and modifying existing ones). A hydride-former is any metal or alloy that is capable of reacting with hydrogen to form a hydride. In it's simplest form this reaction can be represented as follows:

$$M + x/2\ H_2 \longleftrightarrow MH_x$$

where M is the "hydride-former", x is the number of hydrogen atoms, and MH_x is the "hydride". The direction of the reaction is determined by the pressure of the hydrogen gas, the temperature of the reaction, or a combination of both.

Although extensive data now exists on the study of hydrides, with a few exceptions [2,3,4] most of the effort to date has been focussed on FeTi, Mg_2Ni and $LaNi_5$ alloys and their modifications produced by partial substitution of one or both elements of the alloys. This trend is still continuing. Not much work has been reported on the design of hydride-formers from first principles.

The periodic table of elements, given to us neatly arranged by rows and columns, can be a powerful tool in the design of new materials for almost any application. After all, most of the solutions to materials problems are contained somewhere within the periodic table. Having some idea of how individual atoms behave, then by judicious choice of elements, one can formulate and make alloys having the desired properties. For hydrogen storage these properties are; high hydrogen storage capacity, good reversibility for hydrogen absorption/desorption, ease of activation/good kinetics, desired equilibrium pressure, stability in ambient environment, stable to poisoning effects and preferably low cost.

Hydrogen reacts with various elements to form either ionic, metallic, or covalent hydrides [5,6]. The ionic hydrides are formed by the columns IA and IIA of the periodic table, namely the alkali metals and the alkaline earths. Of these, only Mg-hydride, presumed to be of polymeric nature [6], is of interest since it can theoretically store 7.6% by weight hydrogen which can be released below 300 degrees centigrade. The hydrides of the other elements in this group, are either too stable for most practical applications, or very difficult to make.

While most of the elements of columns IIIA to VIIA form metallic hydrides, only when alloyed in appropriate combinations, do they form useful hydrides. In these elements, the bonding of hydrogen now includes d-orbital interactions. It is interesting to note that while hydrides of the alkali and alkaline earths are more dense than the metals, most hydrides of groups IIIA-VIIA are less dense than the corresponding metals, and in most cases the lattice swells as the hydrogen enters [6]. Cr forms endothermic hydride and loses the H at fairly low temperatures, while Mo, W, Mn, Tc, and Re are not known to readily absorb H.

The elements in the VIIIA group (Fe, Co, Ni, Ru, Rh, Pd, Os, Ir, Pt), absorb hydrogen readily, though in some cases at very high temperatures and only when the elements are in finely divided form. Elements of groups IB and IIB (Cu, Ag, Au, Zn, Cd, Hg) do not combine directly with molecular hydrogen.

MATERIALS DESIGN

Background

In a 1974 paper Ovshinsky and Sapru [7], using the periodic table of elements designed "ball and stick" models to explain the electronic properties of amorphous switching and memory materials which were primarily composed of the elements Si, Te, As, and Ge. The periodic table was used to decide how individual atoms in an alloy would prefer to combine in a random arrangement, by just taking into account such factors as bond strengths, bond angles, electronegativity, and the distribution of electrons in the outermost orbitals of the individual atoms. Using colored styrofoam balls for atoms and color-matched pipe cleaners for bonds, and making relatively crude models, these authors were able to successfully explain the electronic behavior of amorphous chalcogenides.

In designing the "ball and stick" models it was assumed that given a choice between two neighbors an atom will prefer to bond to the one with which it can form a stronger bond. The coordination number and the bond angles were adjusted from simple considerations. For example, germanium with it's s2p2 outermost electron configuration was assumed to be mostly tetrahedral with a coordination number of four, and sometimes randomly two-coordinated if the 's' and 'p' character of the electrons remained pure and unhybridized. The tellurium atom with s2p4 configuration was considered to be two-coordinated along with a satisfied 'p' "lone-pair of electrons remaining non-bonded or resulting in weak bonds due to some orbital overlaps with neighboring atoms. Even though the models lacked long-range-order, they were not truly random, since the atoms were not arbitrarily positioned. This created a high degree on "controlled local order". It is the ability to manipulate and tailor-make this controlled local order in a material, whether amorphous of polycrystalline, that allows us to create new and unique families of materials for different applications.

Although this work was unrelated to hydrogen storage, it is mentioned here to illustrate how an intuitive approach resulted in successfully explaining the relationship between structure and properties of amorphous materials, and the power of the period table even when used in it's most simplistic form. An important lesson learned here was that "the key to understanding the properties of solids is in the nature and positioning of their individual atoms". Similar approaches have been used to develop many new families of materials having interesting electronic, optical, catalytic, and other useful properties [8-11]. Not all the materials studied were amorphous, many were multi-component, polycrystalline alloys.

Hydride-Formers for gas phase applications

In the design of hydride-formers, there are two broad and basic requirements. (1) The presence of many catalytically active sites where the hydrogen molecule can split into atomic hydrogen and (2) the availability of a large number of H-storage sites. Both these properties can be designed into the material through structural effects and by careful selection of the alloy composition, both structure and composition being equally important.

An active site can be [12] a dangling bond, an unoccupied bonding orbital with high electron affinity, an occupied bonding orbital with a low ionization potential, location of a high electric field, or imperfections on a heterogeneous surface (such as a dislocation, an impurity, a crystallographic step, etc.). Figure 2, taken from reference [12] shows the active sites on the surface of a crystal. These sites occur only at the surface of a perfect crystal and not in it's bulk. They also give rise to disorder, and many localized energy levels.

Figure 3 is an attempt at illustrating how by going from a two-dimensional surface to a three-dimensional disordered solid, active sites can be increased by several orders of magnitude because these sites now occur throughout the bulk of the solid. This is the structural contribution to the active sites. We need catalytic sites for improved kinetics and H-bonding sites for improved H-storage capacity. The structural, compositional, and configurational disorder that is present in multi-component alloys provides this automatically to some extent. But selective design of the alloys is still necessary.

It is obvious that one needs to start with an element or alloy in which at least one of the components is a stable hydride former. Since the structural integrity of materials is based upon the strong covalent bonding which makes up the backbone of the material, such an element(s) can provide this backbone. All other interactions which result from disorder leading to a spectrum of bond strengths, bond lengths and bond angles, can also be potential sites for catalytic activity, and for H-bonding. For many applications, we need a light weight hydride. FeTi, one of the most well studied hydride-formers is too heavy and gets easily poisoned. In order to design a solid hydride lighter than Ti, let us look at the periodic table. Of the 21 elements lighter than Ti (atomic wt. 22), one is hydrogen and 7 others are gases. We have 13 elements lighter than Ti to work with. Except for a few of these such as Mg, Al, Si, Sc, most others can be easily eliminated. So we have narrowed down to Ti and a few elements lighter than Ti as basic materials.

One can now engineer a material so that the major part of it's matrix consists of one or more of these light-weight elements. Then a small percentage of d-and f-band materials from the transition metals (TM) and / or rare earths (RE) can be added to further increase the density of active sites without significantly adding to the weight of the material. A kind of pin-cushion effect that a TM or a RE atom can produce is illustrated in figure 3 to show how a single atom from this category can have multiple orbital interactions. Many more active sites can also mean less susceptibility to poisoning. This kind of flexibility is not possible when working with stoichiometric crystalline phases.

When selecting an element to introduce or enhance catalysis, the middle part of the TM series (group VIII elements) form the obvious choice. It is well known that catalysts act by chemisorption and in general one expects maximum catalytic activity at some intermediate strength of chemisorption. The chemical reactivity of the TM's decreases as one moves from the left to the right across each period, for example, from Ti to Cu, falling off greatly with Cu (because of the filling of the d-orbitals). Thus the best catalyst are found in the middle of the TM series (Fe, Co, Ni, Ru, Pd, Os, Ir, and Pt).

Governed by such fundamental principles and then introducing disorder to further enhance the probability of H-bonding and catalytic activity, is part of the materials design process. Some examples taken from references 3 & 4 given in Table I illustrate how a light weight element (Mg), is modified by another such element (C, Al) to introduce disorder, and the addition of a TM (Fe) from group VIII to achieve high storage capacity. In some cases the storage capacity is higher than can be attributed to Mg alone. The local chemical environment is modified such that even atoms which do not by themselves form stable hydrides can now contribute to H-storage. A statement taken from one of Ovshinsky's papers [13] applies here very well. This is the concept of TIE (Total Interactive Environment). "....the TIE has a localized geometry encompassing a total chemical site, which, if viewed as composed only of it's parts, would not give the correct picture because the interaction of the parts creates a 'gestaut' different from, and more than the sum of it's parts".

By nature, an amorphous matrix should be less rigid (and thus more easily distortable) than a truly crystalline matrix. It has lower density and hence more volume or open spaces. This means that one can expect less damage during repeated expansion and contraction caused by absorption and desorption of hydrogen. Thus, although a truly amorphous solid with a tailored chemical composition would be the most desirable for the reversible storage of hydrogen, from a practical point of view this is not always possible if one wishes to take into account the cost and ease of manufacture of large quantities of material. Hence, we find that as we go from perfect crystallanity to true amorphicity, most materials of practical interest usually lie somewhere in between. Recent reviews of amorphous hydrides are given by Bowman [14] and Harris & Curtin [15].

Hydride-Formers for Battery Electrode Applications

Although hydrides were originally studied for use in transportation applications, that application will probably be the last one to be practiced on a large scale, because of the complex infrastructure that is needed, and the problem of producing low cost hydrogen from clean energy sources. It is the "non-hydrogen-consuming" uses of hydrides that are of immediate interest. These include rechargeable batteries, air conditioners/heat pumps and temperature sensors. Of these the efforts on hydride battery work have increased exponentially during the past few years [14-17].

Designing a hydride battery electrode is an interdisciplinary project bringing together materials science, electrochemistry, surface science and metallurgy. The two basic requirements for the alloy development namely catalytic and H-storage sites still remain, but now there is the additional requirement that the material be stable in aqueous alkaline environment under the pH-potential conditions of the battery operation. High H-storage capacity will now mean high energy density, good reversibility-long cycle life, good absorption kinetics - high charge rate capability, good desorption kinetics - high discharge rate capability. Additionally, the electrode alloy needs to be tailored for low temperature performance, appropriate desorption pressure (highly disordered materials do not generally have a plateau), and stability to corrosion and oxidation.

During charging, the hydride battery is essentially an electrolyser, while during discharge it is a fuel-cell undergoing H-oxidation, and in between it stores hydrogen in the negative electrode, thus bringing together the properties of alkaline hydrogen evolution, alkaline hydrogen oxidation, and hydrogen storage. Many years of work on the development efficient electrocatalysts and hydride-formers based on the principles discussed in the previous sections, resulted in the development of the Ovonic battery [14]. This is the only hydride battery system known to this author that is designed from first principles and not based on using mischmetal or $LaNi_5$ as a base material.

Here again, light-weight elements such Mg, Si, or Ti can be used to form the "back bone" matrix, and appropriate TM elements introduced for catalysis. For example, Ti as a hydride former and Ni as catalyst makes TiNi a reasonable starting material, particularly since NiTi alloys with proper modification are also good H-evolution and H-oxidation catalysts [18,19]. Further modifications to the alloy include additional elements that form stable hydrides, and those that form stable oxides (using Pourbaix pH/potential diagrams as a guideline).

There are many reasons for working with highly disordered multi-componenet alloys. As discussed earlier, they provide more sites for hydrogen storage and for catalytic activity. The corrosion rate of complex alloys whether due to oxide formation of some other mechanism, will be lower.

When designing a new electrode alloy using the periodic table, we need to ask ourselves questions such as: which elements are hydride formers? which ones are likely to cause an increase of decrease in the hydrogen equilibrium pressure? which ones will cause a decrease in overvoltage? which elements or a combination of elements will cause the formation of a stable oxide? can one control the thickness of this oxide? what will improve the corrosion resistance? what will promote structural integrity? what will improve low temperature performance?

A second set of questions that immediately follow are: how are the above related to bond energies, atomic volumes, electronegativity, oxide properties, catalytic properties, d-and f- orbital availability, weight and cost? do complex oxides improve charge retention? does complex oxide formation mean that several elements are competing for a fixed amount of oxygen and hence the oxide layer is thinner, therefore allowing for both passivation and H-diffusion? Can one element play multiple roles?

In conclusion, although La and mischmetal based hydride electrodes have proved to be promising, I believe that by starting from basic elements and then modifying systematically, one can stay in better control of the composition in terms of future improvements, reproducibility, and addressing any potential problems that may arise.

ACKNOWLEDGEMENTS

The author wishes to thank Raj Sapru, first year engineering student at the University of Michigan, for helping with the figures.

REFERENCES

1. F.E. Lynch, Int. Symposiun on Metal Hydrides, Banff, Canada, Sep. 1990.
2. V.K. Sinha and W.E. Wallace, J. Less-Common Met. 106 (1985), 199.
3. S.R. Ovshinsky, K. Sapru, K. Dec, and K.C. Hong, US patent no. 4,431,561, 1984.
4. K.C. Hong and K. Sapru, Hydrogen Systems vol. I, The Int. Symposium on Hydrogen Systems, Beijing, China, (1985), 487.
5. W.M. Mueller, J.P. Blackledge & G.G. Libowitz, Metal Hydrides, Academic Press, 1968.
6. R.T. Sanderson, Inorganic Chemistry, Reinhold Publishing Corporation, 1967.
7. S.R. Ovshinksy and K. Sapru, 5th Int. Conf. on Amorphous & Liquid Semiconductors, Garmisch-Partenkirchen, FDR, 1974.
8. R. Flasck, M. Izu, K. Sapru, T. Anderson, S.R. Ovshinsky, and H. Fritzsche, 7th Int. Conf. on Amorphous & Liquid Semiconductors, Edinburg, Scotland 1977.
9. S.R. Ovshinsky and K. Sapru, U.S. Patent no. 4,178,415, 1979.
10. S.R. Ovshinsky and K. Sapru, US patent no. 4,544,473, 1985.
11. K. Sapru, S. Venkatesan and P. Pai, Proc. of NACE Conf.

March 1987.
12. S.R. Morrison, The Chemical Physics of Surfaces.
13. S.R. Ovshinsky, Rev Roum Phys Tome 26, nos. 8-9 (1981) 893.
14. R.C. Bowman, Jr. Mat. Sci. Forum vol. 31 (1988) 197.
15. J.H. Harris and W.A. Curtin, Mat. Sci. Forum vol 31 (1988) 229.
16. K. Sapru, K.C. Hong, M. Fetcenko and S. Venkatesan, US patent no. 4,551,400 (1985).
17. M. Kanada and Y. Sato, US patent no. 4,605,603 (1966).
18. M. Yamano, et. al. US patent no. 4,636,445 (1987).
19. M. Ikoma, et. al. US patent no. 4, 837,119 (1989).
20. S.R. Ovshinsky, K. Sapru and A. Reger, US patent no. 4,487,818 (1984).
21. S.R. Ovshinsky, K. Sapru and E. Yee, US patent no. 4,545,883 (1985).

TABLE 1

SOME Mg-BASED HYDROGEN STORAGE ALLOYS

Material Comp. atomic %	Wt. % H	Desorption Temp °C
Mg C O 75.3 7 17.8	8.21	340
Mg Al Fe 89.1 6.5 4.4	6.2	300
Mg Fe Al 60 30 10	5.0	200

FUGURE 1

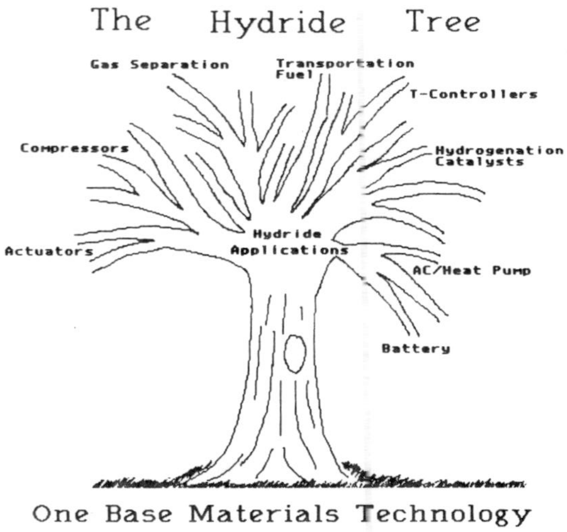

FIGURE 2: ACTIVE SITES ON A CRYSTAL SURFACE FROM REFERENCE 12.

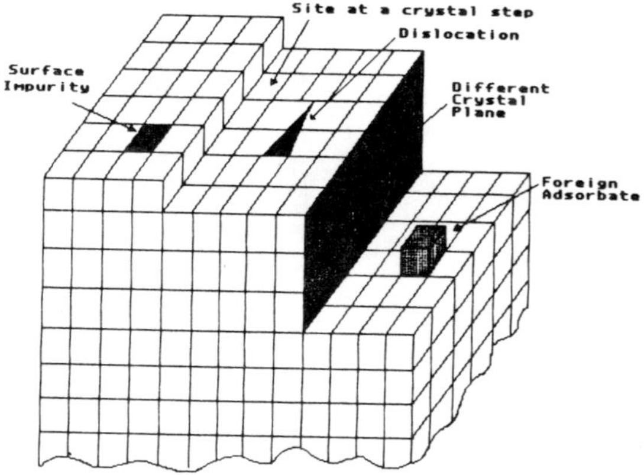

FIGURE 3: SHOWING A DISORDERED MATRIX AND THE PIN-CUSHION EFFECT OF d- AND f- ORBITALS.

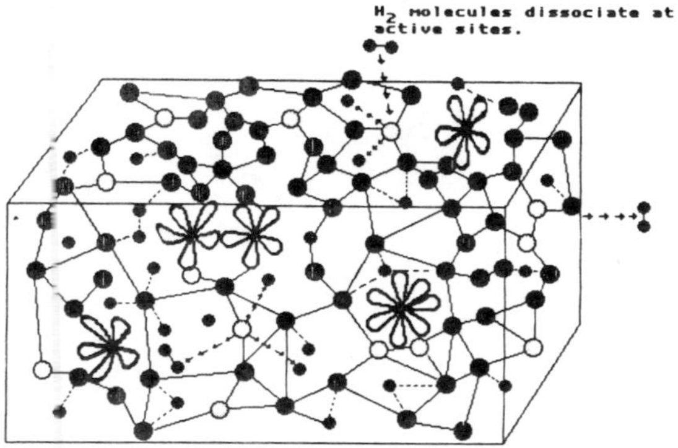

CHAPTER SEVEN

Environment and Safety

GLOBAL WARMING SEA LEVEL RISE AND ITS IMPACT ON MANGROVE ECOSYSTEM IN BANGLADESH

M. Monirul Qader Mirza
Secretary General
and
Subrata Paul
Research Fellow
Centre for Environmental
Studies and Research [CESR]
G.P.O. Box 3290
Dhaka-1000
Bangladesh.

ABSTRACT

It is assumed that the concentration of carbon dioxide in the atmosphere will be doubled between now and 2025 to 2050 and consequently our planet will experience an increase of mean temperature by 3± 1.5°C with predicted sea level rise of 0.3-0-5 m and about 1m by 2050 and 2100 respectively. It is also estimated that by the year 2050 the local rise in sea level in Bangladesh will be between 13 to 209cm. If unabated, the best case would result in a small loss of land, whereas in worst case about 18% of the land would be lost.The problems of climate change and resulting sea level rise are closely entangled with other environmental problems in Bangladesh. The Country ranks fifth in the eastern hemisphere in terms of area under mangrove forest. In South Asia, however,Bangladesh ranks first. The mangrove forests, which form a natural defence against coastal erosion,cyclonic storm and tidal surge, are subject to degradation due to climatic change and consequent sea level rise. This paper will analyse the global warming, sea level rise and its impact on mangrove ecosystem particularly on the Sundarbans forest in the south western corner of Bangladesh.

INTRODUCTION

Around our planet, global warming and climate change is evident now. In the 1980s, the planet experienced six warmest years with 1990 as the most hottest on record. The emission of greenhouse gases is still continuing alarmingly. Although many industrialized countries signed the Montreal Protocol,but still it is uncertain how much emission will be cut off by the end of this century.

Under the present rate of emission , the temparature of the planet may be increased by 1.5-4.5°C over the pre-industrial temperature. Recent analyses indicate that the earth as a whole has warmed by 0.3-0.6°C over the past century. It is expected that this rate will increase significantly,by 0.5-1.0°C per decade through the next few decades, if human activities which emit greenhouse gases continue unabated.

Scientists are not even confident regarding the magnitude of sea level rise (SLR) due to climatic change, however, there is strong consensus that it will happen. Whatever be the magnitude of SLR, the low lying deltas and coastal land will be most vulnerable to the consequent disaster. Undoubtedly, Bangladesh is one of the vulnerable deltas. In Bangladesh, the mangrove ecosystem which is largest in the world will considerably be destroyed which will ultimately generate tremendous unemployment, loss of revenues, closing of industries and displacement of huge population. Moreover, the biological diversity will certainly be threatened.

GREENHOUSE GASES EMISSION

Many different radiative gases can induce tropospheric warming (via greenhouse effect). The important naturally occurring radiative gases in the atmosphere are water vapour (H_2O), carbon dioxide (CO_2), methane (CH_4), Ozone (O_3) and nitrous Oxide (N_2O), CFC-11($CFCl_3$), CFC-12(CF_2Cl_2), methyl chloroform(CH_3CCl_3) etc. These gases absorb the IR radiation emitted by the warm surface of the earth and keep the atmospheric temperatures warmer.

a. Carbon Dioxide (CO_2)

The concentration of carbon dioxide during the period 1000-1800 was found between 270-290 ppmv. It is evident that the concentration was very stagnant. The reason for this constancy has been attributed to the atmosphere-ocean and biota carbon dioxide exchange and the exchange rate is assumed 30 per cent. A precise analysis[1] of a ice core collected from Siple Station, Antarctica indicated that CO_2 started rising around 1800 and had already increased by 15 per cent. Precise direct atmospheric measurements started in 1958, when the level of CO_2 concentration was recorded at 315 ppmv and the rate of increase was 0.6 ppmv per year. The present atmospheric CO_2 level has reached 353 ppmv(1990) with an annual mean growth rate of 1.8 ppmv.

b. Methane (CH_4)

Very significant information have been gathered regarding the concentration of atmospheric CH_4 analysing the ice cores collected from Greenland and Antarctica for the period between 10,000-160,000 years ago. A minimum concentration of 0.35 ppmv was found thereafter and 0.65 ppmv concentration was traced during the glacial and interglacial pransitions(about 15000 and 130,000 years). However, the atmospheric concentration of CH_4 decreased rapidly, prior to, and during the glacial retreat (10000-11000 years ago).

Ice core data indicate that the atmospheric concentrations of CH_4 averaged around 0.8 ppmv between two hundred and two thousand years ago, and increased to 0.9 ppmv one hundred years ago [2]. The atmospheric concentration of methane followed an increasing trend since then. It is correlated with global human population. Over the last 40 years, the atmospheric concentration of CH_4 has been increased by 30%. A global average of 1.51 ppmv methane concentration, recorded in 1978 directly from atmosphere. The current level is about 1.72 ppmv in relation to the atmospheric reservoir of about 4900 Tg of which rate of increase is 0.8-1.0% [3].

c. Nitrous Oxide (N_2O)

From the ice core analysis, the concentration of nitrous oxide before the pre-industrial period was found 288 ppmv, which is known to have increased to a level of 310 ppmv in 1990 following an increasing trend of 0.2-0.3% per year.

Oceans are assumed to be the principal source of Nitrous Oxide but not the dominant one. It has been estimated that ocean sources release 1.4-2.6 Tg per year. IPCC (1990) claims that this estimate is substantially lower than the earlier estimates [4]. The accurate figure of ocean flux could not be determined due to the uncertainties associated with quantifying the gas exchange coefficient and highly ariable partial pressure of N_2O in sea surface water.

d. Halocarbons

Concentration of halocarbons is gradually increasing although its presence was not found during the pre-industrial period. However, the present level(1990) of concentration of CFC-11 and CFC-12 are 280 and 484 pptv respectively. Table-2 presents the Halocarbon concentrations. The concentration of CCl_3F (CFC-11),CCl_2F_2(CFC-12) $C_2Cl_3F_3$(CFC-113) with an exception of CCl_4, have increased more rapidly than the other greenhouse gases in the recent past. The current rate of increase is about 4% per year. The fully halogenated CFCs and CCl_4 are primarily removed by photolysis in the stratosphere, and have atmpospheric life time in excess of 50 years.

THE GREENHOUSE EFFECT AND GLOBAL WARMING

Analysing the surface air temperature record for the Northern and the Southern Hemispheres,scientists find a gradual warming of the atmosphere. During 1860 to 1988 the global net warming is found to be 0.5°C. From the record ,six warmest years in the series have been identified which occurred in 1980s [5][6]. The rank of these years are 1986 ,1980,1981,1983,1987 & 1988.

A relatively stable period is observed during the earliest decades of the record. Since 1860 global temperature rose steadily up until the 1940s with marked warming during the 1920s and 1930s,particularly in the Northern Hemisphere. Temperature then remained relatively constant until beginning of the 1970s. The record also shows that a relatively stable,or slightly decreasing global temperature during the 1940s, 1950s and 1960s was in the Northern Hemisphere when the Southern Hemisphere continued to have a warming trend.

Continuing emissions of CO_2 and other greenhouse gases will cause the atmosphere to future warming. Most atmospheric scientists agree that when the atmospheric concentration of CO_2 reaches approximately 550 ppmv the Earth will be subjected to an average warming of 1.5-4.5°C above pre-industrial temperatures [7].Recent experiments with the most advanced general circulations models of the atmosphere suggest that planet's temperature sensitivity to doubled CO_2 is likely to be in the top half of this range, between 3.0°C. and 4.5°C. If current emission growth trends for all trace gases continue, the combined effects of the most important greenhouse gases could possibly occur as early as 2030[8].

ENVIRONMENTAL SETTING OF MANGROVE FORESTS : THE SUNDARBANS

The predominant feature of the climate in the Sundarbans,as that of Bangladesh as a whole, is the south east monsoon which effectively divides the year into three distinct seasons [9]. The monsoon, from mid June to early October, is characterized by high temperatures, relative humidity, overcast skies and prolonged rain storms which may last for several days at a time. Following the monsoon i.e. Winter from early November to February, is a cool season characterized by low rainfall, low temperature as well as low relative humidity with skies clear. On the other hand, following the cool season a short dry season from mid February to early May, is characterized by a steady rise temperature and violent storms which is followed by rather more settled conditions prior to the onset of the next monsoon.

The average annual rainfall within the forest varies from about 1640 to 2000 mm. It increases from west to east in accordance with the general pattern of rainfall distribution in the country. The rainfall occurs in full force in the months from June to September.

COASTAL MANGROVE RESOURCES OF BANGLADESH

Bangladesh possess one of the largest mangrove ecosystems of the world with three patches of mangrove forests: the Sundarbans,the chakaria Sundarbans and the Coastal mangrove plantation forest. The Sundarbans forest in one patch is the largest mangrove forest in the world having an area of 571,508 hectares, of which 169,908 ha consists of rivers, channels and other water bodies(Figure 1). It alone accounts for about 74 per cent of the total area under reserve forest in Bangladesh. It is located on the west of the Brahmaputra-Meghna delta and extends toward the west beyond the International boundary with India extending from 21°31' to 22°31' N latitude and 89°to 90° E Longitude. The forest is bounded in the east by the Baleswar river, in the west by the international boundary with India which follow the Hariabhanga - Raimangal-Kalindi river system, in the south by the Bay of Bengal and in the north by a complex of small rivers and streams which separate the forest from cultivated land. The forest spreads over the districts of Khulna, Bagerhat and Satkhira.

The Chakaria Sundarbans is a denuded forest and it started to deteriorate in the sixties [10]. The forest is composed of 11 islands in the delta of Matamuhuri river of Cox's Bazar district and extends from 21°36' to 21°45' N latitude and 91°58' to 92°05' E Longitude. This forest used to cover an area of 7489.8 ha reserved and 1012.2 ha protected forest upto 1977 when the process of releasing land from the Forest Department to the Fisheries Department started for shrimp culture presently,this forest covers only 973 ha area.

The coastal mangrove plantation forest was introduced in 1964[10] in the coastal belt of the country stretching from Teknaf in Cox's Bazar to Barisal and Patuakhali over an area of 1215 sq.km.

In Bangladesh, the mangrove ecosystem play a vital role from economic, commercial, aesthetical and ecologic viewpoints. This ecosystem provides the country with a number of services:

a. Food source for offshore fisheries;

b. Essential element for onshore shrimp cultivation ;

c. Coastal protection ;

d. Economic and commercial resources;

e. Recreational resort ; and

f. Habitat for wildlife of national heritage.

The primary productivity of the mangrove forests is manifested in the accumulation of biomass. Shedding leaves of plants after decomposition result in particulate organic matter (POM) and dissolved organic matter (DOM) that enter the estuarine ecosystem. The DOM is consumed by larval and Juvenile organisms that use the mangrove habitat as a nursery and feeding ground, whereas the DOM is consumed by filter feeders (Oysters and mussels)that are common in nearshore environment and that retain the phyto,nauno and macroplankton which need DOM to synthesize organic matter. Besides,the seaward transported organic matter is flocculated by the increasing salinity and this becomes a substrate for benthic feeders and scavengers. The algae, plankton, benthos and other micro fauna are essential features of the mangrove ecosystem and they determine to a large extent the variety,quality and quantity of fish and crustacea.

Cyclonic disaster is a common phenomena in the coastal belt of Bangladesh and it hits almost every year causing huge loss of lives and properties.Like other mangrove ecosystems of theworld, the mangrove forests of Bangladesh, especially the Sundarbans have ensured protection of life and property inland including the cultivated lands from the damage caused by periodic cyclones and tidal surges. For example, a severe cyclonic storm with core wind speed of 160 km/hour with an average storm surges of about 5m sweeped over the greater Khulna district in November 1988, but that cyclonic disaster was not so severe as the cyclonic disaster of Urir Char of 1985 and the unprecedented disaster of April 1991 cyclone where core wind speeds were 154 and 225 km/hour respectively. The Sundarbans served as a great protector against the very severe cyclone of 1988 and that's why the damages were not so colossal. On the other hand, the mangrove trees accelerate the process of siltation by their excellent sediment trapping effect during the ebb period and thus help much to elevate the land.This mangrove is also protecting the Ganges Tidal coastal zone from both landslides and soil erosion and is acting as a buffer against tidal currents, floods and storms.

The directly employment effect is difficult to estimate.Labour in such employment as the Khulna Newsprint Mills and Khulna Hard Board Mill, probably 90-95 per cent of boatmakers in the Khulna, Barisal and Jessore districts,bulk of the labourers engaged in roof thatching, part of the employment in ice plants, fish refrigeration plants and furniture making are all dependent on the products of the Sundarbans.[11] gives a figure of the direct industrial employment generated through the exploitation of forest resources of the Sundarbans(Table 6).

The Sundarbans has spectacular wildlife in greater abundance than any other area in Bangladesh. The mangrove ecosystems of the country support a wide range of mammals,birds,amphibia, reptiles and crustacea. It is estimated that there are 32 mammal species,186 species of birds, 35 reptile species, 8 species of frogs and toads[14] and over 120 species of fish[14] in the Sundarbans. An estimate of the larger terrestrial mammals lists about 80,000 spotted deer(Axis axis),40,000 rhesus macaques (Macaca multta), 20,000 wild bears (Sus Scrofa), 20,000 others (Lutra perspicillata) and 350 tigers (Panthera tigris) (Hendrichs, 1975 and ESCAP, 1987)[14]. Crustaceans account for by far the largest proportion of living creatures, with an estimated 40,000 tons of filders-carbs and 100,000 tons of mud crubi [11].

The Sundarbans forest is endowed by nature with rich fauna and flora. It is a wonderful place for its landscape, geology and wealth of beautiful birds and different kinds of animals, and all these have made the forest most picturesque and fascinating in scenic beauty. The Sundarbans, the paradise for games and game hunters, is a tourist resort of Bangladesh.

GLOBAL WARMING AND SEA LEVEL RISE

Before discussing the future probable change in sea-level let us look in the past.Estimates have been made regarding the changes occurred in the last 100 years. The estimates range from about 0.5mm/yr to 3.0mm/yr,with most lying in the range 1.0-2.0 mm/yr. However, there is no firm evidence of accelerated sea level rise during the century. Eustatic sea-level change are caused by many factors. They are: differences in atmospheric pressure,winds, ocean currents and density of sea water etc.

In the past, rise in sea level occurred largely due to thermal expansion of the oceans; melting of glaciers and the margins of the greenland ice sheets added further rise.There is doubt regarding the contribution of Antarctic ice sheet to the process of sea level rise.[13] finds, "there is no firm basis for supposing that the Antarctic ice sheet has contributed either positively or negatively to past sea level change". Most of the analyses have so far been made support the link between sea level rise and climatic change[14]

Sea level rise for the future has been estimated for different IPCC scenarios. For each scenario, three projections, best-estimate , high and low-were made corresponding to the estimated range of uncertainty in each of the potential contributing factors, and in the climate sensitivity and resulting global warming predictions.

For the "Business-as-Usual"Scenario at year 2030, global mean sea level is 8-29cm higher than today,with a best-estimate of 18cm. At the year 2070, the rise is 21-71cm, with a best-estimate of 44cm.

SEA LEVEL RISE AND BANGLADESH

Sea level rise will certainly affect the low lying alluvial plain of Bengal Delta. However, there is still some uncertainties what will be the magnitude of losses due to sea level rise and resulting inundation.[15] estimated that by the year 2050 the local rise in sea level in Bangladesh will be between 13 and 209 cm. By the year 2100, relative sea-level rise is projected to be between 28 and 447cm. In this paper, we will concentrate our discussion only on the impact of sea level rise in the coastal mangrove ecosystems.

CONCLUDING REMARS

The Sundarbons, is the world's largest estuarine swamp,occupying an area of slightly larger than half a million hectares,accounts for about 74 per cent of the total reserve forest in Bangladesh, is expected to be threatned seriously by the sea level rise due to climatic change. Families of 600,000 people are directly dependent on the Sundarbans. The Forest Department of Bangladesh earns on average 50% of its total revenue from the Sundarbons. Moreover , a number of industries of South-West part of Bangladesh is directly dependent on the supply of raw materials from the Sundarbans acts as a natural barrier against the cyclonic storms and tidal bores of the Bay of Bengal which saved huge lives and properties in the past. The authors estimated that with the 18 cm sea level rise 70 per cent of the mangrove ecosystem of Bangladesh will be destroyed of which major share goes to the Sundarbons.

The Sundarbans support about 25 species of forest flora but Sundri(Heriteria fomes), gewa (Excoeearia agallocha), goran (Ceriops decandra) and keora (Sonnaratia apatala) predominate there [11].The economic significance of the Sundarbans lies in the products that are exploited from the forest. Timber, pulpwood and firewood are the major products. The Khulna Newsprint Mill and the Khulna Hard Board Mill- the only hard board mill of the country are entirely dependent on the forest for their raw materials. Woods of this forest is also used in different match factories and at the same time, many small and cottage industries in the adjacent districts are dependent for their raw materials from the Sundarbans. Fuelwood from the Sundarban is marketed over much of western Bangladesh and also as far as Dhaka both for domestic and industrial use. One major industrial use of it is for brick burning. There are other minor products obtained from the Sundarbans which include golpatta hental, hudo, grass etc. Golpatta palm is of high economic importance, the leaves of which are much in demand for thatching and kutcha fencing (Table 3). The fruits of golpatta are also eaten by the woodcutters in the Sundarbans. Hental and hudo are used under Hatching over which golpatta is used to make the structure strong.

There are four major animal products obtained from the Sundarbans: fishes, honey, beeswax and shells. Fishes constitute a very important animal wealth of the Sundarbans. Over 120 species of fish are known to be caught in this forest. The major fishes caught are prawns and hilsa. In recent years,prawn has become an important commercial fish and exported in increasing number to foreign markets, while hilsa is a very favourite variety demanded all over Bangladesh. During the period of 1983-84 to 1987-88, on an average, fish catching from the forest was about 7700 metric tons. Honey and beeswax are extracted in different quantities in different years. The average for the four years,1983-84 to 1987-88 are about 240 and 60 m tons respectively. Sundarbans honey is in much demand in the country. Both the products, honey and wax,have various uses including that as drug in traditional medicine.Shells are collected throughout Sundarbans to be burnt to produce lime for use with betel leaves or for use in building construction.The forest products of the Sundarbans have a great value in the revenue sector of the country, this forest along is contributing about 50 per cent of the revenue earned by the forestry sector (Table 4).

A considerable number of population is directly dependent on the mangrove Sundarbans, but due to lack of concrete information is difficult to assess the aggregate level. Katebi and Habib (1990) [12] estimated the dependent population numbered to 300,000 on the Sundarbans(Table 5). This may be an underestimate. However,ESCAP (1987)[11] estimated that some 500,000-600,000 people depend directly on this forest for their means of livelihood for at least half of the year.

REFERENCES

1. Neftel,A.,E.Moor,H.Oeschger, and B.Stauffer,1985 :Evidence from polar ice cores for the increase in atmospheric CO_2 in the past two centuries, Nature, 315 , 45-47 &

   Friedli, H.,H.Loetscher,H.Oeschger,U.Siegenthaler, and B. Stauffer, 1986: Ice core record of $^{13}C/^{12}C$ record of atmospheric CO_2 in the past two centuries, Nature,324, 237-238.

2. Craig, J., and C.C.Chou,Methane: 1982: The record in polar ice cores, Geophys. Res. Lett., 9,1221-1224.

3. IPCC,1990, Climate Change; The IPCC Scientific Assessment, Cambridge University Press, New York.p.19.

4. Elkins, J.W.,S.C.Wofsy, M.B.McElroy,C.E.Kolb,and W.A.Kaplan, 1978: Aquatic sources and sinks for nitrous oxide,Nature, 275, 602-606. &

   Cline, J.D., D.P.Wisegraver, and K.Kelly-Hansen, 1987:Nitrous oxide and vertical mixing in the equatorial Pacific during the 1982-1983 El Niño, Deep-Sea Res., 34, 857-353.

5. IPCC, 1990, Climate Change : The IPCC Scientific Assessment, Cambridge University Press, New York, p.207.

6. Sinclar, Jan, 1991: Global Climate Change Has Already Begun, Our Planet, Newsletter ofUNEP,P.14

7. NAS, 1979: Carbon Dioxide and Climate: A Scientific Assessment, National Research Council, Washington, D.C.

8. WMO,1986: Report of the International Conference on the Assessment of the Role of Carbon Dioxide and of Other Greenhouse Gases in Climate Variations and Associated impacts, Report Published by the World Climate Programme, World Meteorological Organisation, Geneva,Switzerland.

9. Chattey, D.R.,F.R.Millor and J.H.Sandom, 1985: A Forest Inventory of the Sundarbans, Bangladesh,Overseas Development Administration Surrey, U.K.p.17-22.

10. Choudhury,M.A.,Quadir, A.D., and Islam, J.1990: Study of Chokoria Sundarbans Using Remote Sensing Techniques, Space Research and Remote Sensing Organisation(SPARRSO),Dhaka, Bangladesh.

11. ESCAP,1987: Coastal Environment Management plan for Bangladesh, Part II, Bangkok, Thailand.

12. Katebi, N.A.M. and Habib, G., 1990: Sundarbans and Forestry, Coastal Area Resource Development and Management, Part II, Dhaka, Bangladesh.

13. IPCC,1990, Climate Change: The IPCC Scientific Assessment, cambridge University Press,New York, p.271.

14. IPCC,1990, Climate Change: The IPCC Scientific Assessment, Cambridge University Press, New York, p.278.

15. Milliman, J.D., Broadus, J.M. and Gable,F., 1989: Environmental and Economic Implications of Rising Sea Level and Subsiding Deltas: The Nile and Bengal Examples, Ambio, Vol.18,No.6 pp.340-345.

TABLE 1 : SUMMARY OF KEY GREENHOUSE GASES INFLUENCED BY HUMAN ACTIVITIES

Parameter	CO_2	CH_4	CFC-11	CFC-12	N_2O
Pre-industrial atmospheric concentration (1750-1800)	280ppmv	0.8ppmv	0	0	288ppbv
Current atmospheric concentration (1990)	353ppmv	1.72ppmv	280pptv	484pptv	310ppbv
Current rate of annual atmospheric accumulation	1.8 ppmv (0.5%)	0.015ppmv (0.9%)	9.5 pptv (4%)	17pptv (4%)	0.8ppbv (0.25%)
Atmospheric lifetime (years)	(50-200)	10	65	130	150

Source :IPCC 1990 [3]

TABLE 2 : HALOCARBON CONCENTRATIONS AND TRENDS (1990)

Halocarbon		Mixing Ratio pptv	Annual Rate of Increase pptv	%	Lifetime Years
CCl_3F	(CFC-11)	280	9.5	4	65
CCl_2F_2	(CFC-12)	484	16.5	4	130
CCl_3F_3	(CFC-13)	5			400
$C_2Cl_3F_3$	(CFC-113)	60	4-5	10	90
C_2ClF_5	(CFC-115)	5			400
CCl_4		146	2.0	1.5	50
$CHClF_2$	(HCFC-22)	122	7	7	15
CH_3Cl		600			1.5
CH_3CCl_3		158	6.0	4	7
$CBrClF_2$	(halon 1211)	1.7	0.2	12	25
$CBrF_3$	(halon 1301)	2.0	0.3	15	110
CH_3Br		10-15			1.5

Source : IPCC (1990) [3] Page : 22.

TABLE 3 : RESOURCES EXTRACTED FROM SUNDARBANS DURING 1983-88

Kind of forest produces	1983-84	1984-85	1985-86	1986-87	1987-88
Round timber excluding Gewa (in m^3)	131380	79154	89149	36491	56790
Gewa (in m^3)	135588	251429	1800844	170736	162099
Misc. fuelwood (in M.ton)	126325	90153	50275	46131	58050
Goran "	56272	77523	69830	70249	69102
Golpatta "	63365	61281	61443	70702	78620
Fish "	9173	8043	8016	6801	6221
Honey "	260	256	225	229	222
Wax "	65	64	56	57	56
Shell "	2430	2675	2657	3184	3599
Grass "	7712	9306	11065	11740	13441
Hental "	6870	8911	5412	6087	7848

Source : [12]

TABLE 4 : COMPARATIVE STATEMENT OF THE REVENUE EARNED BY THE SUNDARBANS FOREST DIVISION AND FOREST SECTOR OF BANGLADESH

Year	Total revenue earned by Forest Department (Million Tk.)	Revenue earned by Sundarbans Forest (Million Tk.)	Percentages
1978	108.11	50.14	46.37
1979	165.50	68.34	41.29
1980	188.00	87.77	46.79
1981	288.00	146.45	50.85
1982	292.62	158.71	54.24
1983	302.62	140.00	46.25
1984	343.77	160.75	46.76
1985	455.65	219.41	48.15
1986	605.87	269.36	44.46
1987	524.11	230.81	44.04
1988	-	282.65	-

Source : [12]

TABLE 5 : NUMBER OF PERSONS DIRECTLY DEPENDENT ON SUNDARBANS

Nature of work	Type of labourer	No.of Labourer	Total population of greater Khulna & Barisal	Total per cent depending on Sundarbans
a. For extraction of timber, fire-wood golpatta and other minor produces and transportation inside and out-side the forest	Bawali	150,000	Greater Khulna =3557,414	
b. Honey and wax extraction	Mowal	8,000	Greater Barisal = 3928,414	
c. Collection of shell and preparation of lime	Jongrakuta	3,000	Total 7485,828	
d. Boat Building and repairs	Carpenter	2,000		4.7%
e. Sawing of timber by sawmill and sawpit	-	1,000		
f. No.of workers working in industries using raw materials from Sundarbans	-	15,000		
g. No.of business-men engaged in forestry business	-	115,000		
h. Other kinds of labourer	-	10,000		
TOTAL		304,000		

Source : [12]

TABLE 6: NO. OF EMPLOYEES IN DIFFERENT INDUSTRIAL SECTORS WHICH ARE DIRECTLY DEPENDENT ON THE PRODUCTS OF THE SUNDARBANS

Industrial Sector	Employees
Forest Department	1038
Saw Mills	1400
Pitsaws	1057
Khulna Newsprint Mill	2250
Hard Board Mill	269
Match Factory	4000

Source : [11]

FIGURE 1 : MAP OF SUNDARBANS

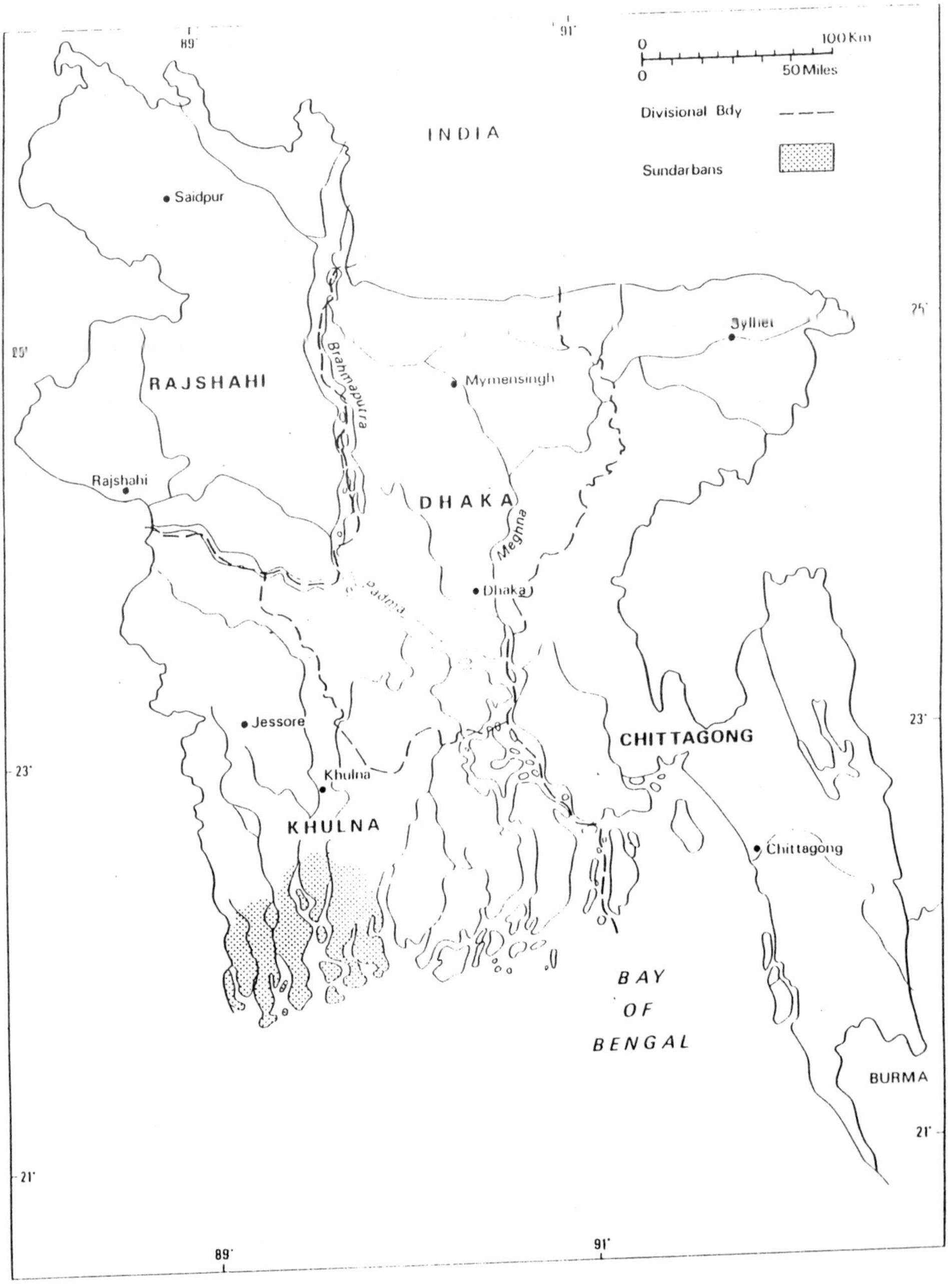

RESEARCH AND DEVELOPMENT PROGRAM RELATED IN MITIGATION OF GLOBAL WARMING TRENDS IN CHINA

Bao Deyou

Chinese Association of Hydrogen Energy
Huayuan Road 3 Haidian, Beijing 100083 China

ABSTRACT
The main reason of global warming is the greenhause Effect gas emission to the atmosphere that is Carbon dioxide (CO2) and other gases. Chinese Scientists at a domestic meeting of "Climate change and environmental Protection" proposed some measures to solve these problems that is :1. Afforestation 2. Carbon dioxide reduction research.3. Hydrogen energy implementation such as hydrogen vehicles. 4. Clean burning coal technology R & D .5. Hydrides technology project ready for fusion reaction and energy conservation.

INTRODUCTION

On the Growing population and CO2 emission and due to unlimited fossil fuel use, the global warming and climate change phenomena is more serious in recent years in worldwide and in China. Chinese government very importance the environmental protection problem, because in our country will have 1.2 billion people in the end of this century.In addition, the what so called "Natural disasters" are more frequently as before, such as Flooding, draught, typhoon, hurrican, etc., the more serious damage of the Bangladesh hurrican recently are almost killed 140 thousand people, and in China, recently serious flooding never before in last 40 years, but in my opinion,this is not a "Natural Disasters" in certain meaning it is a man made disasters and related closely to the climate change and due to the Greenhause Effect and caused by CO2 emission. It is obvious that, decreasing the CO2 emission is a urgent task for all the country in the world, all people and scientists in the world must attach importance this problem.In China in Jan 1991, we have a nationwide Symposium for "Climate Change and Environmeatal Problem Discussion" attended by related scientists such as meteorologist, climtologist Environmentalist, energy experts, chemists and so on, this meeting is hosted by the China Association of Science & Technology and sponser by the State Council. All the participants suggested and proposed a lot of measures and some research and developing project for mitigation of the global warming trends.

MEASURES AND PROJECTS FOR MITIGATION OF GLOBAL WARMING

Afforestation

The main fuel and energy in China in recent years are unavoidable to dependence with coal, according to prediction and statistics, at the end of this century, China will consumer more than 1 billion tons of coal every year, it means will emission more than 3 billion tons CO2 per year, although the CO2 emission per capitation is more less than the developed country but we still must to trying for decreasing the CO2 concentration in atmosphere by every possible means. According to the practicle condition of China today, the afforestation is a best solution for this problem, it is a easily to realize and more effective method. Trees and plants can not only absorb great deal CO2 but also can shelter sand from wind and also can adjust moisture and rainfall and benefits for agriculture production. Under the instruction and initiate by top leader Teng xiaoping, a masses compaign have brocken out in whole country, especially in north China and forming a shelterbelt in north border of China, the massive compaign will continue to the end forever not only in border area but also in cities and townships. Afforestation in China now have been obtained good results, for example, in Beijing in the year before large scale afforestation, there are about 6-7 strong wind and sand weather day every year, but after 1987, only 3-5 strong wind and sand day per year, on same reason, it is also benefits for agriculture and environmental protection and ecology balance.

Carbon Dioxide reduction

Although trees aand plants can absorb large amount of CO2, but the CO2 concentration in atmosphere will still in increasing state, it is due to unlimited burning of fossil fuel and excessive CO2 emission, so that we must try every means to decreasing the CO2 concentration in atmosphere, one of the effect way is the chemical reduction of CO2 and produce methanol, methane, hydrogen and other organic compounds using a catalyst and solar energy as its primary energy in this reaction, if we realized this project, we can turn the damage of CO2 to boon and form a favorable cycling in the environment, it is also an effective method for storage solar energy in the form of chemical products, this reaction system can be express as the following diagram:

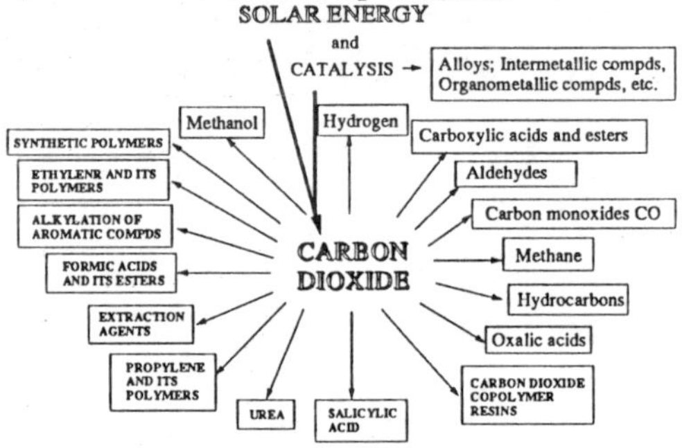

SOLAR ENERGY CO_2
REDUCTION AND ITS PRODUCTS

The main chemical reaction process can express as follows:

$$CO_2 + 2H^+ + 2e^- \rightarrow HCOOH$$
$$CO_2 + 2H^+ + 2e^- \rightarrow CO + H_2O$$
$$\rightarrow H_2 + CO_2$$
$$CO_2 + 4H^+ + 4e^- \rightarrow HCOH + H_2O$$
$$CO_2 + 6H^+ + 6e^- \rightarrow CH_3OH + H_2O$$
$$CO_2 + 8H^+ + 8e^- \rightarrow CH_4 + 2H_2O$$
$$\rightarrow 4H_2 + CO_2$$

and also, CO2 can be produced into some new polymers and plastics as follows:

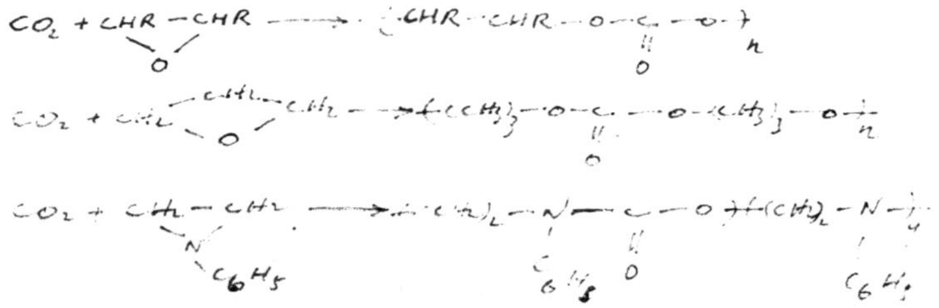

These is polyethlene oxide,polypropyleneoxide and its esters from CO2,they are a kind of transparent and soft polymers or plastics and very useful in daily life. carbon dioxide can also used in artificial rainfull and we can controll the artificial rain in night and sunshine in day, it will benefits for solar energy recovery and application and will promote the solar energy utilization. these three kinds measures could decreasing the CO2 concentration in atmosphere and is a main project for mitigation of global warming trends.

Hydrogen energy implementation such as hydrogen vehicles

Hydrogen vehicles in dual mixed fuel such as gasoline and 5% hydrogen has tested in China more than 3 departments with TiFe hydrides or misch metals hydrides,all the experiments results obtained almost same, it could saving 25--30% gasoline and lower poisoning gases emission, recently, we plan to running regularly in a bus line with a by product hydrogen of an ammonia plant in its purge gas mixture and separated by a hydrides separator. Also in the sodium hydroxide producing factory, a lot of byproduct hydrogen can be used as a fuel, the total amount of such byproduct hydrogen in China will have 1 billion M3 annually,most of this kinds of hydrogen are wasted or not fully used up to Now, a part of the hydrogen are already used in floating glass production and other department.

In a Biology Research Institute , we have developed a hydrogen production by microorganism-bacteria and some of the wasted materials can used as its fermentation raw material,in recent years, they will find out and isolate more effective hydrogen producing bacteria. Photocatalytic spliting of water in sun light and advanced electrolysis of water for producing hydrogen are also in proceeds in China.

Hydride electrode nickel batteries are widely researched in many institutes and universities in China, one of the effective application of this kind of rechargable battery is in traffic vehicles, it is a absolute non-pollution vehicles and with high efficiency in energy consumption, every charging of electricity can drive more than 150 kilometers, it is our first step of research plan. Due to the serious pollution of the traffic vehicles, Chinese scientists always in great effort to research non-pollution vehicles as mentioned above.

We also in research and developing an advanced SPE fuel cell and ready to use in vehicles this kinds of vehicles are more effective than the battery vehicles.

Clean burning coal technology R & D.

It is irreversible that coal will continue to burning for a long time in China, So that we are always attach importance the clean burning technology R & D including low SO2 emission and more perfect burning technology in some large burning coal power generation plant and collective treatment of emission gas of blast furnace, converters and other CO2 containing wasted gases such as ammonia synthetic plant etc.some of the technology are import from foreign country, some of the other project already conducting in research institutes such as gasification of coal and coal liquidfication for producing liquid fuel from coal.

Hydrides technology project ready for fusion reaction and energy conservation

Hydrides technology such as hydride heat pump, hydrides heat storage, hydrides purifier, hydrides hydrogen storage, hydrides refrigerator and hydrides air conditioning etc.. According to the prediction of German scientist Berchner, if all the energy related industry fully utilized the hydrides technology, the energy efficiency of all the economy would increasing 20-25%. In china there is more than 5-8 universities and institutes engaged in research and developing hydride technology mentioned above and have been obtained a great progress, if these project were success, a lot of energy can be saved and also will be benefits environmental protection.Especially in traffic industry ,the pollution problem in urban area will notablly improved.Chinese scientists are now focus their effort in research more high capacity advanced hydrides for further improving the efficiency of hydrogen storage technology.

Certain hydrides can effectively separation the isotopic of hydrogen such as deuterium and tritium, this technology will very useful in fusion reaction process, if fusion reaction would put into parctice and will produce a lot of energy, it is a great contribution to mankind and will completely keep human being from pollution.

REMARKS

Global warming problem is a very complicated problem, it is not only a technical but also a social problem, for example, a very important problem is birth rate control for decreasing population growth rate problem etc.. So that for mitigation of global warming trends must pay attention by all international society and must be a long time effort. All the measures we mentioned above are not enough, at first we must arouse the environmental consciousness of all the people and should gradually increasing the environmental investment including the invest of transfer to realize clean energy especially HYDROGEN ENERGY and must take measures as soon as possible, otherwise it will be have a serious consequences and almost impossible to predict. Chinese scientists with great concern about this problem and determine to make great effort in all sides for mitigation the global warming trends and make benefits for all the mankind.

REFERENCES

1. Proceedings of the symposium of "Climate change and environmtal protection" Jan. 1991 Beijing(in Chinese).
2. Bao Deyou:"Hydrogen and hydride technology activity in China" proceedings of the 8th WHEC Honolulu Hawaii U.S.A. 22-27 July 1990 Vol.2 pp.1081-1084.
3. Bao Deyou "Carbon Dioxide Reduction" Proceedings of the 8th WHEC Honolulu Hawaii in Special Symposia "Global Greenhouse Warming".

SAFETY CONSIDERATIONS IN THE DESIGN OF HYDROGEN-POWERED VEHICLES

James G. Hansel, Glenn W. Mattern, Robert N. Miller

Air Products and Chemicals, Inc.
Allentown, Pennsylvania

for presentation at

The International Association of Hydrogen Energy
and
The American Academy of Science
Technical Seminar

16-18 September 1991

Independence, Missouri

ABSTRACT

There have been several recent and innovative demonstrations of hydrogen fueled road vehicles. Hydrogen's high energy content, renewability, and virtually pollutant-free oxidation properties make it attractive as a fuel to provide mechanical energy using conversion devices such as an engine or a fuel-cell (with an electric motor).

This paper discusses the unique flammability properties of hydrogen as well as its unique migration (leak) properties, and how they influence the design of the fuel storage system and the conversion device, as well as the design of many aspects of the vehicle itself. Materials selection and piping selection are also discussed. These considerations influence not only the design but also the procedures utilized over most aspects of the vehicle's useful life. These include fueling, parked or stored condition, servicing, and scrapping.

Over one-half million operating compressed natural gas vehicles provide an excellent safety learning experience for several aspects of the design of hydrogen vehicles, including crash-worthiness. These are discussed.

Hydrogen suppliers within the industrial gas community and the NASA space shuttle effort have provided valuable hydrogen equipment design and operability experience which can be applied to hydrogen vehicles. The concept of inherent safety can also be invaluable to ensure that hydrogen-powered vehicles are acceptably safe.

INTRODUCTION

Interest in research and/or demonstration vehicles that are powered by hydrogen has grown considerably within the past decade. Applications include automobiles,[1-4] light- and medium-duty trucks,[5] buses,[6] and mining vehicles.[7] Various designs have been considered including feeding on-board hydrogen from compressed gas, liquid, or hydride storage vessels to power an internal combustion engine, or the most recent advance of feeding hydrogen from a hydride storage vessel to a hydrogen/air fuel-cell where the fuel is converted directly to electricity to power an electric motor vehicle.[1] The motivation behind these hydrogen-fueled vehicle programs includes the renewability of hydrogen as an energy source, as well as its zero pollutant emission potential when fuel-cells are used or its very low pollutant emission potential when combustion engines are used. Ironically, the unique oxidative properties of hydrogen, which make it attractive for fuel-cell operation or very lean engine operation, are the same properties that create the risk of fire and/or explosion from hydrogen leaks.

Hydrogen, worldwide production of which exceeds 15 million tons per year, has demonstrated an acceptably safe track record when used as an industrial gas or as a rocket propellant, albeit under carefully controlled conditions using skilled operating and service personnel. However, the data base for operation of hydrogen-powered road vehicles is very limited.

Comparatively, Compressed Natural Gas (CNG)-powered vehicles have accrued an admirable safety record over a relatively large data base. Currently, there are over 600,000 CNG vehicles worldwide operating under a wide variety of conditions. They are generally operated and serviced by less skilled individuals than are hydrogen-powered vehicles.

However, hydrogen is more flammable than LNG or natural gas. Thus, if hydrogen-powered vehicles are to develop into widespread use as fleet vehicles, and perhaps later by the public for private-use vehicles, very careful consideration will have to be placed on their design to ensure that they are acceptably safe. This paper discusses some of the unique properties of hydrogen that affect its safe performance as a fuel in vehicular systems and then some of the key challenges that must be considered in the safe design of hydrogen-powered vehicles.

UNIQUE PROPERTIES OF HYDROGEN

Flammability Properties

Wide Flammability Range

The flammable limits of hydrogen in air range from 4 to 74 vol%, the widest of any common gaseous fuel. By comparison, methane as a typical fuel is 5 to 15 vol%. Thus, in a leak or release situation, if all other considerations are equal, hydrogen fuel is much more likely to be within the flammable range when the hydrogen-air mixture reaches an ignition source. Further, upon ignition, because of this unusually wide flammability range of hydrogen in air, a larger fraction of the accumulated hydrogen will burn.

Low Ignition Energy

The ignition energy of hydrogen is only one-seventeenth that of methane. As a result, low-energy ignition sources, such as ordinary static electricity discharges—which can be generated by simply walking across a carpeted room—are usually more than sufficient to ignite hydrogen. Even the high-flow discharge of hydrogen through a relief valve or burst disc has been known to ignite.

The auto-ignition temperature (AIT) of a hydrogen/air mixture is an additional indicator of flammability, but care must be exercised in using and interpreting published values.[8] For increased safety, a high AIT value is more desirable. The *minimum* AIT, which is often the only AIT value published for a gas, is measured under a "quasi-adiabatic" condition within an enclosed, heated glass vessel. As seen in Table 1, this specific test indicates that the AIT for hydrogen is actually higher, and therefore better than that of propane.

In actual flammability situations, the AIT depends upon how the hydrogen (or any fuel/air mixture) is heated. A number of non-enclosed, non-adiabatic heating situations may occur in a vehicle during normal operation, refueling, or immediately following an accident, such as a hot engine exhaust manifold, a ruptured organic hydride conversion catalyst, or a shorted (overheated) wire, and these should be considered in the flammability evaluation. Table 1 shows data for two AIT test methods —the heated nichrome wire and the heated laminar jet test—that are used to measure the AIT under open, non-adiabatic conditions. Clearly, these data show that the AIT of hydrogen is indeed significantly lower than methane or propane, a comparative disadvantage for the use of hydrogen as a vehicle fuel.

Table 1
Auto Ignition Temperature (AIT) in Air for Three Fuel/Air Mixture Heating Situations

	Heated Glass Vessel* (~ cm. diam.)	Heated Nichrome Wire (1 mm diam.)	Heated Laminar Air Jet (1 cm. diam.)
Hydrogen	520°C	750	640
Methane	630	1220	1040
Propane	450	1050	885

*Minimum AIT

Colorless Flame
When a hydrogen/air mixture is ignited, the flame that is produced during combustion is nearly invisible, although diffraction patterns caused by the heat may be visible under some lighting conditions. The presence of contaminants in the flame may also impart some color (e.g., sodium yellow from ordinary salt). The fact that the hydrogen flame is nearly invisible poses a unique hazard to property and personnel since a fire is often not anticipated. Even if one is looking for a fire, a casual glance over a hydrogen piping system may not necessarily reveal it. Hydrogen fires, occurring in and around the fuel lines and fuel system, as with any fuel, are especially dangerous since the flame and heat of the fire can impinge on and ignite nearby components, thereby escalating the fire situation.

The wide flammability limits, the ease of ignition, as well as the ease of migration of hydrogen, make accidental hydrogen fires more likely relative to other gaseous fuels—but only when there is a release of hydrogen into the air. Hence, a key safety objective is to eliminate the source of hydrogen leakage via design, and if a leak should develop during operation and use, procedures should be established for detection and repair.

Hydrogen is also a simple asphyxiant, displacing the oxygen in air required to support life. Generally, the flammability hazard overshadows the asphyxiation hazard.

Characteristics of Hydrogen Leaks

The relatively small molecular diameter of hydrogen (1.58 Å), compared to all other gaseous fuels, makes fuel lines and systems much more susceptible to hydrogen leakage. Excluding fractures and catastrophic ruptures, common hydrogen leaks fall into two categories: those that involve metal-to-metal seals (e.g., flared or compression joints) and those that involve non-metal seals (e.g., gaskets, O-rings, packings, and pipe-thread compounds). Metal-to-metal joints, once made up properly and leak checked, tend to remain relatively leak free. However, thermal expansion/contraction, vibration, and externally applied forces may eventually cause the joint to leak, but generally the leak seldom becomes catastrophic. On the other hand, non-metal seals may produce sudden catastrophic leaks since it is possible to displace the sealant material with the flowing leaking gas.

Depending upon the gas velocity and the orifice size at the site of the leak, many leaks are probably turbulent, and the turbulent mixing of the ambient temperature hydrogen jet with surrounding air is surprisingly rapid. Thus, within about five diameters of such a leak "hole", air will reach the center line of the jet, and within roughly 500 to 1000 diameters, the jet will essentially attain the composition of air.[9] For example, for a 1/32 in. diameter leak, five diameters is only 0.15 in. and 500 to 1000 diameters is 16 to 31 in. This rapid dilution implies that if the turbulent leak were into the open air, the flammable zone would only exist relatively close to the leak.

The rapid dilution of the jet into the air also implies that the extreme buoyancy of hydrogen relative to air (hydrogen is only 7% as dense as air) rapidly declines with distance away from the leak hole. Under this condition there is a reduced tendency for the leaking hydrogen to continue to rise. Note that these considerations do not apply to hydrogen leaks within confined spaces because within a confined space, due to recirculation, the air being entrained has increasingly higher hydrogen content.

Not all leaks provide turbulent jet mixing. Weeping leaks, in which the hydrogen must follow a tortuous path, have relatively low velocity as they exit into the open air. As a result, jet mixing is poor and diffusion becomes the dominant mode of mixing. Diffusion of hydrogen in air is very rapid relative to some other gaseous fuels, as shown below:[10]

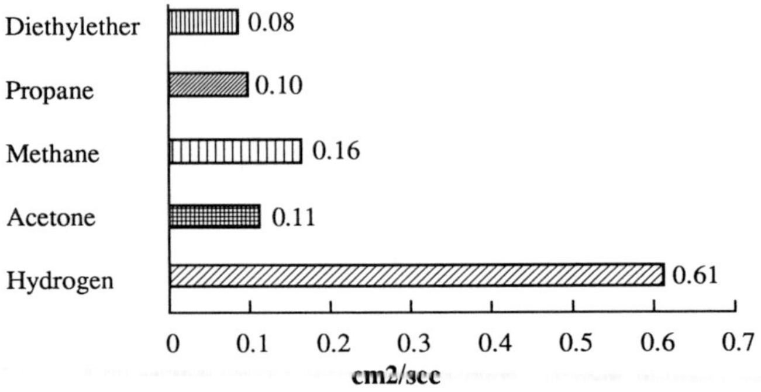

Diffusivity of Gases in Air at Ambient Conditions

If all other factors were equal, this rapid diffusion of hydrogen in air would greatly aid in reducing the risk of hydrogen fires relative to other fuels. Ironically, this enhanced diffusion property of hydrogen relative to other fuels is a net result of its small molecular size—the same property which enhances its ability to leak from systems relatively easily.

Hydrogen buoyancy and diffusion effects in air are typically only important when there is an absence of air currents, such as may occur under the hood or near the fuel line of a parked vehicle. The effects of a slight ambient wind, very slow vehicle motion, or the currents produced by the radiator cooling fan will generally outweigh buoyancy and diffusion effects, and thus further reduce the fire hazard due to a small hydrogen leak.

Hydrogen Embrittlement

In addition to flammability and leak considerations, the issue of hydrogen embrittlement of certain steels, especially ferritic steels, is a major concern in the safety design of hydrogen-fuel systems.[11] The subject is still undergoing a relatively rapid growth in understanding.[13,14] It is known that sensitivity to hydrogen embrittlement is influenced by several parameters, including plastic deformation, cyclic loading, hydrogen purity, temperature, and pressure. Without question, material selection will be an important factor for vehicle applications, and perhaps certification of materials for hydrogen service may be necessary.

SAFETY DESIGN CONSIDERATIONS FOR HYDROGEN-POWERED VEHICLES

On-Board Hydrogen-Fuel Storage Systems

Several methods have been considered for the on-board storage of hydrogen fuel: high-pressure compressed gas stored in cylinders, liquid hydrogen maintained in cryostat storage vessels, or hydrogen in the form of a readily convertible compound, such as a metal hydride or a liquid organic hydride stored in a pressurized tank. Safety design considerations for each system will vary and will depend upon which approach is employed for specific vehicles.

Metal Hydrides

Much recent attention has been given to the use of metal hydrides as a desirable means to store hydrogen fuel on-board a vehicle, and as a result metal hydrides are undergoing development on many fronts.[1,2,7] The inherent safety of very low hydrogen release rates, if the vessel containing the hydride were to rupture, is no doubt a major factor in the development. Another inherent safety feature of the metal hydride storage system is the relatively low pressure of operation—usually no higher than a few hundred psi, and often under 100 psi during operation. Further, the exothermic reaction during hydrogen charging is generally self-limiting as long as the charging pressure is limited. During charging, the surface temperature of the hydride vessel is usually limited to that of hot glycol.

Despite these major advantages of hydride systems for storing hydrogen in vehicles, there are a number of considerations important to the safety design of the hydride container:

- Intended working pressure and relief systems (with redundancy).

- Compatibility of the hydride and hydrogen with the container, and compatibility of the container with surrounding glycol (or other coolant)

- Stress build-up within the hydride powder which, under certain conditions, may crack the container.[15]

- Attrition (break-up) of the hydride particles and potential plugging of relief systems and other piping.

- Quality of welds meeting regulatory codes. Vessel should meet ASME Boiler and Pressure Vessel Code, Section III—Pressure Vessels, or United States Department of Transportation or equivalent specifications.

- Leak and hydrostatic testing following design code requirements.

Liquid Organic Hydride Storage

This relatively new concept for a hydrogen storage system is also undergoing development.[16] One promising system utilizes a catalyst at about 400°C to decompose methylcyclohexane into hydrogen and toluene at a pressure of about 300 psi. Although the pressures are moderate, the high temperature of the catalyst can be an ignition source if the vessel is ruptured. The minimum AIT values for toluene and methylcyclohexane are 480°C and 250°C, respectively.

Compressed Hydrogen Storage

This mode of hydrogen storage is considered based upon the highly developed practice utilized for CNG.[17] Of particular interest in recent developments are lightweight composite vessels. However, for this particular application, the compatibility of hydrogen with the materials of construction must be considered and the vessels must meet U.S. Department of Transportation or equivalent requirements.

The most vulnerable portion of any high pressure vessel is typically the inlet/outlet nozzle connection. It is exceptionally important that this area be protected from collision or other damage. One means is to extend the vessel walls beyond the end of the vessel in order to cover the nozzle connection.

CNG vessels and CNG fuel systems, in general, have demonstrated an admirable safety performance over more than three million vehicle years.[18-21] Part of this success has been the crashworthiness of CNG vessels, especially in comparison to conventional gasoline tanks. The fuel system integrity of CNG vehicles has been excellent.

Liquid Hydrogen Storage

This mode of hydrogen storage is unique in that liquid hydrogen cannot be stored indefinitely. After a period of time, if hydrogen is not utilized fast enough, the heat leak will eventually require venting of some hydrogen. If this were to occur, indoor or otherwise confined-space vehicle parking would not be practical, unless a means were available to temporarily adsorb the gaseous hydrogen. The heat-leak rate depends upon the type of vacuum-jacketed (V-J) insulation construction of the liquid hydrogen vessel, and for a given type of V-J construction, the size of the vessel. Smaller vessels (with a higher surface to volume ratio) provide proportionately higher heat leak rates. For example, an 11,000 gallon, liquid hydrogen V-J (overseas) shipping container of sophisticated construction (including a liquid nitrogen shield), will not vent for over three weeks. However, a less sophisticated 20-gallon liquid hydrogen vehicle fuel-tank will begin to vent within a week if the fuel is not utilized. Current technology provides a boil-off rate from the automobile hydrogen tank of about 1.8% per day.[11] This venting, like any other venting from a vehicle, requires a careful location of the vent.

Commercial liquid hydrogen dewars, trailers, and overseas shipping containers, all employing V-J construction, have accumulated a large data base of safe operation. This is due in part to the use of the V-J vessel-within-a-vessel construction, similar to a thermos bottle, which provides extreme integrity to resist puncture in an accident. The application of this design would be a strong benefit to storing liquid hydrogen in a vehicle.

A number of special circumstances must be considered when sizing the vents and /or relief devices when using this type of V-J constructed tank in a vehicle:[12] If the vacuum fails for any reason (e.g., a random failure of a vacuum fitting or weld or due to a vehicle accident), the hydrogen boil-off will be rapid. Thus, relief devices and vent lines must be appropriate to handle this situation. If, in addition, the V-J constructed fuel tank is engulfed in a fire and the annulus becomes filled with hydrogen, the boil-off rate will increase further, requiring even larger relief devices and vents. Moreover, if the vehicle is on its side such that the relief devices are below the liquid level, the size of the relief devices must be even larger. The latter situation often governs the ultimate design size of the relief device and vent pipe. Finally, redundancy of relief devices must always be considered when using liquid hydrogen.

Liquid hydrogen storage and handling, as well as cold-vapor handling, also requires that attention be paid to cold fracture of carbon steel below -30°C, liquid air formation on the outside of exposed lines or under their insulation, and the possibility that a cryogenic temperature line could cause an adjacent carbon steel component to be cooled below -30°C.

Crashworthiness and Location of the Storage Device

The location of the fuel storage device should follow the practice used for gasoline tanks in recent-model automobiles; that is, to move the tank away from the vehicle perimeter. Hydrogen tank locations in vehicles have been briefly discussed by Pfriem[11] and Quadfleig.[22] Whatever the location, the fuel storage system should be sufficiently anchored to the chassis of the vehicle so as to withstand a force of 8 g's in any direction.[19]

As many of the potential leak sites (e.g., valve packings) as possible should be located outside of a poorly ventilated space. If such sites are located within an unventilated space, then it is advisable to connect them to the exterior via a sheet-metal duct. Also, for example, placing louvers within a trunk lid with corresponding openings in the bottom will provide a natural circulation path for leak flows. On the BMW hydrogen-powered test automobiles, in the event of a crash, the windows, doors, sunroof, and trunk lid open automatically.[2]

Hydrogen Venting and Location

Venting of the stored hydrogen may be necessary during emergency conditions. Generally, emergency venting via relief valves can be connected to a central vent in a safe location away from people. However, burst discs, which would be vented in an extreme emergency, may not always be vented to a central safe location.

The buoyancy of hydrogen relative to ambient air declines with colder hydrogen gas temperatures. This may necessitate a change in vent location for hydrogen gas from liquid hydrogen systems. The buoyancy of hydrogen becomes zero near the boiling point of liquid hydrogen (-253°C at 0 psig). Venting of liquid hydrogen, however, will result in running and flashing liquid which can crack, via cold-fracture, any carbon steel that it contacts.

The natural buoyancy of non-cryogenic hydrogen should always be utilized for safe venting. Generally, the worst-case situation is a parked or stationary vehicle. Hence, the following considerations should be used in the design and location of the vent:[12]

- Whenever practical, vents should be directed straight upward to take advantage of the natural buoyancy of hydrogen. However, provision should be made to ensure that the line is drained of any moisture and rain water to avoid possible plugging of the vent should ambient conditions be below freezing.

- Distance from the vent to (electrical) ignition sources should be maximized.

- Vent locations should be positioned where hydrogen can never enter confined spaces.

- Vents should be kept away from the outside perimeter of the vehicle where individuals could come into close proximity to the vent outlet. Individuals passing casually by the vent outlet may provide an ignition source (static electricity or a lit cigarette), or they may be burned if there is an invisible hydrogen fire at the vent.

Fuel Lines and Hydrogen-Wetted Components

All hydrogen fuel lines and hydrogen wetted components in the vehicle must obviously be compatible with hydrogen. All components and lines should be exterior to the vehicle and should avoid enclosed areas to the extent possible. The propensity for external corrosion (e.g., road salt during winter travel) should be considered when selecting the materials of construction for the external components and lines. The fuel lines should be ductile in order to withstand impact forces without breaking, and vibration from the road or the engine must be dealt with in the design.[12]

The most important concern for the hydrogen fuel-lines is leakage from joints. Designers of industrial hydrogen equipment strive to avoid or at least significantly minimize all joints in the piping.[23] Since serviceability must be considered in the design of a vehicle, it may be necessary that some of the lines will require some joints. Nevertheless, threaded connections and flanges are highly undesirable and must be avoided. Flared joints and certain O-ring-joints are far better, but welded/ brazed lines are by far considered the best for hydrogen use. However, it is important to ensure that the welders are certified and that the welded joint has been properly hydrostatically tested and leak checked. Brazing material should have a melting point over 538°C.[26] Low-melting-temperature metals, plastics, and elastomers should be avoided anywhere within the hydrogen fuel-line system, as they could easily fail in the event of a fire and permit hydrogen to be released.[23]

Valve design should focus on the tightness of shut-off and resistance to packing leakage; positive shut-off or isolation of the hydrogen is essential, and packless valves should be considered wherever possible. Emergency shut-off valves of a "failed-closed" design which can close automatically upon sensing a hydrogen leak or upon sensing a high g-level (i.e., in an accident), require care in determining their location. There should be a minimum of exposed line between the hydrogen source and these valves, and the valve actuator and stem should be located such that they will have a low probability of being damaged in an accident.

Engine Systems

Considerable progress has recently taken place in the development of hydrogen/air spark-ignition or compression/ignition engines.[2,5,22] The use of fuel injection to replace carburetors has significantly reduced the major problem of flashback or back firing of hydrogen in the intake system. Nevertheless, because of the extremely low ignition energy of hydrogen, stray spark-plug discharges, other static discharges, or hot surfaces may still cause unwanted pre-ignition. However, these situations are generally confined to the engine and might not directly affect vehicle safety, unless they are extensive or prolonged.

Avoidance of crank case explosions should also be factored into the engine design.[7] These unwanted occurrences may be caused by the hydrogen/air mixture leaking past the piston rings and igniting via hot surfaces.

The major engine safety concern, however, is leakage of hydrogen from the engine fuel system to the outside air or through the shut-off valves into the engine when the engine is shut-down. This is especially critical because many ignition sources will be near the engine (e.g., the engine ignition system, the alternator, and the starter). Thus, every effort must be made in the design to eliminate internal and external leakage points and to take certain steps to minimize ignition sources.[12,24] Every effort should also be made to use electrically intrinsically safe (e.g., non-arcing) or electrically classified components, especially near areas of potential leaks.

Fuel-Cell Systems

Most of the above safety concerns for the spark-ignition engine also pertain to fuel-cell systems. In particular, for the multi-layer, stack design of a hydrogen/air fuel-cell, there is a significant amount of sealing surface where hydrogen could leak internally or externally. Leakage of hydrogen into the air blower system can lead to undesired consequences so that careful attention must be paid to this potential hazard in the design. Likewise, if hydrogen must be recirculated using an external cooling loop, the potential for leakage from that system must also be considered in the design.

Although spark-ignition engine overheating is handled by established means, fuel-cell overheating is a major concern due to the lack of long-term operating experience. Redundant heat sensors with interlocks and failure indicators could be installed to avoid fuel-cell overheating. However, a challenge will be to properly locate the sensors so that overheating may be sensed at the correct place and time.

Finally, fuel-cell systems will require protection against internal shorting and back-flow of electrical power from any on-board storage batteries.

Hydrogen Leaks Within Heat Exchangers

Leaks of hydrogen into glycol or into any other exchange medium used on the vehicle can lead to a dangerous hydrogen accumulation and/or release. Such a condition could exist in a hydride-unit heat exchanger used for recharging the vehicle or in the heat exchanger used to cool a fuel-cell. Hydrogen sensors with interlocks may be required to sense this dangerous condition. Also, overpressurization of a glycol (or other fluid) circuit due to hydrogen leakage must be countered by suitable relief devices.

Hydrogen Leak Detection

Promising new means of hydrogen leak detection are being defined through the NASA Space Shuttle and the National Aerospace Plane (NASP) programs.[25] For example, a non-intrusive optical leak-imaging system is being developed for manually scanning the shuttle fuel system. Further, the NASP hydrogen fuel system design may incorporate a large number of pinhead-sized hydrogen leak detectors strategically located throughout the fuel system. Both of these systems clearly emphasize the seriousness with which both of these hydrogen fuel users view hydrogen leaks. Those who design hydrogen fuel systems for automobiles or trucks must also consider hydrogen leaks with equal seriousness.

Some of the hydrogen-fueled test vehicles have incorporated multi-point leak detectors which are activated even while the vehicle is inoperative and/or parked.[1,2] One such system[2] provides for the automatic opening of the side windows, the sunroof, and the trunk lid upon detection of hydrogen leakage.

It is important to point out that hydrogen detectors for vehicles should be of a design that will not act themselves as an ignition source for hydrogen. Further, the detectors must be calibrated on a periodic basis, presenting an additional challenge to the vehicle designer.

Safety Considerations in Refueling a Vehicle with Hydrogen

The unique flammability characteristics of hydrogen require extraordinary measures during the refueling operation, as there is a much greater risk of a hydrogen leak during this operation.

- **Avoid static discharge by grounding the vehicle** — During refueling, at least two means of grounding should be utilized. One means is to imbed vertical metal whiskers (approx. 1-ft. in length) in the road at the place of refueling. Another means is to attach a ground wire to the vehicle (using an interlock) prior to refueling. In this case, the ground wire is interfaced to a hydrogen supply interlock. A third means is to incorporate a ground wire directly within the fuel hose designed such that ground contact is made prior to flow or release of any hydrogen.

- **Avoid refueling during the threat of lightning storms.**

- **Prevent air from entering the fill lines and tank** — When refueling with gaseous hydrogen, appropriate steps must be taken to ensure that there is no air within the hose to the vehicle or within the line on the vehicle from the filler connector to the storage device. If this is not done, there is a serious risk of detonation or explosion if the hydrogen/air mixture is ignited. Industrial hydrogen suppliers and consumers use nitrogen purging to safely accomplish transfill of warm gaseous hydrogen or helium purging for cold hydrogen systems, though this approach may not be as practical for vehicle refueling stations. Alternatively, vehicle hydrogen systems could be kept under a small positive pressure of hydrogen, and the filler connection could be equipped with an automatic double shut-off (that is, a shut-off on the hose and on the vehicle) to help maintain an air-free condition.

When refueling with liquid hydrogen, it is likewise imperative to eliminate any air from the refueling system as there is the added problem of air freezing out and entering the hydrogen tank as frozen air. This will cause an extremely dangerous condition. Moreover, liquid hydrogen systems require a hose to return vapor to the refueling station, and this line should also be kept free of air.

A potentially serious hazard would exist if the operator were to add hydrogen to a gaseous hydrogen vessel or a liquid hydrogen tank that actually contained air. This situation could occur either when refueling a "new" tank or when a tank that has been in service has allowed back-diffusion of air following loss of hydrogen via a porous burst disc or a weeping relief valve. If a tank is being filled for the first time or the operator is aware that it contains air, the tank must first be purged with nitrogen and then purged with hydrogen before filling. Helium purge gas is required for liquid hydrogen systems.

One indication that a tank may contain air is that the pressure in the tank is 0 psig. Though this condition does not unequivocally confirm the presence of air in the tank, it is a convenient parameter and can be used as a fail-safe method. One means to reduce the hazard might be to place a pressure switch in the vehicle to sense hydrogen pressure in the tank, with the switch, in turn, wired to an electrical connector on the exterior of the car. If the pressure in the tank dropped below a preset threshold (perhaps 5 psig), an electrical circuit could be interlocked to the fueling system to prohibit refueling of the vehicle.

- **Prevent catastrophic fuel release** — Another potential hazard of the refueling operation concerns the catastrophic rupture of the fuel hose used to transfer the hydrogen to the vehicle. A plausible situation, and one that is common at gasoline service stations, is rupturing the fuel-hose by pulling away from a fuel station with the fuel hose accidently still attached to the vehicle. For hydrogen service, this hazard can be mitigated somewhat by placing an excess-flow valve in the station and a normal check valve on the vehicle.[24] In addition, the excess-flow valve and the normal hydrogen shut-off valves at the station should be positioned and anchored so that they will not be disturbed if the fuel-hose and/or piping are strained or torn apart.

Hydrogen Vehicle Servicing

Proper training of fuel-service and maintenance personnel, availability of suitable and approved repair/test equipment, and proper use of such equipment will provide a significant challenge for the widespread use of hydrogen-fueled vehicles. Servicing CNG-fueled vehicles will undoubtedly provide some lessons on safety training and procedures for use with hydrogen, but the unique flammability properties of hydrogen will require additional considerations.

Industrial gas suppliers that routinely handle hydrogen have developed standard practices for servicing hydrogen systems safely. For example, hydrogen trailers, which require routine service separate from the hydrogen system (such as tires, brakes, lights, etc.) are serviced differently than normal vehicles because of the presence of hydrogen in the trailer. Ignition sources are carefully controlled, and welding and grinding are essentially prohibited during service. Indoor work is restricted. Similar procedures for servicing hydrogen-powered vehicles will have to be established and clearly described in literature that is readily available to both the vehicle service personnel and to the operator. Service personnel will need to be carefully trained and will require right-to-know training for hydrogen hazards.

Servicing the hydrogen system itself will require specialized training, separate labels on the system components, and separate venting and purging equipment within the repair shop. Personnel will have to be adequately trained and periodically retrained in hazard awareness, purging procedures, leak detection, pressure checking, component testing, and replacement. Certification/licensing of a facility and repair technicians will probably ultimately be required.

Scrapping Hydrogen Vehicles

Scrapping a hydrogen-powered vehicle or its fuel components will require warning labels within the vehicle and special procedures, as will be the case for servicing. Most likely, scrapping a hydrogen-powered vehicle will involve disassembly and recovery of some components (e.g., the hydrides, catalysts from fuel-cells, and entire V-J tanks). Thus, warning labels and tags will need to be clearly written, durable, and placed where they can easily be found. As with servicing, scrapping of CNG-fueled vehicles will provide some valuable experience for the hydrogen application.

CONCLUSION

The use of hydrogen-powered vehicles beyond the research and/or demonstration stage will require careful attention to detail with regard to the safety of the fuel system. The elimination and detection of leaks, durability, serviceability, warning labels, and crashworthiness should be the focus of the detailed design. Wherever appropriate, inherent safety should always be included in the design.

Small fleet use would be a logical first trial of hydrogen-powered vehicles because drivers and maintenance technicians could be adequately trained and periodically reinforced. Further, fueling and off-shift parking of fleets would be captive. After the successful completion of a number of small fleet test programs, consideration may be given to use by the public.

REFERENCES

1. Hydrogen Research Progress 1991, *Hydrogen News of the American Academy of Science*, Vol. I, Issue 1 (Jan. 1991).

2. Siuru, W., "R and D in the Fast Lane: BMW's Research of Alternate Power", Mechanical Engineering, Vol. 111, No. 10 (Oct. 1989).

3. Topler, J. and Feuch, T., "Results of a Test Fleet with Metal Hydride Motor Cars", International Symposium on Metal-Hydrogen Systems, Fundamentals and Applicators, Stuttgart, FRG, Sept. 4-9, 1988.

4. Furuhoma, S. and Fukuma, T., "High Output Power Hydrogen Engine with High Pressure Fuel Injection, Hot Surface Ignition, and Turbocharging", Int. Journal of Hydrogen Energy, Vol. 11 (1986).

5. Takiguchi, T., et. al., "Combustion Improvement of Liquid Hydrogen Fueled Engine for Medium-Duty Trucks", SAE Paper 870535 (1987).

6. Peschka, W., "Hydrogen Combustion in Tomorrow's Energy Technology", *Hydrogen Energy Progress VI,* Vienna, Austria (1986).

7. Lynch, F., et. al., "A Hydrogen-Powered Vehicle for Mining", Diesels in Underground Mines, Proceedings: Bureau of Mines Technology Transfer Seminar (1987).

8. Kuchta, J., "Investigation of Fire and Explosion Accidents in the Chemical, Mining, and Fuel-Related Industries - A Manual", Bureau of Mines Bulletin 680 (1985).

9. *Chemical Engineers Handbook*, Perry and Chilton, Eds. Fifth Ed., Pgs. 5-20, McGraw-Hill (1973).

10. *Chemical Engineers Handbook*, Perry and Chilton, Eds. Fifth Ed., Pgs. 3-222, McGraw-Hill (1973).

11. Pfriem, H., "Overview of the Cooperative Program on Hydrogen Storage, Conversion, and Safety of the International Energy Agency", Int. Journal of Hydrogen Energy, Vol. 16, No. 5 (1991).

12. "Standard for Insulated Cargo Tank Specification for Cryogenic Liquids", CGA-341, Compressed Gas Association (1987).

13. Fukuyama, K., et. al., "Fracture Toughness and Fatigue Crack Growth of ANSI 4340 Steel in High Pressure Hydrogen at Room Temperature, Proc 6th Int. Conference Pressure Vessel Tech. ASME, Beijing, Vol. 2 (1988).

14. Deimel, P and Hanisch, C., "Tests on the Steels 15MnNi63 and X56TM in High Pressure Hydrogen Gas of High Purity," Int. Journal of Hydrogen Energy, Vol. 14 (1989).

15. Baker, N. and Lynch, F., "Hydrogen Sorbent Flow Composition and Containment Thereof", U.S. Patent 4,600,525 (1986).

16. Taube, M., et. al., "A Prototype Truck Powered by Hydrogen from Organic Liquid Hydrides", Int. Journal of Hydrogen Energy Vol. 10 (1985).

17. Gordon, R., "Composite Pressure Vessels for Gaseous Hydrogen Powered Vehicles, "Proc. of the 5th World Hydrogen Energy Conf. (1984).

18. American Gas Association Study Reported in *Midwest Engineer* (Dec. 1989).

19. Safety Analysis of Natural Gas Vehicles Transiting Highway Tunnels, Ebasco Services, N.Y. (1989).

20. Krupka, M., et. al., "Gaseous Fuel Safety Assessment for Light Duty Automotive Vehicles", Los Alamos Laboratory, New Mexico (1983).

21. Karim, G., "Some Considerations of the Safety of Methane, (CNG), as an Automotive Fuel - Comparison with Gasoline, Propane, and Hydrogen Operation", SAE Paper 830267 (1983).

22. Quadflieg, H., "From Research to Market Application? Experience with the German Hydrogen Fuel Project", *Hydrogen Energy Progress VI,* Vienna, Austria (1986).

23. "Standard for Hydrogen Piping at Consumer Locations", CGA G-5.4, Compressed Gas Association (1991).

24. NFPA 52, "Standard for Compressed Natural Gas (CNG) Fuel Systems", NFPA (1988).

25. Discussed verbally at NASA Hydrogen Leak Detection Meeting, held June 4-5, 1991

26. NFPA 50A, "Standard for Gaseous Hydrogen Systems at Consumer Sites", NFPA (1989).

27. NFPA 50B, " Liquid Hydrogen at Consumer Sites", NFPA (1989)

PHOTOPYROELECTRIC PD-PVDF SOLID-STATE HYDROGEN SENSOR

Andreas Mandelis and Constantinos Christofides

Photoacoustic and Photothermal Sciences Laboratory
Department of Mechanical Engineering, and
Center for Hydrogen and Electrochemical Studies (CHES)
University of Toronto, Toronto, Ontario M5S 1A4, CANADA

ABSTRACT

A new hydrogen solid-state sensor was made of thin, commercially-available polyvinylidene fluoride (*PVDF*) pyroelectric film, sputter-coated with *Pd*. An intensity-modulated infrared beam produces alternating temperature gradients on the *Pd–PVDF* and on reference *Al–Ni–PVDF* films, which, in turn, generates ac voltages due to the photopyroelectric (*PPE*) effect. Exposure to hydrogen gas in nitrogen was shown to produce an increased differential signal between the *Pd* and reference electrodes. The differential signal was found to be proportional to the square root of the hydrogen partial pressure at very low concentrations (40 to 1000 *ppm*) and to fit the Langmuirian isotherm. Recent results have shown that the thickness of the palladium layer evaporated on the active PVDF film plays an important role in the sensitivity of the device. In an effort to better understand the thermodynamics of the device during the solid-gas exchange interaction, the influence of ambient temperature has been studied.

I. INTRODUCTION

Very recently in a review paper published by Christofides and Mandelis [1] on solid-state hydrogen sensors, a general discussion was presented including a critical comparison of the capabilities and parameters of various conventional sensors. Many of these detectors have been found to be problematic due to noise introduced by temperature fluctuations and especially due to the requirement of high ambient temperature (120 to 240 °C) for their good operation and performance [2]. In 1988 the development of a new photopyroelectric Pd-PVDF (polyvinylidene fluoride) hydrogen detector was undertaken in our laboratory [3-7]. Initial results have been shown to be very promising toward the development of a new type of gas detection and analysis photothermal solid state device, partly based on polymer-metal interface phenomena. The new photopyroelectric (PPE) device not only does it seem to overcome many of the temperature-fluctuation related noise problems, but it also has an easy, simple, and inexpensive fabrication process. It is thought to be the first non-passive device operating on electronic polymer-metal junction phenomena. This device has attracted attention as a new trace hydrogen sensor [8,9] and has substantial potential for extension to the detection of other gases.

The operation of the PPE detector has already been described in the past [3-7] and is based on thermal wave generation in the pyroelectric polymer PVDF, which results in an alternating voltage due to the photopyroelectric effect [10].

The device possesses many advantages such as high detectivity through a broad temperature range, as will be concluded from this work, fast response, high selectivity, durability and reversibility at room temperature. Wide temperature response and high room temperature sensitivity are through to be advantages of our device over the well-known Metal-Oxide-Semiconductor Field Effect Transistor (MOSFET) devices [2]. The work of Lundström *et al.* [11] has definitily established the working detection mechanism of those devices.

This paper deals with the most resent results obtained with the photopyroelectric hydrogen sensor and focuses on three major objectives: (a) hydrogen detection ability (b) device optimization; and (c) physical understanding of the detection mechanism. The necessary instrumentation elements are given in Sec. II. Section III presents some experimental results and discussion, followed by some conclusions in Section IV.

II. INSTRUMENTATION

An overview of the experimental set-up of the PPE sensor is shown in Fig.1. The instrumentation is very similar to the one presented previously [3,4] with some extensions (such as the temperature control system and the capacitance meter) and some minor differences. An infrared laser beam served to produce alternating temperature gradients on the Pd-PVDF and on reference Al-Ni-PVDF films, which, in turn, generated ac voltages due to the photopyroelectric effect. Exposure to hydrogen gas was subsequently shown to produce a controlled differential signal between the *Pd* and reference electrodes; this was attributed to the adsorption, absorption and dissociation of hydrogen molecules on the *Pd* surface.

The signal generation and analysis section of the new set-up of the photopyroelectric detector is also described in Fig. 1. The instrumentation consisted of an infrared laser diode powered by an ac current supply. The intensity-modulated output laser beam was directed to a 3 way model fiber-optic coupler where it was split in three unequal parts. One of the three beams was directed to a photodiode (*PD*), the output of which was then sent to the "monitor" input of the home-made laser current supply for preamplification and feedback control of the laser current, as well as to a miniaturized home-made lock-in. The lock-in was equipped with a ratio option, allowing a normalized demodulated analog function output. A substantial degree of simplification and miniaturization has thus been achieved, with further incorporation of the laser power supply and the laser diode itself soon to be implemented. Our goal is to eventually develop a portable Pd-PVDF PPE sensor head with simple input/output connections to readout voltmeters and alarm circuits via the integrated power supply/detection electronics box. The feedback control consisted of using the optical reference signal to correct for temporal intensity variations in the modulated laser beam. The modulation frequency of the infrared beam was monitored with a frequency meter. The two PPE signals were then connected to the differential lock-in analyzer following two low-noise Ithaco (Model 1201) bandpass-filtered preamplifiers which were referenced by the *ac* laser current supply. The pre-amplifier gains are adjusted to minimize the lock-in output signals at the beginning of the experiments. The use of the lock-in amplifier technique in signal filtering and normalization enables us to obtain stable differential output signals, by adjusting the two gains between 10 to 10000. Providing the input signals from the preamplifier are set to about 10V level, we can achieve $\Delta S \approx 10^{-8}$, where

$$\Delta S = \frac{S_2 - S_1}{S_R} \quad . \tag{1}$$

S_1 and S_2 are the signal outputs from the two channels and S_R is the reference signal. Finally, the home made lock-in amplifier is interfaced through a parallel port with an IBM PS/2 computer for data analysis and detection. For this purpose software capable of data output and data readouts has been developed.

It is important to note that some of the experiments presented below have been performed without a reference PVDF signal. Under normal circumstances there exists a certain amount of dc drift inherent in the signal due to instrumental limitations. This is the reason for which low-drift preamplifiers were used in this study. Out pursuit of minimization of this drift with regard to laser modulation has revealed that a modulation frequency of 40Hz is at least one order of magnitude better than 5Hz in signal stability.

For temperature control, a long heat exchanger has been used in order to ensure that the temperature of the incoming gas flow would be the same as that of the sensor itself. A thermocouple allowing the monitoring of the cell temperature during the experiments with a precision of ±0.2°C has been used. An automatic controller heater (Fenal Model 140) equipped with a second thermocouple has also been used for a temperature stabilization in the range of 20 to 60°C. Lower temperatures have been attained using dry ice (measurements down to -63°C).

III. EXPERIMENTAL RESULTS AND DISCUSSION

Hydrogen Detection Ability

All of the present measurements have been performed at background pressure of 1 *Atm.* Typical experimental results are shown in Figs. 2(a) and 2(b) for various hydrogen concentrations using Pd film of 285 Å thickness (Fig.2a) and 130 Å (Fig.2b), respectively. These figures show details of PPE response as a function of time under the same temperature and flow rate conditions. We plotted the variation of ΔS as a function of time for hydrogen concentrations ranging from 7.5% to 0.075%. The relationship between gas concentration and $Pd-PVDF-PPE$ detector response at saturation ΔS (ΔS_s), is an important parameter for sensor characterization. Figure 3 shows the variation of ΔS_s as a function of hydrogen concentration using *Pd* film of 130 Å thickness. In parentheses one can see the response time at shown hydrogen partial pressure in minutes.

Langmuirian Analysis in Low Concentration Regime (<2000 *ppm*)

Figure 3 shows the variation of the saturated differential signal ΔS_s as a function of hydrogen concentration (or pressure) in the range of 40 to 2000 *ppm* (4 to 200 *Pa*) of hydrogen in nitrogen. This figure will be the main tool for the following semiquantitative analysis based on surface and interface adsorption phenomena. Using the following equation

$$\Delta S_s \approx \Delta S_{max} \left[\frac{K(T)\sqrt{P_{H_2}}}{1+K(T)\sqrt{P_{H_2}}} \right] \qquad (2)$$

on can analyse the PPE response of the device. Figure 3 (solid curve) shows Eq. (2) plotted vs P_{H_2} using $K(T) = 5\times 10^{-3}\ Pa^{-1/2}$. We note an agreement between the Langmuir isotherm and the experimental points especially at low pressures, where the Langmuir behavior is expected to be strictly valid. Thus, the combination of the Langmuir isotherm and the photopyroelectric response gives the possibility of quantitative analysis of the signal response ΔS_s. We did not try to interpret any experimental results for concentrations higher than 2000 *ppm* (200 *Pa*) because the use of the Langmuir isotherm for concentrations over 2 torr (266 *Pa*) is quite problematic. However, it is important to note that Langmuirian behavior has also been reported by other authors up to 2000 ppm [4].

The influence of the Pd-PVDF device history and of the Pd thickness on the PPE Signal

Important "negative learning" concerning the surface of the Pd layer has been achieved in the duration of this work. It has been discovered that the history of the Pd-PVDF active element (Pd deposition process, exposure to environmental conditions) is an important factor affecting performance. For example, a Pd-PVDF film exposed to laboratory air for several months becomes inactive in the presence of H_2, presumably due to oxidation and pollution (impurities) diffused in the Pd layer [12]. Christofides and Mandelis have also shown the influence of oxygen on the Pd surface by using a piezoelectric sensor [13]. However, this phenomenon is reversible and the film can be reactivated after several exposures to pure hydrogen [13]. It was found necessary to keep the Pd film in low vacuum environment in order to protect it from various types of impurities. Other Pd-based sensors present similar long-term poisoning effects and become disabled [2,14]. The definite advantages of Pd-PVDF are that (i) the poisoned active element, Pd-PVDF film, may be changed at essentially no extra cost; and (ii) a sheet of Pd-PVDF capable of yielding over 100 active elements can be easily and effectively stored under low vacuum or a positive N_2 overpressure for long time periods.

The sensitivity of the Pd-PPE device depends strongly on Pd thickness. This work, has unequivocally shown that thick Pd layers (*ca.* 1,500 Å) exhibit a strong response to H_2 gas, contrary to our earlier experience [4]. It is hypothesized that this is the result of improved Pd deposition methods on PVDF [15].

Figure 4 shows the variation of the normalized PPE differential signal ΔS as a function of the logarithm of the Pd thickness L in the case of pure hydrogen at 20°C. The logarithmic scale has been used only for convenience. For these measurements Pd was deposited on pyroelectric PVDF film of 28 μm thickness. In Fig.4 the datum represented by a solid circle well above the straight line (at L=130Å) corresponds to the only film of 52 μm thickness, as expected from capacitive response. Furthermore, it was found that Pd-Cu-PVDF films respond in the same quantitative way as the Pd-PVDF films. This is reasonable, because when *Cu* is used as interfacial electrical conductor between Pd and the PVDF surface, one expects no electric field in that layer:

$$\mathbf{E} = -\,\mathbf{grad}\,\phi(\mathbf{r}) = 0 \tag{3}$$

or

$$\phi(\mathbf{r}) = constant \quad , \tag{4}$$

i.e. there should be no measurable voltage drop across the electrical conductor. This result shows that hydrogen sensing action can be achieved upon Pd deposition, even with previously metallized PVDF and is indicative of the fabrication flexibility built into our sensor.

In terms of signal trends shown in Fig.4, qualitatively similar type of results have been reported by D'Amico *et al.* [16], who performed measurements using a palladium-coated surface acoustic wave (SAW) device. These authors studied the effect of the palladium thickness on the SAW response in a thicker range of Pd (1900 to 7600 Å). Buccur *et al.* [17] have also studied the effect of the Pd thickness on the hydrogen/Pd interface by using a piezoelectric quartz crystal microbalance. These authors have further examined the effect of thickness on the rate of hydrogen desorption. The increase of the photopyroelectric signal with the Pd thickness is consistent with an increased proton charge density in the Pd matrix. A high quality, fast capacitance meter (Boonton 7200) has been used for direct capacitance measurements in the presence and absence of hydrogen. These measurements have shown no measurable (i.e. < 1 pF) capacitance change. It is important to note that, as was shown by Lundström *et al.* [2], their MOSFET-based device presents a capacitive response in the presence of hydrogen gas of the order of 100pF in a hydrogen concentration of 1200 ppm. However, the detection mechanism in the MOSFET device is fundamentally different from the one of the Pd-PVDF device. The capacitance insensitivity of the latter device to [H] presence shows that no depletion layer participates in the PPE response, which is consistent with the hypothesis that only virtual charge (i.e. displacement or polarization charges) on the PVDF side of the interface may be responsible for coupling to the externally introduced charge density on the Pd side of the interface [7]. In order to test the influence of the Pd thickness on the time response of the device we plot in Fig.4 the time response as a function of palladium thickness. There is no significant effect of the Pd thickness on the time response, which remains approximately constant and close to 3 minutes for every thickness and for wide thickness ranges, between 34 and 1588Å. From the known diffusion coefficient of hydrogen in bulk Pd at room temperature ($\approx 2\times10^{-7}\ cm^2\ s^{-1}$) [18], it follows that H atoms diffuse into the bulk a mean distance of 1μm of Pd in 5 seconds [19,20]. This estimate is consistent with Fig.4, which clearly indicates that H diffusion through the Pd layer is not the rate-limiting factor in the kinetics of the Pd-PVDF sensor. Buccur *et al.* [17] have also mentioned that for very small thickness (<5000 Å) the diffusion is very rapid and is not easy to see the difference of time response. An attempt to give quantitative interpretation to the observed rates of adsorption/absorption/desorption based on signal response, by use of the well known Wagner model [21] did not succeed because of our working conditions (STP). Very low pressures ($<10^{-5}$ Torr) will be needed for quantitative kinetic analysis. Comparisons with Wagner's model resulted in slower than predicted rates, which is indicative of the presence and influence of surface impurity molecules controlling the kinetics of the adsorption/absorption/desorption process. Specifically, the kinetic process under STP conditions takes 180 seconds, which is 35 times longer than under clean surface conditions. These observations point to the importance of further experimentation under high vacuum and with molecularly clean surfaces.

The Influence of Ambient Temperature on the PPE sensor

A very important property of the Pd-PVDF sensor proved to be its extraordinary capability to offer unattenuated sensitivity at low temperatures continuously monitored down to -63°C and also qualitatively observed in the liquid nitrogen temperature range. No other H_2 sensor, to our knowledge, has ever been reported with such an ability (or, indeed, with any response at all below 20°C). As such, this property makes the sensor very promising for outdoors detection in cold climates and for fundamental studies of the Hydrogen-Palladium system. Temperature measurements from -63°C to +50°C have been successfully performed, indicating that good detectivity of the sensor is possible throughout a broad range of low temperatures. In fact, due to the high sensitivity of the pyroelectric effect to temperature fluctuations, special attention has been paid to obtaining high stability background signals prior to introducing hydrogen gas in the chamber. In the range of -5 to 15°C achieving stabilization was very different from the rest of the reported temperature range due to some extra problems introduced on account of signal hysteresis [22]. In the absence of hydrogen gas in that temperature range, the background signal depends on the direction of temperature variation. Another important problem with low temperature measurements (below 0°C) was presented by the extreme difficulty to evacuate the hydride from the palladium. Due to this problem low temperature data acquisition had to be preceded by cycling up to room temperature in order to clean the surface (and bulk) of the palladium. Figure 5 shows the variation of the photopyroelectric signal vs. temperature, which was found to satisfy a $T^{-1.4\pm0.2}$ law. This dependence is consistent with a thermostatistical mechanism based on hydrogenic image dipoles performing independent vibrational and librational motions about the polarization field axis of the pyroelectric [7]. Figure 6 shows the variation of the response and recovery time as a function of the working temperature. We note that the response time of Fig. 6(a) depends strongly on temperature: It exhibits a minimum at *ca.* 20°C and is seen to increase with decreasing temperature. Although the presence of the minimum is certainly due to several (more than one) mechanisms, the low temperature portion of the curve is consistent with a gradual "freezing-in" of the PVDF molecular dipole response to the $[H^+]$-induced interface bias. Such a mechanism based on the lowering of the molecular oscillator energy with decreasing temperature would affect (delay) the temporal change of the pyroelectric coefficient in response to the $[H^+]$ field [7]. On the other hand, the [H] concentration in the Pd bulk is expected to build itself up to the thermodynamic equilibrium value more slowly at low T, and this effect can also delay the shift of the Pd metal work function with hydrogen, $\Phi_{Pd}([H],T)$, thus increasing the response time of the device [7,9]. No tentative explanation of the increasing response time at the high-T portion of the curve in Fig.6(a) is possible without further experimentation. No strong dependence on temperature of the recovery time can be seen in Fig.6(b).

CONCLUSIONS

In this paper the development and characterization of a new, sensitive, durable and very economic PPE hydrogen detector has been presented. Our experimental results suggest that the investigated PPE sensor structure can be implemented as a trace hydrogen detector, under ambient and cold conditions or in remote locations without the shortcomings of conventional *dc* pyroelectric sensors. It has been shown that, in terms of sensitivity limits, durability and simplicity, the sensor's characteristics may be at least as good as those of other solid-state hydrogen detectors mentioned in the introduction of this paper. The room temperature operation capability of the new PPE sensor may also be indicative of superior durability over Pd-gate-MOSFET's, the sensitivity and the response time of which at room temperature are inadequate.
The main results of this study can be summarized as follows:

[1] It has been found that the thickness of *Pd* evaporated on the *PVDF* film plays an important role in determining sensitivity and durability, but not in the value of the response time. The observed response times are likely to be surface impurity-controlled and much longer than the one expected from simple gas-clean surface kinetics.

[2] Instrumental hydrogen detection limits were estimated to be 2 *ppm*.

[3] The hydrogen photopyroelectric detector could also become an excellent tool for surface science studies under *UHV*, especially at low temperatures where other sensors do not exhibit good performance. In fact *UHV* studies will also help toward a better understanding of the operating mechanism, because of the highly controlled conditions of the ultrahigh vacuum system.

In this paper we also presented some results toward the optimization and mechanism of the new photopyroelectric hydrogen sensor. Our experimental results suggest that the investigated PPE sensor structure can be implemented as a hydrogen detector, under ambient and low temperature conditions with flexibility in the substrate geometry on which Pd is deposited. It has been shown that this detector presents high potential towards the study of gas-surface interactions especially at low temperatures where conventional sensors present problems or are inoperative.

ACKNOWLEDGMENTS

The support of the Ministry of Energy, Mines, and Resources Canada through a contract to CHES is gratefully acknowledged.

REFERENCES

[1] C. Christofides and A. Mandelis, J. Appl. Phys. **66,** R1-R33 (1990).

[2] K.L. Lundström, A. Armgarth, L-G. Petersson, CRC Critical Reviews in Solid State and Material Sciences **15,** 201 (1989).

[3] A. Mandelis and C. Christofides, Sensors and Actuators, **B2,** 79 (1990).

[4] C. Christofides and A. Mandelis, J. Appl. Phys. **66,** 3975 (1990).

[5] A. Mandelis and C. Christofides, J. Vac. Sci. Technol. **A8,** 3980 (1990).

[6] C. Christofides and A. Mandelis, Int. J. Hyd. Energy (in press)

[7] A. Mandelis and C. Christofides, J. Appl. Phys. (in press)

[8] Sensors Technology, **Vol. 7,** 1 (1990).

[9] A. Balasubramanian, J.J. Santiago-Aviles, and J.N. Zemel, J. Appl. Phys. **69,** 1102 (1991).

[10] A. Mandelis and M.M. Zver, J. Appl. Phys., **57,** 4421 (1985).

[11] K.L. Lundström, A. Armgarth, A. Spetz, and F. Winquist, Sensors and Actuators, **10,** 399 (1986).

[12] M.A. Vannice, J.E. Benson, and M. Boudart, J. Catal. **16,** 348 (1970).

[13] C. Christofides and A. Mandelis, J. Appl. Phys. **66,** 3986 (1989).

[14] S. Abe and T. Hosoya, *Proc. 5th World Hydrogen Energy Conf., Toronto, Ont., Canada, July 15-20,* 1984, **Vol. 4,** pp 1893-1900.

[15] Pennwalt Corp., King of Prussia, PA (private communication)

[16] A. D'Amico, A. Palma, E. Verona, *1982 IEEE Ultrasonics Symposium,* 308 (1982).

[17] R.V. Buccur, V. Mecea, and E. Indrea, J. Less-Common Metals **49,** 147 (1976).

[18] F.A. Lewis, *The Palladium/Hydrogen System* (Academic Press, New York, 1967) p.94.
[19] H. Conrad, G. Ertl, and E.E. Latta, Surf. Sci. **41,** 435 (1974).
[20] D.N. Jewett and A.C. Makrides, *Trans. Faraday Soc.* **61,** 932 (1965).
[21] C. Wagner, Z. Phys. Chem., Abt. A, **159,** 459 (1932).
[22] S.B. Lang, in Temperature its Measurements and Control, Edited by the American Institute of Physics, pp. 1015-1022 (1962).

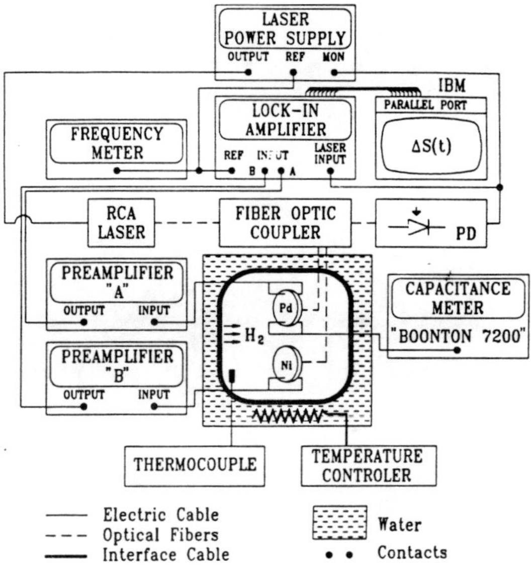

FIGURE 1: SCHEMATIC DIAGRAM OF THE PPE SET-UP.

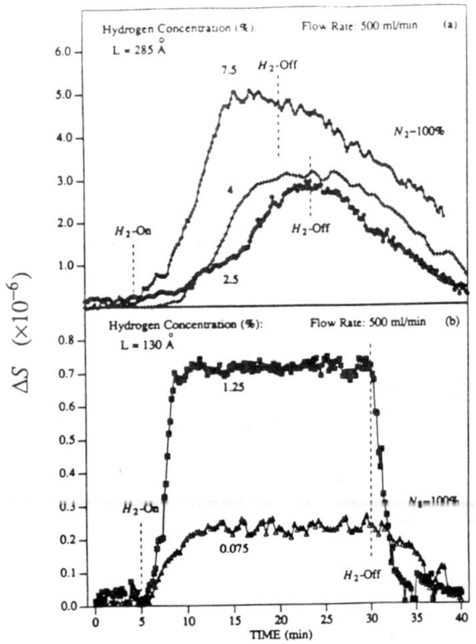

FIGURE 2: PPE RESPONSE AS A FUNCTION OF TIME, FOR VARIOUS H_2 CONCENTRATIONS.

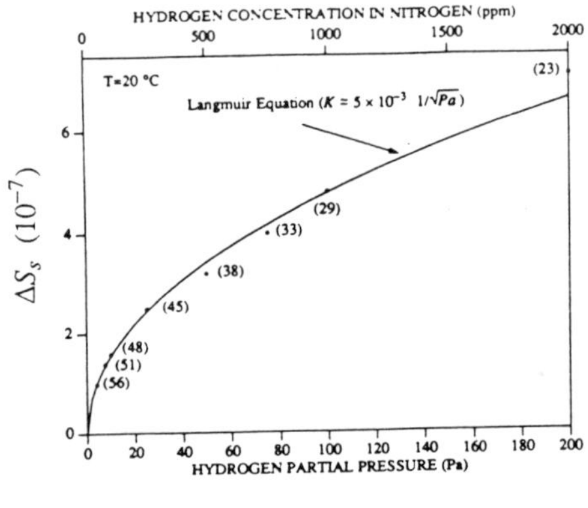

FIGURE 3: ΔS_s AS A FUNCTION OF P_{H_2} (OR CONCENTRATION).

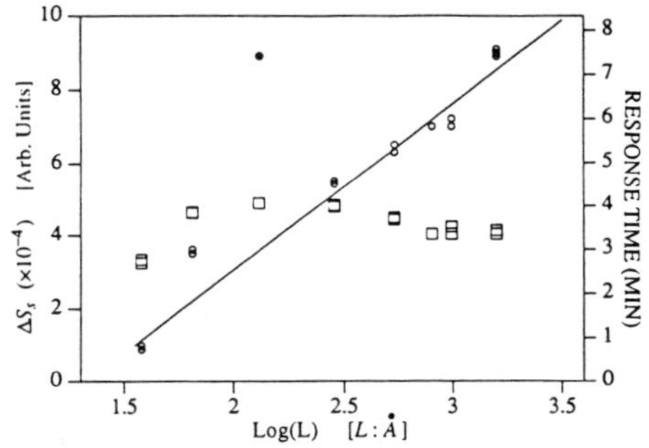

FIGURE 4: PPE SIGNAL ΔS_s (o o o o) AND RESPONSE TIME (□□□□) AS A FUNCTION OF Pd THICKNESS IN THE PRESENCE OF PURE HYDROGEN AT ROOM TEMPERATURE. FLOW RATE: 500 *ml/min* (T = 18 °C and f=32Hz).

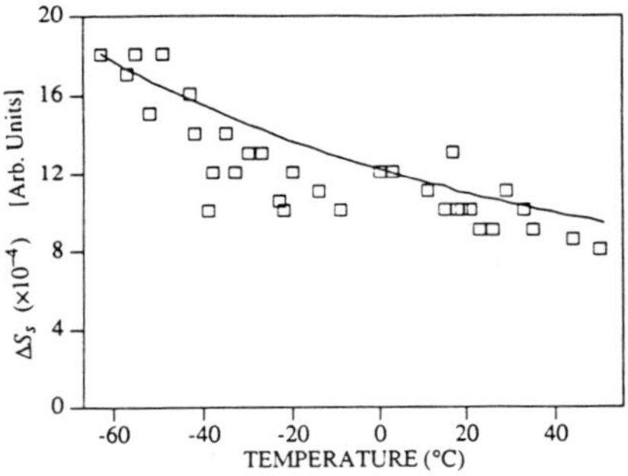

FIGURE 5: PPE SATURATED SIGNAL ΔS_s AS A FUNCTION OF TEMPERATURE IN THE PRESENCE OF PURE HYDROGEN. FLOW RATE: 500 *ml/min* (f=32Hz, and L=1588 Å). SOLID LINE REPRESENTS THE BEST FIT TO THE DATA ($T^{-1.4}$ LAW).

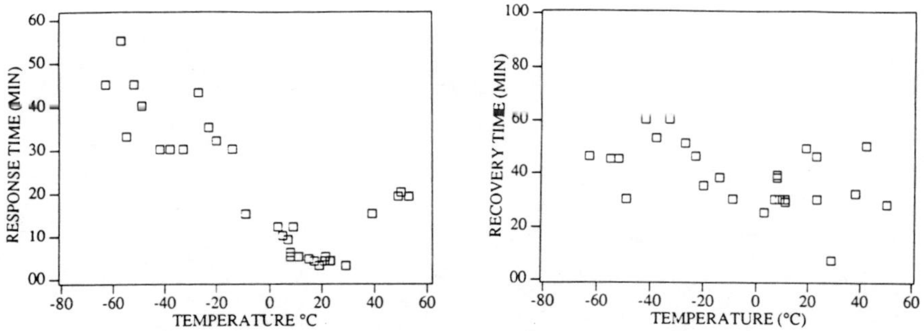

FIGURE 6: PPE RESPONSE (a) AND RECOVERY (b) TIMES AS A FUNCTIONS OF TEMPERATURE IN THE PRESENCE OF PURE HYDROGEN. FLOW RATE: 500 *ml/min* (f=32Hz, L=1588 Å).

KEYNOTE ADDRESS

The Hydrogen Economy: A Strategy for Acceptance

THE HYDROGEN ECONOMY: A STRATEGY FOR ACCEPTANCE

W.M. Hawkins
Senior Advisor
Lockheed Corporation

INTRODUCTION

As some of you may remember from a number of years ago, I entered your world, the world of hydrogen, as an energy medium, by means of the airplane. You have scheduled in this meeting to hear from the real entrepreneur of hydrogen aircraft, Dan Brewer, so I'll limit myself to only a few comments about airplanes. Dan is the real expert. What I have decided to do is to offer a potential course of endeavor that may be useful in bringing about the "Hydrogen Economy" that so many of us have worked for and dreamed about.

As I review my own experience, I find four major subject headings which contain the perceptions, inhibitions, and problems that must be overcome. Included in the list is the promise of the major contribution to our society which any hydrogen economy can make. These four subjects are:

- Safety
- Cost
- Universal use
- Atmospheric cleanliness

As we discuss our future energy efforts we must constantly emphasize these basic fundamentals. They will focus each contribution.

THE COMPETITOR ENERGY SYSTEMS

All of us recognize the convenience, economy, and effectiveness of the energy sources we use today and I'm sure that you agree with me that those are tough competition. To prepare for tonight's discussion I looked up our history in the United States. I found that we were increasing our use of petroleum, natural gas, and coal by a total of 2.6% per year in spite of all our efforts at efficiency improvement and the reduction of energy use. The alarming content of this increase is that the increase in the use of coal, our energy source most potentially destructive to the atmosphere, is 3.4%; petroleum, next most destructive, is 2.5% and natural gas, the most atmospherically benign, is only 1.4%. These three sources of energy comprise nearly 90% of the energy we use with hydro, nuclear, and geothermal and other truly benign sources comprising only 10.7%.

Digging into these statistics further shows that coal and natural gas comprise about 25% each of 1988's total consumption with petroleum approximately 43%. This composite consumption picture is the target for our hydrogen efforts, not forgetting in the process that hydrogen can and must be made, to the extent possible, using nuclear, hydro, wind and other more exotic and non-polluting power sources.

THE CRITICAL CUSTOMER

My candidate as a target for the first use of hydrogen is the private automobile. I recognize that this may be the toughest target of all the potential customers which include industry, the home, electrical generation, and the rest of the transportation industry, trucking, railroads and aircraft. The reason I suggest this target is primarily that this "customer" involves the maximum number of voters, those whose perceptions and habits comprise most of hydrogen's non-technical challenges. If we can successfully compete for this body of customers there is hardly a market we can't enter. Furthermore we will have entered the market (transportation) that uses 26% of our available energy and is probably the most harmful to our atmosphere. It is also, probably, the most technically challenging of all the markets we could target.

HOW TO APPROACH THE AUTOMOBILE

In this audience I'm about to demonstrate the adage "only fools persist where angels fear to tread". I can only plead that I have been exposed to many of your hopes and concepts, your development successes, and the logic behind all your efforts. I've developed my own opinions of what might work and what the market might be in the future. I think it will be different today than several years ago because of the world's final recognition of the fragility of the atmosphere we live in and what we are doing to it with every move we make.

A CANDIDATE APPROACH:

My first blunt belief is that we should not waste time on the conversion of the internal combustion engine and we should concentrate on the fuel cell. The fuel cell can become the core of a remarkably compatible total automotive system since the generation of electrical power opens up the simple 4 wheel drive, permits the use of braking for power generation, and hands us on a silver platter a quiet, benign atmosphere in and outside of the vehicle. Listening to the knowledgeable enthusiasts I've heard the predictions of $500/kW by the year 2000 to be competitive with internal combustion engines by 2020 -- 5 to 7 kW/ft^3 and 15#/kW are also suggested as rational goals. Since the efficiency goals of 40 to 60% appear reachable, the storage efficiency of the fuel can afford to be substantially worse than today's fuel tanks from a weight standpoint, permitting us to consider even today's hydride tanks, should hydrogen be used directly. We can assume, I hope, that this technology will also advance because this method of storing hydrogen appears to me to answer ninety percent of the fears that hydrogen is not safe. Most of us in this room believe that hydrogen is safe -- safer than most any fuel -- if properly designed for and handled, but with the hydrides, I believe we don't have to argue with the people and the press.

I recognize, in this approach, that we have to tackle the toughest technical challenge, but I suggest that this may be necessary to get enough people to individually accept hydrogen and to join with the hydrogen enthusiasts in selling its performance and safety to their neighbors. Trucks and buses would be an easier first step because they are under more technical specialist control. The distribution of fuel may be easier because sources can be more localized and their payload fraction is so large one can afford to have heavier power plants, heavier fuel tanks and volume is not as constrained. But trucks and buses are operated by "Industry", the public acceptance will be mostly through indifference and we will not have created a large public sales force to accelerate progress. This is what I mean when I say we need more voters.

The distribution of hydrogen for individual vehicles is recognized as a major "chicken and egg" barrier demanding large up-front expenditures. We may have to resort to subterfuge and transitory steps, one of which is the use of natural gas or methanol with built-in reformers in each vehicle. Natural gas distribution systems are in place and we're used to the kinds of distribution systems that methanol would use. Once owners discover the operational advantages and quiet of the fuel cell they'll demand more performance and gradually the reformers can become fixed installations delivering hydrogen to automotive hydride tanks. This kind of gradual introduction has the further advantage that it diverts transportation from petroleum to the less polluting natural gas while we sort out how to create hydrogen from a totally non-polluting energy source like nuclear power.

I suggest that there is a further program policy that should be accepted. We should design this car for the luxury market -- the Cadillac, Lincoln, New Yorker, Mercedes, Infiniti customers. The dignity of a quiet vehicle, the prestige of ownership, the initial smallness of the customer base will help to focus design and development efforts and the tolerance of this market to high unit costs will help us over the first step in solving the cost problem. It will also create a prestige target that warms the pockets of those who will want someday to own a vehicle like that -- and they can afford one when we've got our development base secure enough to take the next large production step. If you'll indulge a compulsive designer, I'd like to suggest that these luxury sedans be designed as twin engine vehicles so that there will never be a total breakdown to mar their acceptance. This concept covers the total system including wheel motors, etc.

WHO PAYS FOR ALL THIS?

Our hydrogen economy campaign is in deep serious trouble if we have to depend on the U.S. government to provide the funds. NASA, DOE, DOT, the military services would all benefit immediately if such development actually took place. It would be wasting your time if I listed all the benefits from such a program for any of these agencies. You could do it better than I.

I'm ashamed to say that the cost of any program, if it involves the technical direction and management of any of these government entities, would be disastrously prohibitive -- and technical direction and management is inevitable if we use government money. Thus I conclude that, if it is to be done, industry in some combination using the fundamental knowledge within our universities must somehow step up to the challenge. The amount of money spent in creating a new model of a conventional car would carry us quite a way toward a commercially successful hydrogen fuel cell vehicle system.

The United States still seems to be able to respond to a real crisis in spite of our ascending bureaucratic inertia. Perhaps there is an impending crisis that will keep us participating if not leading. Some nation which understands, respects, and efficiently works with its industry may show us how to do it. Japan, Germany, and perhaps France may see this opportunity more clearly than we do, and will get our competitive juices flowing again.

WHAT THEN?

Next to transportation's use of petroleum, the most concentrated use of energy resources is the electric utilities use of coal. Nearly 70% of this basic source is lost in efficiency of electrical energy conversion and transmission. Perhaps we can sneak into a solution here by means of a transmission system already in existence -- natural gas -- which too may be able to feed a fuel cell through a reformer in the home. The efficiency of the cell should compete very favorably with the loss in electricity conversion and distribution and we'd rid ourselves of 65% of the coal we use. Getting heavy industry to do the same would take care of the rest of the coal.

At this point the distribution system, for all practical purposes (if we did it properly), would be ready for hydrogen when we figured out how to make it at competitive rates which should include a bonus for not destroying the atmosphere.

CONCLUSION

What have I said? I hope it is this:

1. There is a solution to the negative public perception of hazards in the emerging reality of increased safety while using hydrogen. At the core of this is hydride storage.

2. The personal transportation via automobile has the most clear opportunity for bringing voters to the support of a hydrogen economy. This use would reduce atmospheric contamination by approximately 25% all by itself, assuming contamination-free production of hydrogen.

3. The same technology can provide a further gain in home use of electricity and ultimately, heavy industry use of electricity. This would eliminate coal-based energy for all practical purposes with another 20-30% reduction in atmospheric contamination.

4. Aircraft can benefit inordinately since hydrogen as a fuel contains three times the energy per pound as compared to petroleum. When hydrogen is being produced for all the other energy needs aircraft will reap the rewards.

None of this will be easy but I believe that industry and the universities working without government help have the best chance of making progress. My suggestion is to first tackle the luxury automobile with fuel cell application and hydride storage. By this means a critical core of voters will be convinced that a hydrogen economy is the ultimate answer. The possession of such quiet, revolutionary vehicles will whet the appetite of the rest of the voters. Thus all of us can live in a cleaner world.

PROJECT HYDROGEN '91

Conference Participants

Project Hydrogen '91

PARTICIPANTS

Mr. Chris Abbott
Post Office Box 36
Isleton, CA 95641

Mr. Andrew Allman
942 South Carr
Scolalia, MO 65501

Mr. Michael Amelunke
DNR/DE
P.O. Box 176
Jefferson City, MO 65102

Dr. John Amphlett
Dept of Chem. & Chem. Eng.
Royal Military College of Canada
Kingston, Ont. CANADA K7K5LO

Mr. Jan Astwood
Sweetwater H_2 R&D
2211-1 Pine Lane
Saugerties, NY 12477

Mr. Axel Axelsson
3210 Lombardy Rd.
Pasadena, CA 91107

Mrs. Catherine Axelsson
3210 Lombardy Rd.
Pasadena, CA 91107

William A. Ayres
Interchem
8016 Stateline
Leawood, KS 66208

Mr. Giles Babb
Envirnm. Compliance Monitor
5804 Deer Lake Drive
San Antonio, TX 78244

Mr. Micky Beach
Kingsville High School
P.O. Box 7
Kingsville, MO 64061

Dr. Gary Benson
Univ. of WI/Whitewater
2018 Carlson
Whitewater, WI 53190

Mr. Peter Bernhard
Fachhochschule Wiesbaden
Heinrich-Lersch-Strasse
6090 Ruesselsheim, Germany

Ms. Andrea Besses
1109 S. 16th Street
Blue Springs, MO 64015

Mr. Dale Biagi
Longview Community College
500 S.W. Longview Rd.
Lee's Summit, MO 64081

Mr. Chris Boosinger
151 N. Main Street
Willits, CA 95490

Mr. David Booth
Alternative Energy Eng.
Post Office Box 339
Redway, CA 95560

Mr. Leon Bowlin
Blue Springs High School
2000 W. Ashton Drive
Blue Springs, MO 64015

Mr. Bruce Bowman
Delker Corp.
16 Commercial St. P.O. Box 427
Branford, CT 06205

Mr. Larry Boyce
Grandview High School
1510 Clark Road
Blue Springs, MO 64015

Mr. Art Boyt
Crowder College
601 Laclede Ave.
Neosho, MO 64850

Ronald W. Breault
Tecosen
45 First Avenue
Waltham, MA 02254

PARTICIPANTS

Ms. Hiromi Brett
Birmingham Overseas, Inc.
11 Franklin Street
Woodcliff Lake, NJ 07675

Mr. Daniel Brewer
1214 San Julian Drive
Lake San Marcos, CA 92069

Mr. Russell Bright
RR 1, Box 273AA
Kingsville, MO 64061

Mr. Edwin Buckner
EPA Region VII
726 Minnesota Ave.
Kansas City, MO 66101

Mr. Dan Byrne
1008 E. 108th Terrace
Kansas City, MO 64131

Mr. Erik Carlson
1330 S. Wheaton
St. Charles, MO 63301

Mr. John Cash
18014 Dover Fork
Independence, MO 64057

Mr. Jonathan Chance
Avatar Inc.
5 East 41 St., #151
New York, NY 10017

Mr. Dennis Cheney
I.N.E.L.
P.O.Box 1625, 1955 Fremont St.
Idaho Falls, ID 83415-3510

Mr. Bill Christie
Phoenix Litho Inc.
1400 Kansas Ave.
Kansas City, MO 64141

Mr. Robert Clark
324 Kingston
Kansas City, MO 64124

Mr. Stanley Clark
Kansas State Univ.
Agriculture Engineering Dept.
Manhattan, KS 66506

Mr. Phillip Cloud
Nuage Educ Serv/Mobile Museum Co.
1100 S. Bryant
Edmond, OK 73034

Ms. Winifred Colwill
League of Women Voters of MO.
Rt. 11, 1417 Countryshire Dr.
Columbia, MO 65202

Mr. John Cook
Exeter High School
Box F
Exeter, MO 65647

Mr. Tom Corbin
Corbin Bronze LTD
201 Wyandotte
Kansas City, MO 64105

Mr. Scott Courtney
4010 Milton Drive
Independence, MO 64055

Mr. John Craft
The Land Institute
6888 E. McReynolds Rd.
Gypsum, KS 67448

Mr. Robert Crockett
Albany High School
101 West Jefferson St.
Albany, MO 64402

Dr. David Currence
University of MO
208 Agri Engr.
Columbia, MO 65211

Mr. Ryan Dawson
6418 E. 150 St.
Grandview, MO 64030

Ms. Elaine Day
18034 Dover Fork
Independence, MO 64057

PARTICIPANTS

Mr. Robert Dederick
Robert J. Dederick Associates
Post Office Box 367
Port Ewen, NY 12466

Mario L. DeLucia, Mgr,Aerospace
Union Carbide Indust Gases,Inc.
39 Old Ridgebury Road
Danbury, CT 06817-0001

Mr. James Diederich
17300 South River Road
Hartsburg, MO 65039

Ms. Tammy Dillon
4714 NW Riverside St.
Riverside, MO 64150

Mr. Michael Dooley
7351 Crown Park
Belton, MO 64012

James T. Duke
2500 Apache Pass
Altus, OK 73521-1506

Mr. Dennis Duszak
J I Case Company
7 S. 600 County Line Rd.
Hinsdale, IL 60521

Mr. Dan Eberle
Crowder College
601 Laclede Ave.
Neosho, MO 64850

Mr. Herman Everett
NASA
3781 Riverside Dr.
Melbourne, FL 32935

Mr. Jim Everett
Platte County Area VoTech School
P.O. Box 1700
Platte City, MO 64079-1700

Mr. Jerry Feret
Westinghouse Elec. Corp.
Post Office Box 10864
Pittsburgh, PA 15236

Mr. Jeffrey Fisher
Transit Advan Consortium
44 N. Bedford St., #1
Arlington, VA 22201

Mr. Jim Flory
Delker Corp.
1957 S. Stacey
Wichita, KS 67207

Eric P. Fogg
City of Lansing
2655 Ingall Road
Leslie, MI 49251

Dr. T. Foley
British Gas
Tech. Transfer 326 High Holborn
London WC1V 7PT ENGLAND

Mr. Henry Fuller
27470 Hwy 74
Paris, CA 92570

Mrs. Jean Fuller
27470 Hwy 74
Paris, CA 92570

Dr. Shoichi Furuhama
Musashi Instit of Technology
1-28-1 Tamazutsumi Setagaya-ku
Tokyo 158 JAPAN

Mr. Joe Galbraith
1000 Stafford Dr.
Pleasant Hill, MO 64080

Mr. Chris Gannan
Pattonsburg R-II
323 West 4th
Pattonsburg, MO 64670

Mr. Richard Garver
Poplar Bluff Sr. High
1025 Ivadean
Poplar Bluff, MO 63901

Mr. Jim Gerdom
Design Factory
P.O. Box 14037
Lenexa, KS 66215

PARTICIPANTS

Mr. Tony Goen
American Hydrogen Assoc.
219 S. Siestaln, Suite 101
Tempe, Arizona 85281

Ms. Lisa Goodwin
616 S. Maple
Lathrop, MO 64465-0432

Mr. John Gotthold
1401 Pointe Claire Ct.
Sunnyvale, CA 94087

Mr. John Gray
Black & Veatch Engr.
39 W. 7th Street
Kansas City, MO 64114

Ms. Cynnie Green
Southwest High School
6512 Wornall Rd.
Kansas City, MO 64113

Mr. Claud Greenlee
St. Charles High School
Kings Highway at Waverly
St. Charles, MO 63801

Mr. George Gregory
5436 Willow Lane
Vermilion, OH 44089

Mr. Wyman Grindstaff
Southwest MO State University
Chemistry Department
Springfield, MO 65804

M. K. Gupta
Macromolecular Research Centre
Rani Durgavati University
Jabalpur-482001, M.P., INDIA

Mr. David Haenke
Ecol Soc Proj/Tides Found
Rt. 1, Box 20
Newburg, MO 65550

James G. Hansel
Air Products & Chemicals, Inc.
7201 Hamilton Boulevard
Allentown, PA 18195

Mr. Hiro Hara
ICA
8 N. Main St., Suite 222
West Hartford, CT 06107

Mr. Troy Harvey
Electronic Design Labs
3 Great Pasture Rd.
Redding, CT 06896

Mr. John Hawkins
Penn Valley Community College
3201 Southwest Trafficway
Kansas City, MO 64111

Dr. Willis Hawkins
Lockheed, CA
4500 Park Granada Blvd.
Calabasas, CA 91399

Mr. Herb Hayden
American H_2 Assoc.
219 S. Siesta Ln., Suite 101
Tempe, Arizona 85281

Ms. Nancy Hazard
Northeast Sustainable Energy Assn
23 Ames Street
Greenfield, MA 01301

Mr. Ehbort Heath
Penn Valley Comm College
39909 E. Truman Road
Oak Grove, MO 64075

Mr. Harold Hennen
N.W. Elect. Power Inc.
RR3
Gallatin, MO 64640

Mrs. Marge Hennen
RR3
Gallatin, MO 64640

Mr. Del Hennings
Excelsior Springs High School
Lynn Rd.
Excelsior Springs, MO 64024

PARTICIPANTS

Ms. Jamie Hillmer
Rt. 3 Box 506
Harrisonville, MO 64701

Ms. Carolyn Hilton
229 S.W. Pinetree Lane
Lee's Summit, MO 64063

Mr. Shinichi Hirano, Sr. Res. Eng
Mazda R&D of N. America
1203 Woodridge Avenue
Ann Arbor, MI 48105

Dr. I. Hodara
Weizmann Instit. of Science
Solar Research Facilities
Rehovot 76100 ISRAEL

Ms. Rebecca Hodges
23901 Cowherd Rd.
Lee's Summit, MO 64064

Mr. W.J. Hoenigmann
AIRCO, H_2 Oper. Mgr.
575 Mountain Avenue
Murray Hill, NJ 07974

William C. Hopkins III
Ouray Cottage Motel
Post Office Box 486
Ouray, CO 81427

Tey Say Howe
Petronas Menara Dayabumi
Kuala Lumpur MALAYSIA

Ms. JoEllen Hudson
314 E. Bridlespur Dr.
Kansas City, MO 64114

Dr. Orlo Hudson
Advantage Hytech
8705 E. 63rd. Tfkwy
Kansas City, MO 64133-4703

Mr. Jimmy Hurt
111 E. Forest Lane
Excelsior Springs, MO 64024

Mr. Paul Inman
718 Newton
Kansas City, MO 64125

Mr. Dwight Iott
Rt. 1 Box 547
Warrensburg, MO 64093

Mr. Bob Jackson
MO/DNR
Box 176
Jefferson City, MO 65101

Mr. Michael Jackson
4545 Agnes
Kansas City, MO 64130

Mr. Thomas Jackson
Electrical Heating Equip. Co.
Post Office Box 392
Conshohocken, PA 19428-0392

Mr. Richard Jervis
Elect. Heating & Equip.
Post Office Box 392
Conshocken, PA 19428-0392

Mr. David Katz
Alterenative Energy Eng.
Post Office Box 339
Redway, CA 95560

Donald W. Kent, Jr.
Electrical Heating Equip. Co.
Post Office Box 392
Conshohocken, PA 19428-0392

Mr. Joe King
Coriolis Assoc.
123 W. 8th. St.
Lawrence, KS 66044

Mr. Katuyoshi Koyanagi
Musashi Instit. of Tech.
1-28-1 Tamazutsumi
Setagaya-ku Tokyo 15B JAPAN

Mr. Duanne Lamiman
Duanne Lamiman & Assoc.
10003 West 95th
Overland Park, KS 66212

PARTICIPANTS

Mr. Robert Landwehr Jr.
311 E. Lincoln
Carrollton, MO 64633

Dr. Peter Lehman
Humboldt State Univ.
Engineering Depart.
Arcata, CA 95521

Ms. Carolyn Leuthold
League of Women Voters of MO.
2000 Valley View Dr.
Columbia, Mo 65202

Mr. Michael Lightrain
Post Office Box 69
Willits, CA 95490

Mr. David Litchfield
MO/DNR
Post Office Box 176
Jefferson City, MO 65102

Mr. Zack Lyon
Design Dimension
1418 Glenwood Avenue
Raleigh, NC 27605

Mr. Dave Macey
MO Public Service
10700 E. 350 Hwy
Kansas City, MO 64138

Prof. Andreas Mandelis
University of Toronto
5 King's College Rd. Dept of Mech Engn
Toronto, Ont. CANADA M5S1A4

Mr. Ambrose Manikowski
Lockheed Missiles and Space
P.O. Box 3504
Sunnyvale, CA 94088-3504

Mr. Leonard Marianowski
Institute of Gas Tech.
3424 S. State Street
Chicago, IL 60616

Dr. Whit Marks
Halim International
1300 S. Bryant Ave
Edmund, OK 73034

Mr. Michael Marshall
U.S. EPA
726 Minnesota Av.
Kansas City, KS 66101

Mr. Andrew Matson
19016 E. 37th Terr. #11
Independence, MO 64057

Dr. Marleta Mattos
R. Matias Aires, 268, Apt.72
01309 Sao Paulo-SP BRAZIL

Mr. Roy McAlister
American Hydrogen Ass'n
219 S. Siesta, Suite 101
Tempe, AZ 85281

Dr. Emmet McCord
University of Missouri
Rt 1 Kenney Drive
Cassville, MO 65625

Ms. Kristin McGovern
908 Timber Oak Dr.
Blue Springs, MO 64015

Mr. William Meurer
Green Motor Works
5414 Cahuenga Blvd.
N. Hollywood, CA 91601

Frank D. Middleton
Post Office Box 1256
Cartersville, GA 30120

Dr. Robert Miller
Air Products & Chem., Inc.
7201 Hamilton Boulevard
Allentown, PA 18195

Mr. Robert Monconduit
GAN Capitalisation
Tour Gan 6 eme Etage
Paris Ladefense Cedex FRANCE

PARTICIPANTS

Dr. Nandor Nemeth
EMR
172 Crestview Road
Ottawa, Ont. CANADA K1H5G2

Mr. Mark Neuhaus
Solar Electric
116 4th. Street
Santa Rosa, CA 95401

Mr. James Nichols
Nichols Company
3906 Waddell Street
Kansas City, MO 64111

Mr. Joe Nichols
1803 Woodland Blvd.
Kansas City, KS 66106

Mr. Brock Nicholson
US Environ. Protect. Agency
US EPA, OAQPS/AQMD (MD-15)
RTP, NC 27711

Lars Henrik Nielsen
Riso Nt'll Laboratory
Post Office Box 49
DK-4000 Roskilde DENMARK

Mr. Frederick Nuessle
1935 Berryman Street
Berkley, CA 94709

Mr. Patrick O'Halloran
9414 Jarboe
Kansas City, MO 64114

Mr. Jim Ohi
Solar Energy Research Instit.
1617 Cole Boulevard
Golden, CO 80401

Mr. Mike Owens
1702 S. 14th Ave.
Ozark, MO 65721

Mr. Scott Pattinson
#1140, 1304 15 Ave. S.W.
Calgary, Alberta CANADA T36OX7

Mr. Hugh Pettit III
Post Office Box 1256
Cartersville, GA 30120

Dr. Thomas Pirotte
Spotlight Mktg, Inc.
4609 Seagraves
Joplin, MO 64804

Mr. Justin Pirtle
Rt. 2 Box 196
Lathrop, MO 64465

Mr. Timothy Pitts
Calhoun High School
Box 53
Ionia, MO 65335

Ms. Jackie Player
470 NE Colbern Road
Lee's Summit, MO 64063

Mr. Michell Pogue
2605 Shady Ln.
Poplar Bluff, MO 63901

Ms. Jenessa Pollard
608 Mark St.
Lathrop, MO 64465

Ms. Gina Potee
909 N. 13th
Blue Springs, MO 64015

Mr. Jason Praski
TransGas Limited
Post Office Box 67
Regina, Saski CANADA S4P2Z5

Dr. Keith Prater
Ballard Power Systems
980 W. 1st Street, Unit 107
N. Vancouver, B.C. CANADA V7P3N4

Mr. Kenneth Press
State Controllers Office
7280 Hillside Avenue #207
Hollywood, CA 90046

PARTICIPANTS

Ms. Susan Pryor
Lathrop School
Rt. 2 Box 261
Holt, MO 64048

Walter R. Pyle
Chevron Resh. & Tech. Corp.
100 Chevron Way
Richmond, CA 94802

Mr. John Rainbolt
536 Nature Ln.
Arcata, CA 95521

Mr. Ravinder Rana
Southwest High School
6512 Wornall Rd.
Kansas City, MO 64113

Mr. Jim Reilly
Brookhaven Nat'l Laboratories
Dept of Appl. Sci., Bldg. 815
Upton, New York 11973

Mr. Ken Riead
Independence Power & Light
Post Office Box 1019
Independence, MO 64051

Dr. Vernon Roan
Cntr for Adv. Studies in Engineering
3950 RCA Boulevard, Suite 5003
Palm Beach Gardens, FL 33410

Mr. Scott Rogers
104 N. E. Street
Albany, MO 64402

Mr. Sam Romano
Georgetown University
Car Barn, Suite 223
Washington, DC 20057

Mr. Denver Rosberg
HCR 1, Box 96
Underwood, ND 58576

Ms. Charla Rudisill
US Environ. Protect. Agency
US EPA, OAQPS/AQMD (MD-15)
RTP, NC 27711

Mr. Philip Salem
27 E. 22nd
Edmond, OK 73013

Dr. Krishna Sapru
TIPS, Inc.
1924 Smallbrook
Troy, MI 48098

Mr. William Sayer
Cactus Services, Inc.
200 84th Street
Virginia Beach, VA 23451

Mr. Brandon Scheuerman
Rt. 1 Box 74
Exeter, MO 65647

Mr. Fred Schieszer
Central Missouri State University
Depart. of Power & Transportation
Warrensburg, MO 64093

Mr. Helmut Schiller
Hydrogen Service
Scholzenviertel 7
D6140 Bensheim-Zell, W. GERMANY

M.Sc Lotte Schleisner
Riso Nat'l Laboratory
Post Office Box 49
DK-4000 Roskilde DENMARK

Mr. Fred Schwartz
Mid America Regional Council
300 Rivergate Ctr 600 Broadway
Kansas City, MO 64105

Dr. David Scott
University of Victoria
Post Office Box 3055
Victoria, B.C. CANADA V8W3P6

Mr. Darin Selby
P.O. Box 283
Tecopa, CA 92389

Mr. Ethan Silva
E.A. Silva Assoc.
499 Embarcadero
Oakland, CA 94606

PARTICIPANTS

Mr. Jerry Spector
8514 De Soto Avenue
Canoga Park, CA 91304

Ms. Dorothy St.Clair
Ozark High School
P.O. Box 166
Ozark, MO 65721

Mr. Newton Stafford
2405 Nabal Street
Escondido, CA 92025

Dr. Dieter Staschewski
Karlsruhe Nuclear Research Center
P.O. Box 3640, D-7500 Karlsruhe
F.R. Germany

Mr. Richard Stern
U.S. EPA (MO-63)
Research Triangle Park, NC 27711

Ms. Heather Stevens
Lathrop High School
RR1 Box 132
Lathrop, MO 64465

Mr. Todd Strother
Lathrop Highschool
612 Center Street
Lathrop, MO 64465

Mr. Masayuki Sukigara
Asahi Chemical Ind. America
350 5th Ave., Suite 7412
New York, New York 10118

Dr. William Summers
Westinghouse Electric Corp
Adv. Energy Sys. P.O. Box 10864
Pittsburgh, PA 15236

David H. Swan
Texas A&M University
238 Wisenbaker Eng. Research Cntr
College Station, TX 77843-3402

Mr. M. Swarnkar
Macromolecular Research Centre
Rani Durgavati University
Jabalpur-482001, M.P., INDIA

Mr. David Tabb
800 N. 19th
Blue Springs, MO 64015

Mr. Rod Talley
07 E. N. Second St.
Wright City, MO 63390

Ms. Karin Tarc
11112 Applewood Dr.
Kansas City, MO 64134

Mrs. Delores Tarman
1913 Forest Road
Edmond, OK 73034

Mr. Joe Tarman
1913 Forest Road
Edmond, OK 73034

Mr. Elroy Thieszen
Thieszen Irrigation Co. Inc.
RR1, Box 39
O'Neill, NE 68763

Mr. Bill Thompson
R.R. 3
Pattonsburg, MO 64670

Mr. D. Tillmetz
DORNIER
Post Office Box 1420
D-7990 Friedrichshafen GERMANY

Mr. Harry Toll
Wright City R-II High School
P.O. Box 198
Wright City, MO 63390

Mr. Don Trott
5 S. 680 Heather Ct.
Naperville, IL 60540

Ms. Teresa Tschirhart
416 Ralston
Raytown, MO 64133

Dr. Yoshiharu Tsujikawa
Univ. of Osaka Prefecture
4-804 Mozu-umemachi
Sakai 591, JAPAN

PARTICIPANTS

Dr. John Turner
Solar Energy Research Instit.
1617 Cole Boulevard
Golden, CO 80401-3393

Ms. Kathleen VanHorn
LHS
Box 336
Lathrop, MO 64465

Dr. Nejat Veziroglu
Intn'l Assoc of H_2 Energy
Post Office Box 248294
Coral Gables, FL 33124

Mr. Demetri Wagner
American Hydrogen Assoc.
200 Concord Dr.
Sedona, AZ 86366

Mr. William Ward
U.S. EPA
726 Minnesota Ave.
Kansas City, KS 66101

Mr. Dave Warrender
2685 Burnside Road
Sedaftopol, CA 95472

Dr. Harry Watson
Univ. of Melbourne
Dept Mechanical Eng.
Parkville, Vic 3052 AUSTRALIA

Ms. Linda Wheeler
Greenside Ventures
3517 Virginia
Kansas City, MO 64109

Mr. Bengt Wikstrom
Wiknell
Post Office Box 691
2510G Helsingborg SWEDEN

Dr. Robert Wills
SkyLine Engineering
Box 134
Temple, NH 03084

Professor Roland Winston
Dept. of Physics, Univ. of Chicago
5640 South Ellis Avenue
Chicago, IL 60637

Mr. John Wynhausen
1524 South 11th
Lincoln, NE 68502

Mr. Omar Yaakub
Laclede Gas Company
720 Olive Street-Room 1401
St. Louis, MO 63101

Mr. Al-Saedi Yaseen
KACST
KACST Solor Pro. P.O. Box 6068
Riyadh, Saudi Arabia 11442

Ms. Peg Young
Penn Valley Comm. College
4505 W. 74th. Place
Prairie Village, KS 66206

Mr. Terry Young
Middle TN State Univ
Route #4 Box 196
Woodbury, TN 37190